FUNDAMENTALS OF SYSTEMS ANALYSIS

TO THE STUDENT: A STUDENT WORKBOOK for this textbook is available through your college bookstore under the title *Student Workbook to accompany Fundamentals of Systems Analysis, 2 ed prepared by Warren D. Stallings, Jr.* The student workbook can help you with course material by acting as a tutorial, review and study aid. If the Student Workbook is not in stock, ask the bookstore manager to order a copy for you.

SECOND EDITION

FUNDAMENTALS OF SYSTEMS ANALYSIS

Jerry FitzGerald, Ph.D.
Jerry FitzGerald & Associates

Ardra F. FitzGerald
SRI International

Warren D. Stallings, Jr.
Coopers & Lybrand

John Wiley & Sons
New York • Chichester • Brisbane • Toronto • Singapore

Library of Congress Cataloging in Publication Data:

FitzGerald, Jerry.
    Fundamentals of systems analysis.

    Includes bibliographical references and index.
    1. Business—Data processing. 2. System analysis.
I. FitzGerald, Ardra F., 1938–   joint author.
II. Stallings, Warren D., joint author. III. Title.
HF5548.2.F476 1981   658.4'032   80-11769
ISBN 0-471-04968-9

Printed in the United States of America

20 19 18 17 16 15 14 13 12

ABOUT THE AUTHORS

DR. JERRY FITZGERALD is the principal in Jerry FitzGerald & Associates, a management consulting firm located in Redwood City, California. He has extensive experience in, and currently consults in, the areas of systems analysis, data processing security, EDP auditing, computer control systems, and data communications/tele-processing. Prior to consulting, Dr. FitzGerald was an associate professor in the California State University system, a senior management consultant for an international consulting firm, and a systems analyst at a major medical center and a computer manufacturer. He received his Ph.D. in business from the Claremont Graduate School, has an M.B.A., a bachelor's degree in industrial engineering, is a certified data processing auditor (CDPA), and has a Certificate in Data Processing (CDP). In addition to three books and numerous articles, he received the Joseph J. Wasserman Award in 1980 for outstanding contributions to the field of auditing computerized systems and security controls for computers. This is an annual award given by the EDP Auditors Association to the person who made the most outstanding contributions in the area of EDP auditing, control, and security. Currently, Dr. FitzGerald manages his own firm and devotes approximately one-half of his time to consulting in the above-mentioned areas, and the other half of his time to conducting various seminars and other instructional courses.

ARDRA FITZGERALD is a senior information specialist in the Library of SRI International located at Menlo Park, California. Her areas of expertise are technical literature searching, designing computerized bibliographic systems and, especially, computerized bibliographic data base searching. In this position, a systems approach to problem solving and research is utilized in an international consulting environment. Mrs. FitzGerald specializes in research in the areas of management, computers, accounting, systems analysis, engineering technology, and the social sciences. She uses more than 200 different data bases in a wide variety of subject areas. Before joining SRI International, she was a reference librarian and bibliographic specialist in business and engineering in the California State University system. She also had total responsibility for the management of a technical library for an aerospace firm. In addition to a bachelor's degree in sociology, she has both an M.B.A. and a master's degree in librarianship. She is an active member of the American Society for Information Science and the Special Libraries Association.

v

WARREN STALLINGS is a manager for the international public accounting firm of Coopers & Lybrand. His foremost areas of expertise are in applying audit and control theory to accounting systems that have been computerized and in analyzing the framework of controls within the data processing environment. He has been involved in the development of EDP audit curricula at the university level, and teaches EDP audit theory and mechanics within Coopers & Lybrand. Prior to joining the firm, Mr. Stallings helped to build and also administered the EDP audit functions for a major national bank and a West Coast electrical utility. In addition, he has provided management consulting services in the areas of finance, data processing, and organizational development to the health care, government, and educational industries. His academic background includes a B.S. in information systems and an M.B.A. from California State Polytechnic University (Pomona, California). Mr. Stallings holds the Certificate in Data Processing (CDP), is a certified data processing auditor (CDPA), and is a certified public accountant (CPA).

PREFACE

Systems analysis is a course that must be taught in every Data Processing/Computer Science curriculum. Because the development or use of computer systems is basic to *all* business functions in today's environment, it also should be a basic university-wide course that cuts across disciplinary lines.

The primary objective of this book is to teach the first course in systems analysis and design. It covers all the general concepts and problem-solving steps of systems analysis. Current supplemental periodical articles should be required to gain the in-depth knowledge for some of the more computer-related areas that are introduced and discussed within this text.

The book is very control-oriented and it is also people-oriented. It discusses the involvement of both controls and people in the systems study, systems operation, and in the use of computerized systems.

The book is divided into three parts. These three parts are the Preliminaries, The System Study Itself, and The Tools of Systems Analysis.

Part One. Its discussion delineates a thorough introduction to systems analysis. It covers how to conduct a feasibility study; whether that feasibility study is to try to decide whether to develop a new system, or whether it is to purchase/lease a new computer system (Chapters 1–2). In this second edition, we have added sections such as the emergence of the software engineering discipline, as well as numerous changes throughout these two chapters.

Part Two. This is the real heart of this textbook and it thoroughly discusses the ten steps of systems analysis (Chapters 3–12). There is an entire chapter on each of the ten steps of systems analysis, starting with the problem definition, on through system design, and finally implementation. In this second edition we have made most of our changes in this part of the book. Two chapters have been combined into one step that involves gathering background information on the area under study and understanding the interactions between the various areas that will be using the system. There also is an expanded part on defining input and output requirements for on-line systems. There is a major new section on structured system design, as well as on-line input and output design considerations. There is an entirely new chapter on designing system controls into computerized systems (this chapter carries over to Appendix II on design controls), as well as many other changes throughout these ten basic chapters.

Part Three. This part of the book covers all of the tools of systems analysis such as charting, forms design, records retention, report analysis, procedure writing, techniques for the systems manager, and the research needs of the systems analyst

(Chapters 13–19). In the second edition we have put much of our attention on changes throughout these seven chapters; although there have been some major new sections such as: HIPO diagrams, micrographics, reviewing management reporting systems, evaluating and selecting packaged software, an introduction to data base structures, and computerized literature searching in order to meet the research needs of the analyst.

Finally, at the end of each chapter there are questions that pertain to the material covered in the chapter. Also at the end of each chapter, there are short situation cases for the student to complete. These cases depict situations that pertain only to the specific chapter involved. The student should be able to read the chapter and its related case, and then derive answers to the case questions using only the information contained in that chapter. Each case should be discussed further in the classroom because the case questions point out only the most blatant mistakes made by the systems analyst in the case. Both the situation cases and the end-of-chapter questions have been revised as a part of the second edition. In addition, a new appendix has been added on controls that can be used as a handy reference guide. The teacher's manual has been appropriately updated and includes the following sections for each chapter:

Lecture outline

Answers to the questions at the end of each chapter

Sample test questions including:

Fill in

True-False

Multiple-Choice

Essay

Matching

Situation case answers

Situation cases were not prepared for Chapter 19 because the scope of the material does not call for them.

We feel that this edition provides students with a solid foundation in the fundamentals of systems analysis and equips them to deal better with the demands of today's business environment and the need for effective controls.

The authors are indebted to their reviewers Deane Carter (Colorado State College), William Cotterman (Georgia State University), and Daniel Couger (University of Colorado).

Jerry FitzGerald
Ardra F. FitzGerald
Warren D. Stallings, Jr.

CONTENTS

PART TWO
THE SYSTEM
STUDY ITSELF

PART THREE
THE TOOLS
OF SYSTEMS ANALYSIS

CONTENTS

PART ONE

THE
PRELIMINARIES

**Part One of this book is concerned with items
that are preliminary to a full systems study.
These preliminaries include the orientation to
systems analysis and a chapter on the function
and performance of the Feasibility Study.**

CHAPTER 1

INTRODUCTION TO SYSTEMS ANALYSIS

Systems abound in nature and appear to be an essential characteristic in all things. No approach to organization could be more natural or rational than a systems approach. This chapter contains a basic explanation of the various parts of systems analysis, including the systems situation, definitions, the systems analyst, relationship of the systems department to the rest of the organization, systems department functions, philosophies of systems design, desirable characteristics of a system, consideration of alternatives, the area under study, and documentation of the system. In addition this chapter introduces the concept of software engineering.

THE SYSTEMS SITUATION

Human organization can be viewed as a network of interdependent systems designed to perform the activities vital to human existence. If we had a static world and perfect systems, this book could end right here. But, of course, we live in an ever-changing world where "perfection" is a vague and moving target.

Our formal organizations, our businesses, government, hospitals, and so on, require a framework of systems and procedures to guide them in their day-to-day operations. In short, we live in a world of systems, whether they are biological or man-made.

Many resources may be combined for the effective operation of a system. But the most important ingredient is people—the people who operate the system and the people who use its output. Unless the system has the support of its operators, it will not work as intended. And even when it has user support, it will collapse if it fails to satisfy the needs of its users.

One of the resources of today's system is data processing. In order to meet the challenge of a world of changing business conditions, the systems analyst has learned to rely heavily on data processing as a resource for more effective and efficient handling of business data. In the past, some companies experienced significant difficulties because of ineffective use of this vital resource. In these cases, however, the fault often was with the people who determined how the data processing resource was to be used. If they set unrealistic goals, had an inadequate understanding of what was needed, or could not obtain cooperation, their efforts were doomed to failure.

Obviously, then, the human factor is a vital one in systems design. The men and women who design our systems must design them for human operation and benefit.

Every manager is responsible for systems because having systems is basic to being a manager. Each manager in an organization employs systems to carry out organizational objectives. And within each system there are usually sets of subsystems called procedures to guide the manager's team in its efforts.

The systems and procedures found in today's organizations are usually based upon a complex collection of facts, opinions, and ideas concerning the organization's objectives. In many cases, however, the systems and procedures have not been well defined or documented, and represent a collection of informal working practices. The complicated and multipurpose orientation of today's business firm makes it difficult for its members to all agree on how things should be done. Thus we may encounter departments working at cross purposes, or managers who are unable to cooperate with each other's approach to things. Unity of purpose becomes more difficult to achieve because effective communication deteriorates as the number of organization members increases. For an organization to survive, it must learn to deal with a changing environment effectively and efficiently. To accomplish the making of decisions in an uncertain environment, the firm's framework of systems and procedures must be remodeled, redefined, or tailored on an ongoing basis.

It often falls upon the person who is charged with designing a new system in the firm to perform the work in such a way as to redefine and refine the firm's objectives, sometimes on a very broad scale.[1] Fact must be sorted out from opinion and each used in its place. The vital must be separated from the trivial. Most important, the analyst must continue to ask "What are we trying to accomplish?" Only when this question is answered will a systems design effort make sense. Systems should be realistically aimed at the achievement of legitimate objectives of the organization, or they become wasted resources. In the following chapters we will discuss in detail the methods the analyst uses to develop realistic goal-directed systems and procedures.

DEFINITION OF SYSTEMS AND PROCEDURES

A *system* can be defined as a network of interrelated procedures that are joined together to perform an activity or to accomplish a specific objective. It is, in effect, all the ingredients which make up the whole. And a *procedure* is a precise series of step-by-step instructions that explain

1. What is to be done.
2. Who will do it.
3. When it will be done.
4. How it will be done.

The procedures tell how the ingredients are made into the whole. Systems are often classified into the following two categories.

A *closed system* is one which automatically controls or modifies its own operation by responding to data generated by the system itself. For example, high-speed printers used with computer systems usually have a switch that senses whether there is paper in the printer. If the paper runs out, the switch signals the system to stop printing.

An *open system* is one which does not provide for its own control or modification. It does not supervise itself so it needs to be supervised by people. For example, if the high-speed printer used with computer systems did *not* have a switch to sense whether paper is in the printer, then a person would have to notice when the paper runs out and signal the system (push a switch) to stop printing.

Another common example of this concept is the household furnace. One with a *closed system* (see Figure 1-1) has a thermostat which automatically switches the furnace on when the temperature goes below a certain degree or off when the temperature goes above another degree. By contrast, an *open system* furnace (see Figure 1-2) is a wall heater which an individual switches on when cold or off when warm. The former (closed system) is mechanically controlled on an automatic basis, while the latter (open system) is controlled by an individual as circumstances suggest.

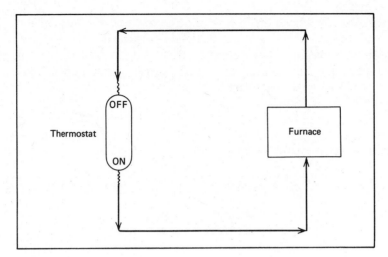

FIGURE 1-1   A closed system with a thermostat that automatically switches the furnace ON and OFF.

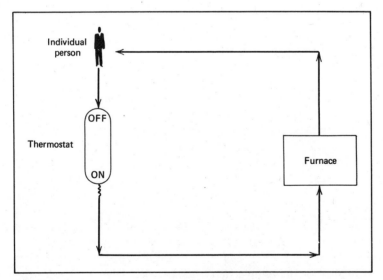

FIGURE 1-2   An open system showing a person that switches the furnace ON and OFF.

In order to apply the "open" and "closed" system theory to computer-based systems, consider the following alternative approaches that could be taken in designing portions of an inventory system. In an open system (see Figure 1-3), the normal sequence of events might include

- Customer order for parts received.
- Computer-based system processes order
  –updates inventory master file.
  –produces report of parts on hand.
- Clerk reviews report and decides whether additional parts should be ordered.
- If more parts are needed, clerk prepares a purchase order for additional parts.

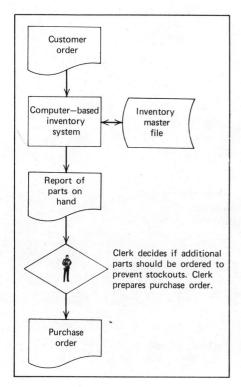

FIGURE 1-3   An open computer-based inventory control system showing a clerk deciding if additional parts should be ordered to prevent stock-outs.

In a closed system (see Figure 1-4), the normal sequence of events might include

- Customer order for parts received.
- Computer-based system processes order
  —updates inventory master file.
  —determines if current inventory level is below minimum level that should be maintained.
  —Automatically produces purchase order for parts if needed.

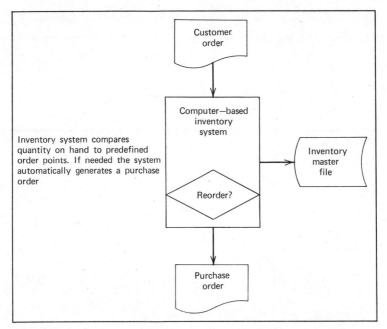

FIGURE 1-4   A closed computer-based inventory control system where the computer program determines if additional parts should be ordered to prevent stock-outs.

In the open system example, the control over inventory level is manual, based upon the clerk's review of the report of parts on hand. In the closed system example, the computer-based system automatically controls the inventory level by referring to a predefined reorder point for each part.

The difference in principle is the important point here. A systems analyst should consider the opportunities offered by each operation of a system for application of the "closed" technique, and should also understand that the "open" variety is appropriate to many operations.

For example, some operations in a system may have little bearing on the probability of retaining a customer's continued goodwill and business, such as the operation that routes the customer's order to one clerk or another for processing. The customer does not care which of 15 clerks fills the order, so the operation that makes the choice can be a closed type of decision generated somehow by the mechanics of the system on the basis of appropriate criteria. On the other hand, the operation that provides the customer with a special discount or other preferential treatment may be one which should have the benefit of human judgment added. It is probably an operation that should be supervised by responsible employees of the firm using the open system category.

Fully closed systems, the kind we all look forward to, are still rarities. In the previous example, a fully closed furnace would be one in which the temperature never varied above or below a set comfort zone. This comfort zone would be determined at the time of installation when a program would be written instructing the computer to maintain temperature within the specified range. Progress made in the past ten years indicates a high probability that in the future we will be able to rely significantly upon the computer for many of the decisions now supplied by people. The systems analyst plays a vital and exciting role in this transition.

DEFINITION OF SYSTEMS ANALYSIS

The *systems analysis* approach to a problem differs from a trial-and-error approach. The trial-and-error approach involves identifying a number of potential solutions to the problem and then testing each randomly until one alternative appears to provide an acceptable solution. In the systems analysis approach, all influences and constraints are identified and evaluated in terms of their impact on the various "decision points" in the system. A decision point is that point in a system at which some person or automatic mechanism must react to input data and make a decision. A decision point may be seen in the previous Figures 1-1 and 1-2. In Figure 1-1 the thermostat is a decision point. For example, the thermostat makes the decision whether the furnace should be turned on or turned off depending on the temperature setting of the thermostat and the temperature in the room surrounding the thermostat. In Figure 1-2 the individual makes the decision whether the furnace should be turned on or turned off. In this case the thermostat functions only as a switch and not as the "decision point." Therefore in the systems analysis approach a system is designed around the various "decision points." The system itself conveys the information or material between the different decision points and it determines the criteria with which to make the decision. Figure 1-5 shows a system decision point. The decision point is whether it is too hot or too cold, regardless of whether that decision is made by an individual or made automatically by a thermostat.

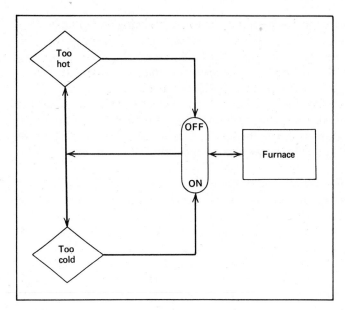

FIGURE 1-5   System decision point.

A decision point also is clearly reflected in Figures 1-3 and 1-4. In Figure 1-3 the decision point is represented by the clerk's manual review of the report of parts on hand to decide if additional parts must be ordered. In Figure 1-4, the decision point is embedded within the inventory system itself which makes the decision automatically on whether to order additional parts.

As another example, a hospital's X-ray processing system would be designed around its key decision point: the doctor's evaluation of the exposed film and the alternative actions available to the doctor at that point.

To properly understand a decision point in one system it is usually necessary to understand how other systems within (or even outside) the organization interact with and affect the decision point. As we will see, many outside influences can converge on a single decision point, which is often the reason for its existence. The decision point is a main focal point in the system. The systems analyst works to identify all such decision points. The analyst also attempts to ascertain their significance in relation to the objectives of the system before beginning any efforts to improve upon the current system.

In summary, a PROCEDURE is a set of step-by-step instructions that explain how a SYSTEM operates. Within the system there are a number of DECISION POINTS at which either one action or some other action is taken as the information or material flows through the system from start to finish. In order to be effective, the analyst must be able to understand and analyze the existing system and procedures and design new ones when required.

THE SYSTEMS ANALYST

The systems analyst is a methods person who can start with a complex problem, break it down logically, and identify the reasonable solutions. The analyst can study an ailing system and come up with superior alternatives. Or given some number of objectives, the analyst can devise systematic means of attaining them. The systems analyst views a systems situation in terms of its scope, objectives, and the organizational framework.

For example, the scope of a systems project is the range of the project. In other words, the range is the area or the breadth that the study will encompass. The objectives of a systems project are whatever the analyst is trying to accomplish with the new systems design. The analyst also views a system in terms of the information with which that system operates. It is very important that a systems analyst completely understands the information that is used in the current system and the information that may be used in the new proposed system.

Another very important item for an analyst to consider in viewing a systems situation is where people are involved in the operation of a system and what they do. The basic ingredient of a system is people. If the analyst overlooks people, the system will not be as efficient as it might. The equipment employed in a system, the forms and reports used in a system, and the responsibilities assigned to each department involved in the system are all important items to a systems analyst. An analyst who overlooks any of the above items may be jeopardizing the possibility of a good system design. The primary responsibility of the systems analyst, therefore, is not only to develop systems that meet the objectives and goals of the entire company, but also to meet the objectives and goals of the individual departments that are involved with the specific system being designed. A good systems solution meets all of its objectives. An excellent systems solution meets all of its objectives and also is tailored to both fit comfortably within the organization's framework and to take advantage of existing human resources.

An analyst can work with manual or computer-based systems techniques to creatively modify the status quo. Later, in Chapter 18, we will discuss the details of the analyst's job description.

Although it is common for a systems analyst to know computer programming, one usually does *not* do the programming work on a system. Rather, an analyst works with programmers who are assigned the specific programming tasks in the project. The systems analyst is more like a general manager who determines the design of the overall system, obtains the necessary technical help—programmers, forms specialists, equipment engineers—and follows the system through design, implementation, follow-up, and re-evaluation.

The analyst needs some knowledge of computer programming techniques in order to communicate effectively with the programmers; an analyst needs to be able to understand why certain things can or cannot be programmed. On the other hand, the analyst also may need to understand something of the nature of the work being performed by the people for whom a new system is being designed. The analyst is often the only true liaison between two areas. Each area speaks its own language and has its own techniques, with the analyst serving as both interpreter and mediator. As such, the analyst must be both open-minded and willing to learn new things. And, most important of all, the analyst should have the ability to deal with people at all levels of the organization.

RELATIONSHIP OF THE SYSTEMS DEPARTMENT TO THE REST OF THE ORGANIZATION

In today's world the systems department is usually more closely related to the use of computers than to manual systems methods; however, systems analysts do plenty of both. The success of a computer-based system depends largely upon the concise definition and documentation of the relationship between it and the manual systems with which it must interface.

If the firm has a computer, the systems department may or may not have the responsibility for its use and management. There is no hard and fast rule about which department shall control the computer. In some firms the EDP (Electronic Data Processing) operation is a separate department and is considered a service organization operating for the benefit of all other departments. The systems department may operate separately on the same theory, or it may be an adjunct operation of the EDP function. Still other firms have the EDP operation under the accounting department, the industrial engineering department, or as an arm of administration. The organizational location of the EDP function could probably be traced in many cases to the first department that showed management it had a need for a computer. The need for a systems department has often not been determined until after an EDP function has gone awry. For this reason, independent systems departments are still relatively new.

Whether the systems department exists as a separate entity or as an activity within another department, it is a staff activity that renders service to all other departments. It advises and assists, rather than directs.

The systems department can become a powerful influence in the firm because of its comprehensive involvement in most areas of the organization. It often holds the magic power of the computer, and is "in-the-know" about most long-range plans. The department's activities cut across organizational boundaries, because systems do. It cuts through interdepartmental networks, and involves itself in the detailed mechanics of the firm's operations. It provides management with the real-world means of implementing its plans and dreams. It also serves as a mechanism that seeks to ensure the effective use of the EDP resource and provides feedback to management on the current and potential impact systems have on the organization.

Nevertheless, the systems department remains a staff function and can design and implement systems changes only when it has authorization from line managers.

FUNCTIONS OF THE SYSTEMS DEPARTMENT

The principal responsibility of the systems department is *systems design.* But to effectively perform this broad function it should have centralized control, or at least strong influence, in the following areas.

1. Forms design and control.
2. Procedure writing and procedure manual control.
3. Records management.
4. Report control.
5. Office layout.
6. Work simplification studies.

A centrally coordinated systems network has many advantages for the firm, as we shall see later. If the systems department is unable to oversee the above functions, many of these advantages are lost.

As we noted, the primary responsibility of the systems department is *systems design.* And, as systems people, we desire a systematic approach to things. Therefore the following general outline has been developed as an overall description of the system development process. It will guide the analyst, in a systematic fashion, through all the phases of a systems study.

1. Define the problem (Chapter 3).
2. Prepare an outline of the systems study (Chapter 4).
3. Obtain general background information on the areas to be studied, and understand the interactions between the areas being studied (Chapter 5).
4. Understand the existing system (Chapter 6).
5. Define the system requirements (Chapter 7).
6. Design the new system (Chapter 8).

7. Design new system controls (Chapter 9).

8. Prepare economic cost comparisons (Chapter 10).

9. Sell the new system to management (Chapter 11).

10. Provide implementation, follow-up, and re-evaluation (Chapter 12).

The analyst can use this sequence, as amplified in subsequent chapters, as a comprehensive guide during any systems study. It is a system that works—one that has been tested successfully and in the heated pace of industry.

PHILOSOPHIES OF SYSTEMS DESIGN

The aim of a system is coordination of managerial efforts toward the goals of the firm. The systems study must go beyond the simple paper work studies and simple procedures. It must go far enough to include the philosophy, the objectives, the policies, the interactions, and the management thinking within the firm. It should harmonize with the style of the individual firm. By observing all aspects of the firm's organizational structure, objectives, other systems, and legal requirements, the analyst is more likely to design a system that will be accepted and used.

A system has as its objective the coordination of actions involving people, equipment, time, and money; and it should produce results such as these:

1. The right information furnished to the right people, at the right time, and at the right cost.

2. Decrease in uncertainty and improvement of decision quality.

3. Increased capacity to process present and future volumes of work.

4. Ability to perform profitable work that was previously impossible.

5. Increased productivity of employees and capital, and reduced costs.

A system may be viewed in terms of information flowing between departments. The information is continuously being recorded, processed, summarized, used, stored, and discarded. This information flow is the life blood of an administrative system. The information is necessary for making decisions and plans. It is necessary for initiating and directing actions as well as for comparing results against plans. Or information may be an objective in itself (for example, the information compiled for reports to the government).

It is important that we clarify the difference between data and information. *Data* is a general term used to denote any or all facts, numbers, letters, and symbols that refer to or describe an object, idea, condition, or other factors—such as a name, address, telephone number, or a list of such items. *Information* is a meaningful assembly of data, telling something about the data relationships—such as how many names are in the list, or how many persons live in a certain area, and so on. In essence, information can be derived from data only to the extent that the data are consistent, accurate, timely, economically feasible, and relevant to the subject under consideration. *Data processing* usually refers to the conversion of data to information.

Every systems operation consists of input, processing, and output. An *input* is the energizing element that puts the system into operation. The next activity is *processing*, which transforms the input into an output. This processing activity may be performed by an individual, by computer, or both. The *output* is the purpose for which the system was organized. Figure 1-6 shows this process, which is called the Input/Output Cycle.

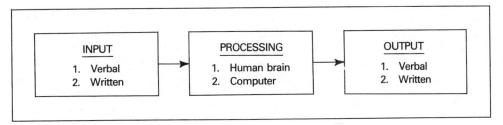

FIGURE 1-6   Basic input/output cycle.

DESIRABLE CHARACTERISTICS OF A SYSTEM

The system should provide information that is Consistent, Accurate, Timely, Economically feasible, and Relevant. These characteristics *CATER* to the needs of the organization in which the system is used. Consider their importance in the information system of the firm, for example.

Other characteristics of a good system are that it

1. Establishes standards.
2. Specifies each area's responsibility.
3. Delineates actions and decisions.
4. Is easily understandable.
5. Provides criteria with which to judge its own performance.
6. Identifies the decision points.

A system must be acceptable to the organization's managers. Equally as important, it must be accepted by the people engaged in its operation. As already mentioned, the system must harmonize with company style, policy, organization structure, and external requirements imposed by custom or law. The system must also be compatible with other systems in the firm in order for it to be accepted and used.

A well-planned system has a foundation of sound procedures, and is designed and assembled in such a way that the system can be easily adjusted to changing conditions. Flexibility should be built into a good system because its environment is dynamic and not static.

SYSTEM CHANGE AND CONVERSION ALTERNATIVES

At this point we can discuss three of the more common alternatives open to a firm with regard to changes in its systems (see Figure 1-7).

1. *No change;* keep using the existing system. This alternative is usually chosen when the cost of a new or modified system will exceed the benefits. Or sometimes powerful forces within the organization are unable to agree on the desirability of a new system. Rather than risk conflict, no change is made. Of

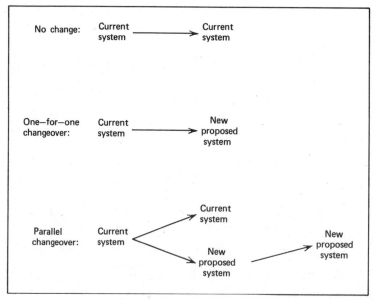

FIGURE 1-7   Three alternatives when changing systems.

course we could list many other reasons for choosing the no-change alternative.

2. In the *one-for-one changeover* method, the existing system is completely replaced with some other system. This method abruptly stops the use of the old system and simultaneously puts the new system into operation.

3. The *parallel changeover* method allows both the existing system and the proposed system to operate simultaneously until the new system has proven its reliability. This method is appropriate when installing a new computer in a computer-based information system, or when replacing one computer system with another. It may also be used when replacing a manual system with a computerized system.

WHAT IS THE "AREA UNDER STUDY"?

The systems study may involve only a small functional area in one department, or it may involve the entire organization. Figure 1-8 shows five levels that may comprise the "area under study." It should be noted that this is only one possible approach to defining the area under study and you should tailor it to fit your own organization. However, we have found it to be a logical and easy-to-use method because it more or less parallels traditional organizational lines. We will refer to these five levels throughout the text.

Level One

A problem that involves the entire organization—all its divisions and locations. Decisions on this level can have great economic and directional weight. The stakes are

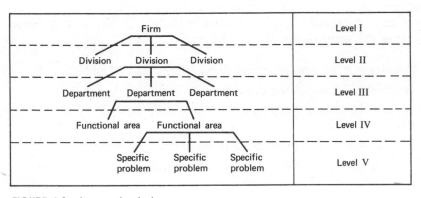

FIGURE 1-8 Area under study.

high. Management personnel require superior quality information. All decision points require critical treatment. Systems analysis at this level requires the use of most or all of the techniques detailed in Chapters 3 through 12.

Level Two

A problem that involves one division of the firm; a little less demanding, but conceptually similar to level one. Much integration is required between the division level and the firm level.

Level Three

A problem that involves departmental interaction within a firm. This is the middle management area where most systems studies are performed. Information is usually more technical and specialized here than at the division or firm level. The tasks of planning, organizing, and controlling are very detailed at the lower levels of the organization.

Level Four

A problem that involves the functions within one department. The activity to be studied is easily defined and understood. Only a small close-knit group is involved. The problems are usually easy to identify but difficult to resolve. Individual personalities and human relations play a big role at this level.

Level Five

A problem that involves one specific problem within a function of a department. This is usually the smallest area studied in an organization. The problem is a specific one and is usually lacking in specific data. Only a limited number of people are involved.

The analyst must be conscious of the level which a study will affect because the nature of the objectives and tasks will be different for each level. The analyst should always be aware of the scope of the "area under study."

Level I: A study that involves the entire firm—all its divisions and locations.

Level II: A study that involves one division of the firm.

Level III: A study that involves departmental interaction within the firm or division.

Level IV: A study that involves the functions within one department.

Level V: A study that involves one specific problem within a function of a department.

DOCUMENTATION OF THE SYSTEM

Documentation provides the necessary means of coordinating the procedures, programming, and other operations involved in the system. Proper system documentation refers to a thorough written description of all the component parts and operations of the system. The documentation usually describes the forms used, the personnel required to run the system, the equipment needed, and the sequence of operations from input to output. System documentation should be compiled as the system is being designed, and revised as the system is changed and perfected. Computer programs also must be thoroughly documented.

There are six principal reasons why documentation of a system is of vital importance.

1. Projects that are postponed may be difficult to restart unless there is adequate documentation stating the problem, the systems objective and scope, and the degree of completion.
2. When no documentation of previous work exists, a complete restudy or a complete reprogramming effort may have to be undertaken just to obtain an understanding of the prior effort, before making a change.
3. If documentation is inadequate, conversions from one system to another can result in unnecessary delay and additional cost.
4. When file layouts are changed, the system documentation may not accurately reflect the relationships between all the various file formats.
5. Without proper documentation, effective communication of the who, what, where, when, and how of the system is virtually impossible.
6. Without adequate documentation, auditability and control will be hard to achieve.

THE EMERGING SOFTWARE ENGINEERING DISCIPLINE

In the past decade, the development of computer software has been plagued with many symptomatic problems such as

- Late delivery of software.
- Excessive costs for design, development, and implementation.
- Failure to satisfy original specifications.
- Difficulty in maintaining existing software.

In many cases these problems resulted because the analyst (designer or software engineer) failed to establish and apply a framework of basic design procedures and techniques. This approach is common to all engineering disciplines and greatly improves the quality, consistency, and efficiency of the resulting products. The techniques required to support the design, development, and implementation of sophisticated computer-based information systems have undergone much redefinition and refinement over the years. The result has been the emergence of a new technical discipline often referred to as *software engineering.* One of the most descriptive and accurate definitions of software engineering was provided by F. L. Bauer of Munich, Germany.[2] In his definition he states that software engineering involves the establishment and use of sound engineering principles (methods) in order to economically obtain software that is reliable and works on real machines.

The data processing industry has begun to realize that software engineering is not just a collection of tools and techniques. Because of the growing complexity of the software products being produced and the changes in software development technology, software engineering has begun to come into its own as a true engineering discipline. As a discipline, software engineering will draw many of its principles from the areas of management science, engineering fundamentals, computer science, physical science, and business communication skills.

Software engineering can be differentiated from other engineering disciplines by one major facet. Most branches of engineering deal with visible and tangible materials. The chemical engineer, industrial engineer, aerospace engineer, civil engineer, mechanical engineer, and electrical engineer all deal with material products. Of the classical engineering disciplines, electrical engineering is the most abstract since electricity is not a tangible entity. However, through the use of technologically sophisticated devices electricity exhibits characteristics that can be measured materially. Software, on the other hand, is considered to be intangible in every aspect. Software usually is defined as a set of programs, procedures, and the associated documentation concerned with the operation of a data processing system. With the large number of programming languages in use today, and the large number of approaches that can be taken in logical program design, software can be more accurately described as an "intangible product."

The technical aspects of software engineering are beyond the scope of this text; however, many of the related areas will be touched upon as the system design process is examined.

ORGANIZATION OF THE BOOK

The remainder of this text is arranged to illustrate the overall steps in the system analysis and design process. It is graphically shown in Figure 1-9.

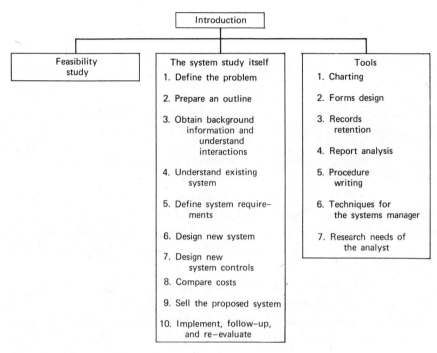

FIGURE 1-9 The systems design process.

This process can be viewed as three different areas of activity. The first area is the feasibility study. A feasibility study may or may not be performed, depending upon the size of the study, the cost involved in the new system, and the speed with which management requires the new system. Feasibility studies are usually reserved for large costly systems that may require a long time for implementation. The second area is the system study itself. The ten steps of a systems study are shown in Figure 1-9. These ten steps are covered in Chapters 3 through 12 of this textbook. The third area involves the tools of systems analysis. These tools also are shown in Figure 1-9 and are thoroughly covered in Chapters 13 through 19.

SELECTIONS FOR FURTHER STUDY

In order to have a successful system, it must be acceptable to the people who operate it and to the people who use it. Since the people involved can cause the success or failure of a new system, the systems analyst would be wise to read further on organizations.

The works of Kast and Rosenzweig, Baker, DuBrin, Etzioni, or Mott might be useful here.

To gain further insight into the working vocabulary of the systems analyst, other texts on systems analysis also should be examined. These might include those by Gore, Optner, Lazzaro, or Harpool. Trends in systems and designing for tomorrow's organizations are treated by Bensoussan and Jun, while Gall stresses how systems fail.

The following selected readings reflect what has been discussed in this chapter. This procedure will be followed throughout the text.

1. ABRAHAMSSON, BENGT. *Bureaucracy or Participation: The Logic of Organization.* Beverly Hills, Calif.: Sage Publications, Inc., 1977.

2. BAKER, FRANK, ed. *Organizational Systems: General Systems Approaches to Complex Organizations.* Homewood, Ill.: Richard D. Irwin, Inc., 1973.

3. BENSOUSSAN A., ed. *New Trends in Systems Analysis: International Symposium, Versailles, France, Dec. 13–17, 1976.* New York: Springer-Verlag New York, Inc., 1977.

4. BERNE, ERIC. *Structure and Dynamics of Organizations and Groups.* Westminister, Md.: Ballantine Books, Inc., 1973.

5. BINGHAM, JOHN, and GARTH W. DAVIES. *A Handbook of Systems Analysis, 2nd edition.* New York: Halsted Press, 1978.

6. CARLSEN, ROBERT D., and JAMES A. LEWIS. *The Systems Analysis Workbook: A Complete Guide to Project Implementation and Control, 2nd edition.* Englewood Cliffs, N.J.: Prentice-Hall, Inc., 1979.

7. COUGER, J. DANIEL, and ROBERT W. KNAPP. *System Analysis Techniques.* New York: John Wiley & Sons, Inc., 1974.

8. DRABEK, THOMAS E., and J. EUGENE HAAS. *Understanding Complex Organizations.* Dubuque, Iowa: William C. Brown Company, 1974.

9. DUBRIN, ANDREW J. *Fundamentals of Organizational Behavior: An Applied Perspective, 2nd edition.* Elmsford, N.Y.: Pergamon Press, Inc., 1978.

10. ETZIONI, AMITAI. *Comparative Analysis of Complex Organizations, revised edition.* Riverside, N.J.: Free Press, 1975.

11. GALL, JOHN. *Systemantics: How Systems Work and Especially How They Fail.* Scranton, Pa.: Times Books Division of the New York Times, 1977.

12. GORE, MARVIN R., and JOHN W. STUBBE. *Elements of Systems for Business Data Processing.* Dubuque, Iowa: William C. Brown Company, 1975.

13. HALL, RICHARD S. *Organizations: Structure and Process, 2nd edition.* Englewood Cliffs, N.J.: Prentice-Hall, Inc., 1977.

14. HARPOOL, JACK D. *Business Data Systems: A Practical Guide.* Dubuque, Iowa: William C. Brown Company, 1978.

15. JUN, JONG S., and WILLIAM B. STORM. *Tomorrow's Organizations: Challenges and Strategies.* Glenview, Ill.: Scott, Foresman & Company, 1973.

16. KAST, FREMONT, and JAMES ROSENZWEIG. *Organization and Management, 3rd revised edition.* New York: McGraw-Hill Book Company, 1979.

17. LAZZARO, VICTOR. *Systems and Procedures: A Handbook for Business and Industry, 2nd edition.* Englewood Cliffs, N.J.: Prentice-Hall, Inc., 1968.

18. LUCAS, HENRY C., JR. *The Analysis, Design, and Implementation of Information Systems.* New York: McGraw-Hill Book Company, 1976.

19. MAURER, JOHN G. *Readings in Organization Theory: Open System Approaches.* Westminister, Md.: Random House, Inc., 1971.

20. MORRILL, CHESTER, JR., ed. *Systems and Procedures Including Office Management: Information Sources.* Detroit, Mich.: Gale Research Company, 1967.

21. MOTT, PAUL E. *The Characteristics of Effective Organizations.* New York: Harper & Row Publishers, Inc., 1972.

22. OPTNER, STANFORD L. *Systems Analysis for Business Management, 3rd edition.* Englewood Cliffs, N.J.: Prentice-Hall, Inc., 1974.

23. TOU, JULIUS T. *Software Engineering.* New York: Academic Press, Inc., 1970 (in two volumes).

24. WILLOUGHBY, THEODORE C., and JAMES A. SENN. *Business Systems.* Cleveland, Ohio: Association for Systems Management, 1975.

QUESTIONS

1. What is one of the most vital factors to consider in systems design?
2. What is the difference between a system and a procedure?
3. Define a closed system.
4. What is the difference between the systems analysis approach and the trial-and-error approach?
5. Define a decision point.
6. What is the primary responsibility of the systems analyst?
7. Name four areas the systems department should have centralized control over, or at least exert a strong influence upon.
8. How does *information* differ from *data*?

   9. Name four characteristics of a good system.
  10. Name the parts of the organization to which the five levels of the "area under study" pertain.
  11. Name three reasons why documentation of a system is of vital importance.
  12. What are the eleven steps of "the systems study itself"?

SITUATION CASES

CASE 1-1

The controller of the Graham Water Company noted an increase in overdue account balances resulting in the automatic suspension of delivery. He requested that a systems analyst be assigned to investigate this situation.

The Graham Water Company specializes in delivering bottled water to Northern California communities and businesses twice a month. Several of their biggest clients are retirement communities where the schedule calls for the bottled water to be delivered on the first and third Monday of each month. The setup of new accounts and the monitoring of account status is the responsibility of the Credit Department. Current payment policy requires that customers pay either by mail or to the delivery-man on the first delivery of the following month. When a residential customer's account balance exceeds $35.00, his account is flagged as overdue and a notice is sent requesting payment. When the balance of an overdue residential account exceeds $50.00, it has been company policy to refuse delivery on the first scheduled date of the month unless the customer makes full payment of his liability at that time.

In order to monitor account balances the Credit Department reviews a weekly report provided by the Accounts Receivable Section of the Accounting Department. Payments received are processed manually by the Accounts Receivable Section. Payments and deliveries are posted to customer ledger cards and a running balance is maintained. Because of the sheer volume of activity, at times, the posting of payments gets backlogged. As a result, the Credit Department may not always receive current account balance information upon which to make collection decisions. The Credit Department, after reviewing account balances and identifying those with balances exceeding $50.00, prepares a collection sheet which is given to the deliveryman. The deliverymen have reported an increasing number of problems with the retirement communities. Apparently the timing of the receipt of pension checks and the backlog of payment processing have created a problem, particularly for those people receiving delivery (and payment requests) on Mondays. In addition, the current collection sheet form does not allow for partial payments.

QUESTIONS

1. What functions of the Systems Department may be performed by the systems analyst to satisfy the controller's request?
2. What level of study is the analyst likely to be involved with?

CASE 1-2

Ms. Evans, a former computer programmer, had just been transferred to the Systems Department at Westlake & Company. Her first assignment as a full-fledged systems analyst was to perform a full systems study involving the automation of the accounts payable function. Ms. Evans decided to outline the entire sequence of events that would take place during the systems study in order to get a better feel for the size and scope of the effort required.

As an analyst, Ms. Evans developed a plan for carrying out the systems study in the following manner. She decided the first step should be to define the nature and extent of the problem relative to the accounts payable system as concisely as possible. Next she would prepare an outline of the specific steps to be taken in performing the assigned systems study. Ms. Evans felt it would then be important to obtain some general background information on the Accounts Payable Department. Next the analyst would study the existing system until she obtained a clear understanding. The system requirements for the new proposed system would be concisely defined. This would facilitate her understanding of what the new system would have to do in order to meet the current and future requirements of the company. Next Ms. Evans planned to draw together the problem definition, her understanding of the existing system and the definition of the new system requirements as a basis for designing the new system.

The design would be used as a basis for presenting the new system to management for their evaluation and, hopefully, approval to proceed with development. Ms. Evans felt it would be wise to prepare a formal system design report for management to review. This was to be followed with a verbal presentation of the system to management so that questions could be answered and potential problems identified. Finally, with management's approval, the analyst would proceed with the development of the new system. Ms. Evans realized that after installation, it would probably be necessary to do some follow-up work and some re-evaluation before the project was entirely completed.

QUESTIONS

1. Evaluate the sequence of events Ms. Evans planned to follow in performing the assigned systems study.
2. What *level* was the area under study in this case?

25

FOOTNOTES

1. Firms have goals and objectives, and policies to be followed in order to achieve these goals and objectives. The systems analyst writes procedures (step-by-step instructions) that detail the activities required to carry out the policies. A network of such procedures constitutes a system.

2. F. L. Bauer, "Software Engineering," in *Information Processing 71* (Amsterdam: North Holland Publishing Co., 1972), p. 530.

CHAPTER 2

FEASIBILITY STUDY

In this chapter we will discuss a vital planning tool which is available to management and the systems analyst: THE FEASIBILITY STUDY. It is a planning method that stresses a "look before you leap" approach to any important project.

WHAT IS A FEASIBILITY STUDY?

A feasibility study can be undertaken with the objective of determining the possibility or probability of improving an existing system at a reasonable cost. It also can be used by management as a tool to evaluate the feasibility and probable cost of developing a new system which may not have been needed in the past, but which is now required to deal effectively with the changing demands of the business environment. It refines the statement of the problem and breaks it down into workable projects. It isolates problem factors for analysis.

A feasibility study group may recommend an analysis of the present company procedures or it may recommend project abandonment. It may examine various types of computer equipment and recommend for or against the purchase or rental of such. A feasibility study may estimate the cost of a specific undertaking such as job retraining, or cost of converting from a manual to a computer-based system.

WHY CONDUCT A FEASIBILITY STUDY?

A feasibility study in some form should always be conducted prior to any commitment to a large or long-term investment or change. The impact of proposed major changes must be weighed carefully because there is usually a great deal at stake. The consequences of a major change are not all felt at once. Instead, these consequences tend to unfold, for good or ill, over some period of time after the commitment is made. The impact of unwise capital expenditures, for example, can be disastrous. A firm buys another plant, only to realize later that the plant is not adaptable to the work that management thought could be done there. Such a consequence might have been foreseen if a formal feasibility study had been carried out in order to delve into the details.

Of course, our focus is systems, an area rich in the opportunities offered by the application of feasibility studies. The failure of the plant in our example above might easily have been caused by its systems inflexibility, or by a prohibitive cost required to install needed equipment and procedural routines. The right forms, the right computer system, the right personnel, and so on, are all things that must be anticipated *before* an irreversible commitment is made—or the firm may find itself in serious functional and economic trouble. Consider the airline which commits itself to a computerized ticket reservation system without a feasibility study appropriate to the size and importance of the investment. This is a dangerous road to travel. The surprises are certain to include many unhappy ones; perhaps, for example, the system will quickly become inadequate to process increased business volume. At this point the system becomes "infeasible," and the airline probably loses its competitive ability to instantly confirm plane reservations. There may be no economical solution to such a problem and the firm may fail as a result. The feasibility study offers management the opportunity to deal with such contingencies in advance.

A well-done feasibility study enables the firm to avoid six common mistakes often made in project work.

1. *Lack of top management support.* Top management has to understand and support subordinate managers in their efforts to improve the firm's operations. Feasibility studies also get subordinate managers directly involved in exploring and designing the systems they will have to live with in the future. Such involvement results in increased conscientiousness, which in turn enables top management to have more confidence in subordinates' plans. The result will be top management support for such plans.

2. *Failure to clearly specify problems and objectives.* The feasibility study can be directed toward defining the problems and objectives involved in a project, after management has given the group some understanding of what they would like to accomplish.

3. *Overoptimism.* A feasibility study can be conducted in an objective realistic manner to prevent overoptimistic forecasts. The study should be conservative in its estimates of improved operations, reduced costs, and so on, to ensure that all the firm's future surprises with a new system are happy ones.

4. *Estimation errors.* It is very easy to underestimate the time and money involved in the following areas:
   a. impact on the company structure;
   b. employees' resistance to change;
   c. difficulty of retraining personnel;
   d. system development and implementation;
   e. computer program debugging and running.

5. *The crash project.* Many managers do not realize the magnitude of work involved in developing new systems. Crash projects usually involve changing too quickly. A feasibility study might determine that a present system, with all its inadequacies, is superior to a crash project—assuming, of course, that the feasibility study itself is not run as a crash project.

6. *The hardware approach.* Firms have been known to get a computer first and then decide on how to use it. A feasibility study can identify, in advance, the uses to which the computer will be put and can identify the best computer for the job *before* any irreversible commitments are made.

TWO TYPES OF FEASIBILITY STUDIES

First, let us briefly examine the nature and purpose of two types of feasibility studies, then go on to describe just how each is carried out. For purposes of comparison (later in the text) we have labeled these Type A and Type B. There are, in fact, many variations, for there are few absolutes in systems analysis. The same analyst may perform two very similar analyses for two different companies in two entirely different ways. The differences may be caused by any number of factors, generally related in one way or another to the *people* with whom the analyst deals and the organizational framework in which the system must operate.

The *Type A* feasibility study is a study that either recommends for or recommends against a full system study. Its primary objective is to report on the time and costs involved in undertaking a full systems study, and to estimate the benefits that a full systems study would offer in terms of a new system. A Type A feasibility study is usually reserved for a Level One or Level Two problem, involving the entire firm or a major division of the firm.

The luxury of performing a Type A feasibility study should be reserved only for those projects involving extremely high cost, long development time, or those projects upon which the survival of the firm depends.

A *Type B* feasibility study is entirely different from Type A. It is a study to determine the feasibility of using a computer. A Type B study identifies the use to which a computer will be put and then identifies the best computer for the job. The study can involve the feasibility of upgrading to a larger computer, or the evaluation of a purchase or rental of a computer for the very first time.

HOW TO CONDUCT EACH TYPE OF FEASIBILITY STUDY

A feasibility study can be viewed as a "mini-systems study" because it is preliminary to and smaller than a full systems study. Any feasibility study (Type A or Type B) can be expanded to a full systems study if the estimated benefits are sufficient to warrant the systems study. If the feasibility study is expanded to a full systems study, the analyst will be expected to carry out the ten steps presented in Chapters 3 through 12 of this book. These are the chapters that explain, in detail, the procedures to be followed through to completion of a systems study.

Conducting a Type A Study

The first order of business in conducting a Type A study is to determine the nature and extent of the problem or problems as accurately as possible (see also Chapter 3). In defining the problem determine the

1. Subject, the central theme or topic of the study.
2. Scope, the range the study will encompass.
3. Objectives, what you will be trying to accomplish.

Arrive at some agreement, with the managers who will be directly affected by the study, on the subject, scope, and objectives.

Develop a To Do list that will serve as an outline or blueprint of what is going to be done in order to carry out the feasibility study.

The following is a recommended sequence of steps the analyst might follow when conducting a Type A feasibility study. The analyst should conduct a Type A study keeping flexibility in mind so that the sequence of steps can be changed if it becomes necessary.

1. Interview key personnel to get facts about the problems they are having, or the changes and improvements they would like to see. Try to put a finger on the *real* problem and its causes. This can often be very difficult since it is natural for people in certain positions to become quite defensive if they feel you are prying. Although it is often not possible, it is good to try not to criticize a person directly. Try, instead, to focus on events rather than on personalities. Of course, all criticism should be as constructive as possible. Also, try not to assume excessive authority or status. You are there to help the people, not hurt them—and you must show them by your manner that you are trustworthy of their time and confidence. It is best to interview the key line managers first. If a line manager has staff assistants, interview them next; and, finally, interview the supervisors and clerical employees. It is often best to work from the top down; but the order can be changed to fit the situation. As in all systems work, flexibility is the key word here. This order is not always possible, nor is it always desirable. In some cases, it is best to interview a key manager's staff assistant first in order to prepare for the interview with the manager. (Chapters 5 and 6 detail many of the things for which the analyst should stay alert during these interviews and techniques to be used; but of course the specific information sought must be dictated by the objectives of the feasibility study.)

2. Study any written procedures that exist relative to the subject of the study *before interviewing the clerical employees.* This not only will save a lot of time in learning the mechanics of the operations involved, but also will minimize interference with the firm's routine operations. Ask the area's manager about any such written procedures during your interview.

3. Try to learn the informal (unwritten) procedures while interviewing and observing the work flows. (You should develop a flowchart of these work flows

during this phase.) But, be careful, because informal procedures involve personalities, and the quickest way to loose cooperation is to get people upset with your tactics. Again, the analyst should try not to continually criticize PEOPLE. You are there to help, *not* to enhance your own status or feeling of importance. Be objective, but friendly. Never carry an attitude of superiority with you since it will show for what it is, and may wreck the project. Remember you probably will be dealing with these people for a while, and their cooperation will be *needed*. Start selling yourself as a helpful, friendly person right from the beginning since getting cooperation is half the battle in many projects.

4. Not only must the problem be defined, but its source must be determined. Redefine the problem in light of the facts obtained from this preliminary review. Reestablish the scope and objectives if necessary, and perhaps switch from a feasibility study to a systems study if the problem is found to be different from originally thought.

   Now organize your thinking about the subject, scope, objectives, and other conclusions, so an accurate picture can be presented to the line managers when you meet with them again. Remember they will be looking for your clear understanding of the situation, and will be making decisions about whether they can trust your interpretation and recommendations.

5. Meet again with the involved line managers and go over the findings and conclusions. Give them a rough estimate of solutions, timetables, benefits, and costs. Be sure to call their attention to any changes you think are necessary in the subject, scope, or other objectives. This step is to keep management informed and to get their tentative agreement and approval on progress to date. One sure way to hurt the success of any project is to change the definition, subject, scope, or objectives and not let management know until late in the study. Try not to spring too many surprises on line managers!

6. With management's concurrence, finish the study by analyzing and estimating the costs of performing a full systems study. Study the manpower requirements and the time that will be required to develop a new system or to modify the existing system. Be sure to include an estimate of the costs of any materials needed in the new system, such as desks, office space, calculators, supplies, and so forth (see Chapter 10). Also try to estimate the cost of operating the new system in terms of salaries, computer rental (Type B feasibility study required), or operating costs. Match all the costs against the potential benefits to be gained from the new system and present it to management in a feasibility study report. Recommend for or against undertaking a full systems study. (The details of the Feasibility Study Written Report are covered later in this chapter.)

It will sometimes happen that the feasibility study will actually turn out to be a full systems study because the original problem was underemphasized. Your conclusions plus the cost and benefit estimates then become the *actual solution* to the problem and not a recommendation for a full systems study to solve the problem.

Conducting a Type B Study

This is the most technically complex of all feasibility studies for its object is to determine the feasibility of installing a computer or upgrading the present computer to a more effective system. A Type B feasibility study closely approximates a full systems study. This is especially so if the study being performed is for a first-time installation. The plan detailed below encompasses the considerations required, whether the purpose is a new computer evaluation or a change in the existing computer system. The plan is easily adjusted to fit either case.

The job of planning for and converting to a computer system involves a large amount of preinstallation planning. The most obvious, but often overlooked, step is to establish a data processing organization (EDP operations). A strong manager, with data processing experience, should be put in charge as early as possible (before the computer arrives).

In the Type B feasibility study, probably the most important single item is the careful selection of a seasoned computer systems manager—one who can guide the firm around the many pitfalls that exist in computer planning. The manager selected should have experience with the conversion to equipment similar to the equipment appropriate for anticipated applications, and also should have experience managing the operation of such equipment over a span of time sufficient to have become familiar with the numerous inevitable problems that will occur in these operations. The firm should consider the salary of such a person as part of the cost of computerizing.

The first major item is to decide what applications are to be put on the computer. This really involves a study to determine priorities, that is, which applications should be put on the computer first, second, and so on. Economic return, internal structure between managers, and survival of the firm may be some of the criteria used to determine which applications get programmed first.

After deciding upon the applications, you must decide on the computer hardware itself. Choose the hardware to fit the applications, and not vice versa. Get modularity (see Glossary) and equipment that can provide the needed programming languages and other software.

Call in various computer manufacturers and have them evaluate the systems requirements and applications. This is done by way of the "Request for Price Quotations" (RPQ's). By soliciting proposals from vendors in this manner, you are assured of responses based on a uniform format and with similar assumptions. The manufacturers can then make proposals containing

1. Descriptions of equipment: speed, modularity, capacity.
2. Cost of equipment: purchase, lease, multiple shift cost, peripheral cost.
3. Maintenance costs and test time (which is time that the computer is not available).
4. Installation requirements and cost.
5. Manufacturers' assistance, for example, software, application programming, and training.
6. Bench mark tests: Test run one of the applications on the recommended computer to see how long it takes to run.
7. Delivery schedules: Consider imposing a penalty clause for late delivery.
8. Software: Is the software currently available?

In selecting the final equipment, you must evaluate the computer manufacturer's proposal (RPQ) as to

1. The appropriateness of this computer to the desired applications.
2. The speed of processing the applications during bench mark tests.
3. The potential growth of the firm. Does the proposed equipment have sufficient modularity and capacity?
4. System flexibility. Will it be able to handle probable future applications?
5. Availability of the programming languages the firm needs now, and will need in the future.
6. Software (computer control programs). Is it completely debugged and working satisfactorily?
7. Will the manufacturer assist in training and programming? At what cost?
8. Availability of service in case of breakdown.

Once the computer is *selected,* programming can begin. The programmers can use a similar computer at the manufacturer's plant, at another firm, or at a service bureau to test the programs they write. The point is that programming should begin *before* the arrival of the new computer.

Initial selection and education of personnel should begin immediately after the specific computer is chosen. Relevant past experience is extremely important. Some computer manufacturers offer instruction classes for data processing personnel, and teach everything from basic computer concepts to high-level technical familiarity with the specific equipment being installed.

Many fine texts are available which can be used to supplement the on-the-job training given by experienced personnel (hired to run the operation).

There will be the general task of educating the personnel in the other departments that are affected by the computer as to

1. New procedures and control practices.
2. Precise schedules for data input.
3. Changes in the arrangement of data on forms, in files, and so forth.
4. Adding or deleting data in the computerized system.
5. Use of the computer's output.
6. The reasons for all the changes.

The next item in the feasibility study is to plan how the conversion is to be handled after the computer arrives. A *parallel conversion* is one in which the old system is operated simultaneously with the new system for at least one cycle using current data. If one is satisfied that the new computer system is performing satisfactorily, the old system can be discontinued. Parallel conversions can be costly, but they can save the firm the disastrous results of a computerized system that fails, loses vital data, and causes serious economic problems.

A *one-for-one changeover* is a simultaneous installation of the new system and removal of the old system. This may take place at one time or it might be phased-in logically. One method of protecting the firm against loss of data during the conversion is to use emulation. Emulators process the old programs by the old procedure using current data, that is, emulators enable a new computer to execute instructions from a previous computer. It should be noted here that not all computers can use emulators. In addition, there are other disadvantages to be considered: the manufacturer must supply the emulator, and if it is a "hard wired" emulator the normal batch stream of jobs is interrupted. Also, a "software" emulator lowers the throughput of the batch job stream. Whether one chooses a parallel conversion (run both systems simultaneously) or a one-for-one conversion with emulation, the point is that both systems or emulation are continued until everyone is satisfied that the new programs are accurate and that the old system data and the new system data can be reconciled.

The next planning phase is the physical installation planning.

1. Decide on the location of the computer.
2. Consider special electrical circuits or plumbing.
3. Check air conditioning duct work, water supply, and capacity required. Often the normal building air conditioning is *not* sufficient.
4. Check floors: loading capacity and the need for raised floors.
5. Decide on floor space partitioning: computer room, offices and soundproof areas for the keypunch machines.

6. Provide storage facilities for disks, tapes, program documentation, and other typical items.
7. Check miscellaneous items, for example, furniture, telephones, and electric wall sockets.

Computer operator training must be completed before the computer is installed. A computer library should also be established to control the movement of magnetic tapes and disk packs and to protect them from fire, humidity, and contamination. Layout flowcharts, PERT charts, and Gantt charts will help in controlling the location, time, and sequence of these events (charting is covered in Chapter 13).

FEASIBILITY STUDY WRITTEN REPORT

The feasibility study written report is the documentation of the feasibility study. It is the medium by which you tell management what the problem is, what you have found its causes to be, and what you have to offer in the way of recommendations. The report should

1. Define the problem in such a way that it clearly demonstrates your understanding of the problem.
2. Clearly describe the subject and scope. List the areas included *and excluded.* State the level of detail pursued (see Chapter 3), and explain modifications to the original study plan.
3. State the objectives and whether they were met or not. Are the objectives technically feasible? Are they economically feasible? And are they operationally feasible? Will the objectives, if met, result in a solution or computer installation which will fit within the current organization structure or will reorganization be needed?
4. Point out special attention areas such as unusual situations or interrelations between problems.
5. Describe the entire system or study in detail. (This is the body of the report. It is here that you must put your thoughts across.) You will want to include a description of both the existing system and the proposed system of the Type B feasibility study.
6. List all of the economic cost comparisons and benefits (see Chapter 10).
7. State recommendations clearly. Explain the logic behind the recommendations.

8. Provide a suggested time table for implementation of the recommendations, with appropriate milestones identified.

9. Have an appendix that includes any pertinent flowcharts, graphs, pictures, floor plans, or layouts not used elsewhere in the report.

PROCEED . . . IF REQUESTED

If management accepts the recommendations, then you should proceed to carry them out according to the principles discussed earlier in this chapter.

1. Type A: perform a full systems study.
2. Type B: implement the new computer system.
3. Drop the whole project if you have recommended *against* further study—and if management agrees with you.

SELECTIONS FOR FURTHER STUDY

The parameters of a feasibility study may vary tremendously from one job to another. Since this is true, it is difficult to pinpoint precise library tools. Oddly enough, there still is not much written that concentrates specifically on the feasibility study. One older volume, *EDP Feasibility Analysis* by Massey and Heptonstall may still be located in some libraries. Clifton's *Project Feasibility Analysis* is one of the few newer books on the subject. When searching for information on feasibility analysis, one normally must take a roundabout approach by looking under such subjects as cost-effectiveness, electronic data processing, computer installation, or some other topic related to profitability of new projects.

If it is a Type A study, your library usage may be heavy. You may, for example, have to locate information on the industry as a whole. If in involves a technique, a thorough search of the patent literature may be in order. Or, if a change in production methods seems to be needed, you may have to find out what other firms have tried and whether the methods tried were successful or not.

A Type B study is almost certain to involve a literature survey of some sort. You may need to find general books on computers if your company is considering its first computer installation. These might include the works of Sanders, Condon, Stimler, or Kanter. Two other competitive publications, the Auerbach series and the Datapro series, can provide much useful information to the data processing systems analyst. Each service provides subscriptions to series on such topics as minicomputers, terminals, peripherals, and other computer hardware subjects. Each also provides special reports such as Datapro's *How To Select and Install a Small Business Computer, How To Analyze Your Data Communications Needs,* or *All About Remote Batch Termi-nals.* Auerbach publishes similar reports such as *Computer Performance Evaluation, Designing Transactions and Controls, Guidelines for the Project Manager in the Development of an On-Line Order Entry System,* or *54 Ways To Reduce DP Costs.* The Auerbach and Datapro series provide the analyst with information on what hardware is available, what purpose it serves, and how much it costs.

Trade literature, which is usually housed either in the corporate library or in the purchasing department, also may be consulted. The term "trade literature" as used here refers to the manuals and sales brochures issued by manufacturers. It may or may not include specifications. Some trade literature may be your best source of information. On the other hand, there is the type that is designed to sell a product, and you need to be critical in examining and using these materials since the manufacturers may oversell their products.

Since cost estimating is a major part of any feasibility study, the library can be especially helpful if it has a representative collection of the many books published on business economics, cost accounting, cost control, and cost effectiveness. In an industrial situation, Maynard's *Industrial Engineering Handbook* is particularly useful.

1. AUERBACH PUBLISHERS, INC. *Auerbach Computer Technology Reports* (formerly *Auerbach Standard EDP Reports*). Pensauken, N.J.: Auerbach Publishers, Inc., 1962– . (Updated monthly.)

2. CLIFTON, DAVID S., JR., and DAVID E. FYFFE. *Project Feasibility Analysis: A Guide to Profitable New Ventures.* New York: John Wiley and Sons, Inc., 1977.

3. CONDON, ROBERT J. *Data Processing Systems, Analysis, & Design, 2nd edition.* Reston, Va.: Reston Publishing Company, Inc., 1978.

4. DATAPRO RESEARCH CORPORATION. *Datapro Research Reports* series. Delran, N.J.: Datapro Research Corporation, 1969– . (Updated monthly.)

5. DATAPRO RESEARCH CORPORATION. *All About Remote Batch Terminals.* Delran, N.J.: Datapro Research Corporation, November 1976.

6. DATAPRO RESEARCH CORPORATION. *How To Analyze Your Data Communication Needs.* Delran, N.J.: Datapro Research Corporation, October 1975.

7. DATAPRO RESEARCH CORPORATION. *How To Select and Install a Small Business Computer.* Delran, N.J.: Datapro Research Corporation, April 1977.

8. DUGGAR, JOHN. *Computer Performance Evaluation.* Philadelphia, Pa.: Auerbach Publishers, Inc., 1977.

9. GITOMER, JERRY. *Designing Transactions and Controls.* Philadelphia, Pa.: Auerbach Publishers, Inc., 1977.

10. GITOMER, JERRY. *54 Ways to Reduce DP Costs.* Philadelphia, Pa.: Auerbach Publishers, Inc., 1976.

11. HARPOOL, JACK D. *Business Data Systems: A Practical Guide.* Dubuque, Iowa: William C. Brown Company, 1978.

12. KANTER, JEROME. *Management Guide to Computer System Selection and Use.* Englewood Cliffs, N.J.: Prentice-Hall, Inc., 1970.

13. KANTER, JEROME. *Management-Oriented Management Information Systems, 2nd edition.* Englewood Cliffs, N.J.: Prentice-Hall, Inc., 1977.

14. LETTEER, CHARLES E. *Guidelines for the Project Manager in the Development of an On-Line Order Entry System.* Philadelphia, Pa.: Auerbach Publishers, Inc., 1975.

15. LUCAS, HENRY C., JR. *The Analysis, Design, and Implementation of Information Systems.* New York: McGraw-Hill Book Company, 1976.

16. LUCAS, HENRY C., JR. *Computer-Based Information Systems in Organizations.* Palo Alto, Calif.: Science Research Associates, 1972.

17. MASSEY, L. DANIEL, and JOHN HEPTONSTALL. *EDP Feasibility Analysis.* Braintree, Mass.: D.H. Mark Publishing Company, 1968.

18. MAYNARD, HAROLD B. *Industrial Engineering Handbook, 3rd edition.* New York: McGraw-Hill Book Company, 1971.

19. SANDERS, DONALD H. *Computer Essentials for Business.* New York: McGraw-Hill Book Company, 1978.

20. SANDERS, DONALD H. *Computers in Business: An Introduction, 3rd edition.* New York: McGraw-Hill Book Company, 1975.

21. STIMLER, SAUL. *Data Processing Systems: Their Performance Evaluation, Measurement & Improvement.* Moorestown, N.J.: Stimler Associates, 1974.

QUESTIONS

1. Name two alternative objectives of the feasibility study.
2. When should a feasibility study be conducted?
3. What is the primary objective of a Type A feasibility study?
4. What two things does a Type B feasibility study identify?
5. When a computer installation is being implemented, personnel in other de-

partments that will be affected by the computer need to be educated. Name three areas that should be covered in this task.

6. Define one-for-one changeover.
7. Name four factors to consider in the physical installation planning for a computer.
8. Identify four items that should be covered in the feasibility study written report.

SITUATION CASES

CASE 2-1

When the plant vice-president authorized a feasibility study for the installation of automatic data processing equipment, it was decided that only a computer capable of meeting both the scientific and commercial interests of the plant should be considered.

A data processing committee, composed of the comptroller, an assistant to the comptroller, and the assistant manager of systems design, was appointed to handle the responsibility of completing the feasibility study. Each of the firm's six departments designated one or two people to work with the committee. A scientific subcommittee composed of a solid-state physicist, a mathematician, and an operations researcher was formed to advise the data processing committee on scientific requirements.

Initially a **Type A** feasibility study was augmented. First, the **subject** of this study was the use of electronic data processing equipment at this plant. Second, the **scope** of this study was to encompass two major areas.

A. Scientific Areas. The scientific applications should include the solution of problems in molecular theory and solid-state physics, calculations on problems resulting from the use of a nuclear reactor, and the feasibility of using mathematical relationships to correlate physical properties of metals with process histories.

B. Business Areas. (1) The initial business applications are in the areas of production control, cost accounting, top management reporting, material inventory control, and two related systems, payroll and personnel records. (2) Subsequent areas include the balance of top management reporting, aspects of general accounting not included in the initial application, fixed asset accounting, and additional budgeting systems. (3) Other applications include material and time standards and literature searching in connection with scientific reporting.

Third, the *objectives* which were to be achieved were stated.

A. Scientific areas.

1. To enable the scientific staff to solve problems too difficult to resolve without computer facilities.

2. To permit more rapid problem solving and more efficient utilization of scientific personnel.

3. To permit the rapid tabulation and analysis of large volumes of scientific data.

4. To provide researchers more time to keep abreast of the latest scientific developments, by allowing the computer to do the time consuming data manipulations.

B. Business areas.

1. To make available new management control data as well as improved processing in automatic billing, payables, and disbursements.

2. To reduce the cost of data processing through reduction in personnel cost and through processing more data per unit of time and per dollar of cost.

3. To improve management control through timely and accurate reporting.

4. To provide increased flexibility in modifying EDP procedures.

5. To provide the capability of expansion with a relatively small increase in cost.

6. To further integrate the system through use of a computer.

7. To provide the ability to solve mathematical or logical sequences too complex or lengthy for other types of data processing.

After the Type A study was properly completed, it was decided to go to a computerized system, and a Type B study was initiated.

Two applications are to be put on the computer in the scientific and business areas.

A. Scientific area.

- Elimination of the time spent on repetitive and monotonous computations.

- Increasing the researcher's ability in certain areas, by making available to him programs and solutions developed by competent persons with similar interests in other organizations.

B. Business area.

- Internal and external reporting may be derived simultaneously (in other words *parallel conversion* will take place) from the integrated files established.

- Management reports will be available at the required time and all duplication of effort will be eliminated.

- The frequency of issuance of management reports will be as desired by management.
- Manual methods will be reduced to only those procedures which must be accomplished by human beings.

Selection of the computer hardware itself was decided upon. Invitations to bid were sent to six manufacturers.

The model selected met all required specifications and was capable of accomplishing economically and efficiently both the business and scientific data processing requirements. The selected system was extremely flexible and could readily expand or contract to meet changing requirements. All critical reporting requirements could be accomplished on a one-shift basis. Expected delivery time was within acceptable limits even in a "worst case" situation.

Selection and education of personnel began immediately after the computer was selected. It included organization, staffing and training.

- The Automatic Data Processing System Organization was to be established as a branch of the comptroller's office.
- Technical training of personnel would be accomplished by the computer manufacturer at the plant and at the manufacturer's facilities.
- Upon completion of the necessary training, personnel of this installation would begin programming the selected applications.
- As each application was completed, the program was to be tested and debugged at the manufacturer's service center.

The site preparation phase was completed next.

- The proposed site for the computer operation was located on the first floor of the main administration building.
- The plant facilities department prepared a cost estimate for the inside site preparation including the air conditioning system, sprinkler system, plumbing, raised flooring, painting, heat absorbing glass, excavation, and electrical wiring.

QUESTIONS

1. What evidence was there that top management supported the feasibility studies?
2. What factors in the case justified doing both a Type A and Type B feasibility study?
3. Identify any important items the Type B feasibility study did not include.

The Billing Department of a company has been slow in preparing the bills for mailing to its customers. The company's customers are beginning to complain because they are not able to meet the discount periods. The discount period begins when the goods are delivered, not when the bill is received. The sales staff believe that the delay in billing, which results in the lost discounts, is causing them to lose sales.

The Systems Department has been asked to perform a Type A feasibility study. The analyst who is assigned to this study will recommend for or against a full system study. The Systems Department manager assigned this study to a junior analyst who had been with the company for about one year. In assigning the study the manager explained that the subject was the preparation of the bills and that the scope of the study would cover the Billing Department. He also defined the objective of the study to be a recommendation as to whether or not a full systems study should be conducted.

The analyst decided that she would first make a "To Do" list to aid in conducting the study. Her list looked like this.

1. Interview clerical help in the Billing Department.

2. Interview key management in the Billing Department.

3. Read the written procedures.

4. Summarize all information into a report.

5. Discuss summary with Billing Department management.

6. Make cost comparisons.

7. Submit feasibility study report to Systems Department.

Then the analyst proceeded to the Billing Department. She interviewed the clerical employees who were involved with the billing procedure. From what she was told in these interviews, she learned how the actual billing procedures operated.

The next step the analyst took was to interview the management of the Billing Department. The information that she received through these interviews told her how the managers thought the billing was being handled.

The analyst then returned to her desk and began to read the procedure manual on the billing procedure. The manual presented the way that the billing operation was formally intended to operate.

A summary of the interviews and procedure manual was the next thing that the analyst made. The summary was then discussed by the analyst and the Billing Department management. The Billing Department management gave their concurrence for the analyst to make any cost comparisons required.

After making the cost comparisons the analyst added to the report her conclusions and recommendations. The feasibility study written report was then turned in to the Systems Department manager.

QUESTIONS

1. Identify the major mistakes the analyst made while conducting her study.
2. What else should the analyst have done?

CASE 2-3

Con-Can is one of the leading producers of metal containers. Over the years the company has kept up with their competitors in all aspects of operations.

During the last few years the paperwork connected with running the operation has increased. This has been caused by an increase in sales volume and also by new government regulations that require more current records to be kept. In order to meet these requirements the company considered installing a computer at their Los Angeles branch.

Because of the high cost of installing and operating the equipment, top management decided to have a system analyst perform a Type B feasibility study.

On February 4, the system analyst began his study. First he called a meeting of the various department supervisors who would be directly affected by the change from a manual system to a computer system. The departments represented were Sales, Accounting, Production Control, and Personnel.

The meeting was held very informally. Each supervisor was asked to express what they felt the computer's most important function should be. After the meeting the analyst determined the following priorities based on the supervisors' comments. The highest priority item to be run on the system would be payroll followed by inventory control. Other items would be added later as needed.

Second, the analyst suggested that Personnel begin taking applications for a data processing manager of the new department. It also was suggested that the selected applicant should have some practical experience with computers in addition to being able to manage people.

The third step carried out was to begin educating the personnel in the various departments affected by the new system. Group meetings were held in each department to explain the various changes that were to be made. Items discussed were

1. New procedures that would be needed.
2. Precise schedules for inputting data.
3. How the data on forms would be rearranged.
4. How a person would make corrections to data already in the system.
5. How the computer printouts would be read.
6. Why the changeover was necessary.

The fourth step was to decide if the computer should be rented, purchased, or leased. Before asking for bids from the different computer companies, top management and the analyst devised a list of requirements and applications that the firm had in mind for the new equipment.

Each bidder was required to evaluate its equipment in relation to Con-Can's requirements and applications. The manufacturers' evaluations were to be returned to the firm accompanied with their bid. After reviewing the bids and evaluations, it was decided to lease the ABC 300. This decision was based primarily on the financial aspects of the equipment, rather than its ability to meet the requirements of Con-Can's users.

Two weeks after Con-Can began taking applications for manager of the new department, Personnel hired Tom Snodgrass. Mr. Snodgrass had received his B.S. in data processing from NBA University and had been working as a program troubleshooter for Peanuts, Inc. during the last four years.

With the impending arrival of the computer, two final conditions needed to be met. First, the layout of the computer room needed to be finished. The raised floor, storage library, and room dividers had to be installed. Second, since Con-Can now knew the applications and the language used by the computer, they had to begin preparing programs for the new system.

QUESTIONS

1. What other factors should have entered into the equipment decision, besides cost?
2. Identify any additional areas that should have been considered in conducting the study.

THE SYSTEM STUDY ITSELF

Part Two of this book is devoted to a step-by-step approach to conducting a full systems study. These steps depict the overall development of the system design process. It is divided into ten separate steps or phases.

CHAPTER 3

PROBLEM DEFINITION

The object of this chapter is to teach a systems analyst how to define a problem. Where problems come from, how to define them, and how to present the findings to management are covered. A successful analyst will always start by defining the problem at hand prior to working toward its solution.

HOW DO YOU KNOW YOU HAVE A PROBLEM?

The word "problem" is defined as "a question proposed for solution or consideration." All too often a firm finds out it has a problem only when things start going wrong. No one knows the problem exists until something fails or someone brings the problem to the attention of management.

But what often appears to be the problem itself will only be a *symptom* of the *real* problem. To be successful, the analyst must be able to tell the difference between a symptom and a problem. A symptom is best described as a noticeable condition caused by the problem. For example, it might seem that the problem is not enough office space for the employees. But while defining the problem more exactly, one might learn that the *real* problem is the absence of a records disposal system. In this case, the lack of office space is a symptom; the lack of a records disposal system is the problem. In many cases the symptoms of a problem gradually become evident. Because they develop slowly and are difficult to spot, the problem and symptoms may not be connected until the situation has deteriorated noticeably.

You may have heard the old saying that a problem defined is half solved. In systems work one learns the truth and value of this saying many times over.

THE PROBLEM-REPORTING MACHINERY

The problem-reporting machinery is an expression used to describe the method by which the systems analyst learns of problems. Problem information in a business firm often flows in the manner shown in Figure 3-1.

The problem-reporting machinery may involve a verbal message given to a systems analyst, or it may be submitted on a written form similar to the one shown in Figure 3-2.

The systems analyst often faces a constant barrage of problem reports from both internal and external sources. In general these sources are

From the external environment	*From the internal environment*
Management consultants	Data processing
Professional associations	Financial records
State and federal agencies	Management
Crediting agencies	Informal organization
Community relations	Formal line organization
Outside auditors	Employees
Competition	Company auditors
Customers	Systems department
Unions	Budgets
Regulatory agencies	Compliance auditors

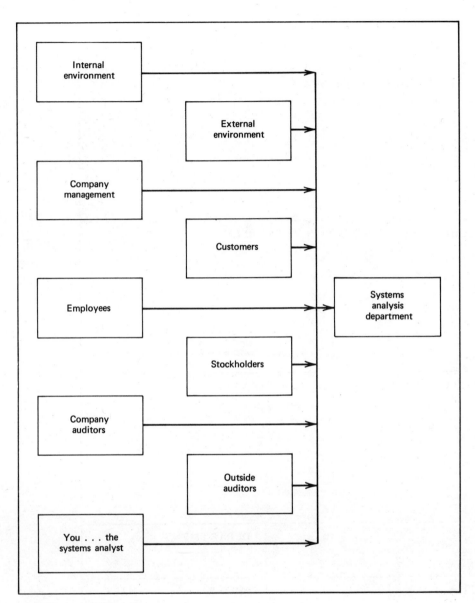

FIGURE 3-1 Problem flow.

PROBLEM REPORT FORM

STATEMENT OF THE PROBLEM:

INCIDENTS SURROUNDING THE PROBLEM:

WHY YOU ARE REPORTING THIS PROBLEM:

NAME	DEPARTMENT	
TITLE	PHONE	DATE

FIGURE 3-2 Typical problem report form.

WHERE PROBLEM SIGNALS COME FROM

The systems department must be sensitive to changes in the firm's operating environment, so it can anticipate problems and deal with them early. A systems department that is right on target will sense problems even before they are reported officially. The alert systems department will constantly monitor and review activities throughout the firm, looking and listening for problem signals. For example,

Activities to monitor/review

Relocation of work areas

New equipment installation and use, such as computers

New system implementation and use

Product change

Management policy decision

Customer, supplier, or employee feedback

Employee morale

Project budgets

Number of people required for tasks

Problem signals

Work or processing is too slow

Too many people required for a task

Too few people performing a task

Indirect problem reports for managers

Delays in new equipment installation and use

Delays in new system implementation and use

Customer, supplier, or employee complaints

Dwindling profits or market share

Low morale evidenced by high employee turnover

Budget slippage on projects

DEVELOP A TO DO LIST

It is often useful for the analyst to develop a habit of utilizing a To Do list to use as an aid in task planning. The To Do list is a common planning method which is nothing more than a variation of the old shopping list. It simply helps the analyst plan in an organized manner. Although there are many variations, two types of To Do lists are especially useful.

The first type is a list of all the tasks to be done tomorrow. It is prepared late in the work day. Any tasks that went unfinished today are the items that should come first on tomorrow's To Do list. Then, any new tasks are added to the list. Each day a new To Do list is prepared for the following day.

The second type of To Do list is an outline of how you are going to carry out the project at hand—in this case, the problem definition study. List in order the tasks that have to be accomplished and how they are going to be done.

Following are some of the areas that are most important to the problem definition study To Do list.

1. Have preliminary discussions with management and attempt to identify problem areas.

2. Study written procedures and attempt to identify procedural problems.

3. Observe the current system.

4. Interview the personnel involved and attempt to identify organizational problems.

5. Gather other data, for example, facts and figures.

6. Evaluate the findings.

7. Come to a conclusion on the definition of the problem; decide not only what the problem is, but also what it is not.

8. Rediscuss the findings with management.

9. Issue Problem Definition Written Report.

These major headings will be discussed more completely in the next section "Define the Problem . . . Subject . . . Scope . . . Objectives." The list should have subheadings and sublists which specify the particular departments, people, records, equipment, and so on, that you intend to see. The To Do list is the analyst's written plan of action. The better the plan, the more likely the analyst is to arrive at a correct problem definition.

The To Do list can be changed to the situation or to the personal liking of the analyst. In any event, if the analyst approaches people in an efficient and organized manner, it will be recognized, and the needed cooperation might be more easily forthcoming.

DEFINE THE PROBLEM . . . SUBJECT . . . SCOPE . . . OBJECTIVES

Let us suppose that you as an analyst have detected the symptoms of a problem, and are ready to begin the all-important phase of concise problem definition. In this phase you must define three important things.

First, define the *subject*. The subject is the topic or central theme of the problem study. When the subject is clearly defined, you automatically have a title for the problem definition report. For example, the subject of a problem definition study might be "employee payroll complaints."

Second, define the *scope*. The scope is the area or range that the study will encompass. Sometimes it is limited by time, dollar resources, or organizational boundaries. It is always limited by the subject. If the subject is not adhered to, your frame of reference will be lost and you will be unable to focus on the problem area indicated by the subject. For example, if the subject is "employee payroll complaints," then the scope should include only the relevant activities that seem to be involved. It should not be allowed to wander into other unrelated complaints.

Third, define the *objectives*. The objectives are the things you will be trying to accomplish. In our example, the subject is "employee payroll complaints." The scope includes payroll-related complaints only. The objectives, then, must fit the subject and scope. The objectives should not be too ambitious nor, on the other hand, too limited. In short, they must express exactly what things you intend to accomplish in the problem definition study. Two objectives in our "employee payroll complaints" study might be

a. Review timekeeping system methods and forms for adequacy.

b. Determine whether check writing system (manual or computerized) is up-to-date relative to workload and techniques.

In order to define the subject, scope, and objectives, hold preliminary discussions with whoever raised the problem (interviewing will be discussed in Chapter 6) and with other key management personnel involved with the problem. Get their interpretations of what the problem is, what needs to be done, and what the subject, scope, and objectives should be. Work to arrive at common definitions of the subject, scope, and objectives.

Read any pertinent written procedures that relate to the subject of the problem definition study. Ask the key management personnel which written procedures pertain to the study.

Go out and observe the system in action. Talk with the employees who operate it. Interview the personnel who are directly affected by the problem. Try to get as many opinions as possible. Study any written procedures that are available. Gather any other data such as quantities, times, amounts, numbers of failures, and so on, that may be significant. Throughout this process, any important information should be recorded for future reference. These notes might include names of people who appear most knowledgeable and from whom further help might be expected, anything historically significant such as other approaches that were tried unsuccessfully, or any procedures that appear to be unnecessary or outmoded.

After all the preliminary data has been gathered, analyze the findings in order to arrive at a precise definition of the problem. Then go back and again talk with the people who raised the problem or the key management of the area in which you are working to advise them of the findings. The point is that you should discuss the findings with the proper management *before* issuing the final problem definition report. The evaluation and design approach is further detailed in "Pattern of Systems Design . . . a Step-by-Step Approach" in Chapter 8.

VERBAL PRESENTATION

After you have successfully defined the problem, management may decide that a verbal presentation of the general findings is appropriate. If so, the following discussion may help you be more effective in getting your ideas across.

The first step before making a verbal presentation is to make an outline of the talk. This outline should follow the outline of the written report you intend to make on the problem study. Remember that management does not have time for lengthy and dull presentations, so make the talk short and meaningful.

Summarize how you came to be aware of the problem, its subject, scope, objectives, and finally the definition of the problem. Outline the departments and/or important people interviewed (so management will feel free to point out anyone missed). As precisely as possible, note the findings, both pro and con. Describe any proposed changes and how successful or unsuccessful you feel they would be and why. You will probably want to include known cost factors. Any impending changes of standards, for example, also should be noted to give management a chance to postpone a decision until the needed information arrives. You also may point out whether any need exists for outside consultants. Above all, demonstrate a clear understanding of everything you are talking about. They will be looking for it. A more detailed approach to verbal presentations is included in Chapter 11.

PROBLEM DEFINITION WRITTEN REPORT

The written problem definition report is a short report that sets the stage for an advanced feasibility study or a major systems study. It is mandatory that this written report on the problem study be made, not only for communication now, but also for future use in further studies. Again we must stress that the analyst should demonstrate a clear understanding of the problem. Failure to do so will cause management to lose confidence in the analyst's ability to handle the situation.

The outline for the verbal problem definition report may be the basis for the written problem definition report.

The written problem definition report outline will vary with the situation, but in general it should contain the following.

1. Introduce the problem: subject, scope, objectives.
2. Explain any modifications to the original plan.
3. Indicate which areas of business were included or excluded and describe the level of detail pursued in the study (see next section).
4. Clearly define the problem.
5. State which objectives were met and which ones were not, and why not.
6. Point out interrelation between problems, or any unique situations.
7. Make recommendations.
8. Explain your logic if need be.
9. Use any charts, graphs, pictures, floor plans, or layouts required.

OTHER CONSIDERATIONS

There are many pitfalls that can be the undoing of a systems analyst. Of these, the following four tend to cause the most damage.

1. Improper problem definition by management accompanied by the lack of a problem redefinition by the analyst. (The world's best solution to the wrong problem has a net value of zero.) It is in the problem definition/redefinition phase that the analyst and management agree to agree on the subject, scope, and objectives.
2. Excessive ambition, leading to the consideration of the analyst's favorite alternatives only. The analyst tries to force a favorite solution as the last word on the situation.
3. A solution oriented more toward the technical peculiarities of a computer than to the objectives of the people who will use the system. If a system fits the computer but not the users or the organizational framework, the system will not be the solution to your problem. It will be a new problem.
4. Straying from the objectives of the problem study and becoming preoccupied with techniques or equipment for their own sake. Do first things first. Stick to the objectives.

There are also three general *levels of detail* which the analyst may pursue in a problem definition study, feasibility study, or systems study. We call them

1. *Prob-view.* Only general purpose items are documented, such as major decision points, key volumes or quantities, and critical events.

2. *Feas-view.* A more specific, detailed documentation of all decision points, most volumes or quantities, and important events.

3. *Sys-view.* The most extensive level of documentation. It includes all decision points, all volumes or quantities, all events, and any other incidents that appear relevant to the system study.

The three levels of detail listed above can be applied to a problem definition study, a feasibility study, or a major systems study. It is just a matter of *how much detail* is required for the study you are doing. For example, programming can be done easily from a sys-view level systems study, but it cannot be done from a prob-view level systems study without further investigation by the systems analyst or the programmer.

SELECTIONS FOR FURTHER STUDY

When all around you are griping, you *know* there is a problem! But seeing the problem and solving it are two different things. The solution to some problems is obvious; in others, even finding out what the problem is, is difficult. Creative thinking can help you do a better job. Reading some books on problem solving or creativity will show you how others have dealt successfully with the same problems you face as an analyst.

Other subjects in this chapter also relate to personal improvement. If you have always relied on memory to tell you when to get things done, then you will not be used to planning by list. Books on personal planning may be of assistance in this area. If you have never made a *formal* verbal presentation, you may be in for some surprises. Even the most confident person may freeze when called upon to "make a speech." If your presentation is not well organized or if you do not have "stage presence," your case may be lost.

1. ADAMS, JAMES L. *Conceptual Blockbusting: A Guide to Better Ideas.* San Francisco, Calif.: W.H. Freeman & Company, 1974.

2. BLISS, EDWIN C. *Getting Things Done.* Des Plaines, Ill.: Bantam Books, Inc., 1978.

3. BROWN, LEROY. *How To Make a Good Speech.* New York: Frederick Fell, Inc., 1964.

4. DEBONO, EDWARD. *Five Day Course in Thinking.* New York: Basic Books, Inc., 1967.

5. DEBONO, EDWARD. *New Think: The Use of Lateral Thinking in the Generation of New Ideas.* New York: Basic Books, Inc., 1968.

6. FELDMAN, EDWIN B. *How To Use Your Time To Get Things Done.* New York: Frederick Fell, Inc., 1968.

7. FRIANT, RAY J., JR. *Preparing Effective Presentations, revised edition.* New York: Pilot Books, 1978.

8. GREGORY, S.A. *Creativity and Innovation in Engineering.* Levittown, N.Y.: Transatlantic Arts, Inc., 1973.

9. HARPOOL, JACK D. *Business Data Systems: A Practical Guide.* Dubuque, Iowa: William C. Brown Company, 1978.

10. HAYNES, JUDY L. *Organizing a Speech: A Programmed Guide.* Englewood Cliffs, N.J.: Prentice-Hall, Inc., 1973.

11. HEGARTY, EDWARD J. *How To Write a Speech.* New York: McGraw-Hill Book Company, 1951.

12. HIMMELBLAU, D.M. *Decomposition of Large Scale Problems.* New York: Elsevier-North Holland Publishing Company, 1973.

13. JACKSON, K.F. *The Art of Solving Problems.* New York: St. Martin's Press, Inc., 1975.

14. JENSEN, RANDALL W., and CHARLES C. TONIES. *Software Engineering.* Englewood Cliffs, N.J.: Prentice-Hall, Inc., 1979.

15. KAUFMAN, ROGER. *Identifying and Solving Problems: A Systems Approach.* La Jolla, Calif.: University Associates, 1976.

16. LAIRD, DONALD A. *Techniques for Getting Things Done.* New York: McGraw-Hill Book Company, 1947.

17. MAIER, NORMAN R. *Problem Solving and Creativity in Individuals.* Belmont, Calif.: Brooks-Cole Publishing Company, 1970.

18. MICKEN, RALPH A. *Speaking for Results.* Boston: Houghton Mifflin Company, 1958.

19. MORGAN, JOHN S. *Improving Your Creativity on the Job.* New York: American Management Association, 1968.

20. NIZER, LOUIS. *Thinking on Your Feet.* New York: Liveright Publishing Corporation, 1963.

21. ROBINSON, DAVID. *Writing Reports for Management Decisions.* Columbus, Ohio: Charles E. Merrill Publishing Company, 1969.

22. SAMSON, R.W. *Problem Solving Improvement.* New York: McGraw-Hill Book Company, 1970.

23. SANDERS, DONALD H. *Computers in Business.* New York: McGraw-Hill Book Company, 1975.

24. SHEFTER, HARRY. *How To Prepare Talks and Oral Reports.* New York: Pocket Books, Inc., 1971.

25. STEELE, FRITZ. *Consulting for Organizational Change.* Amherst, Mass.: University of Massachusetts Press, 1975.

26. TARR, G. *Management of Problem-Solving: Results from Productive Thinking.* New York: Halsted Press, 1973.

27. YECK, JOHN D. *How To Get Profitable Ideas for Creative Problem Solving.* New York: McGraw-Hill Book Company, 1965.

QUESTIONS

1. What is the difference between a problem and a symptom?
2. Name four sources of problem reports from both the internal and external environments.
3. Name five problem signals (symptoms).
4. Describe the two types of To Do lists.
5. When you are conducting a problem definition study, what key things must be defined?
6. What is the first step that must be taken before making a verbal presentation?
7. Name four points that should, in general, be contained in the outline for the written problem definition report.
8. How does the *prob-view* differ from the *sys-view*?

SITUATION CASES

CASE 3-1

ABC Corporation manufactures a broad line of children's toys. As the result of a recent study, the production manager became aware that the level of factory output had dropped to a level 25% below last year's monthly average. The production manager requested that a systems analyst be assigned to study the situation and identify the problem(s). Because of the upcoming holiday season when peak demand for toys is experienced, the manager requested that the analyst complete the work within three weeks.

The analyst first visited each of the production departments including (1) Raw Materials, (2) Manufacturing, (3) Inspection, (4) Assembly, and (5) Packaging. In each department, the analyst interviewed the department supervisor and other key employees.

During the review of the Assembly Department, the analyst noted that although the supervisor felt there were no problems, several employees indicated that conditions were deplorable and work had slowed down considerably. As a result, the analyst decided to spend some additional time studying this area and did not plan to waste time looking at the Packaging Department.

At the end of the analyst's review of the Assembly Department, he concluded that the supervisor was not relating well to his employees and that the lighting and ventilation in the area were unsuitable. In an effort to speed up the solution of the problems noted, the analyst confronted the supervisor with his findings. The analyst and the supervisor engaged in a verbal battle, whereupon the production manager was asked to mediate. When asked for an explanation, the analyst reviewed his findings. The Assembly Department supervisor challenged the analyst's facts and stated that the real problem was in the Packaging Department where the equipment was old and where breakdowns were occurring more frequently.

QUESTIONS

1. Discuss the scope of the analyst's work as it relates to problem definition activities.
2. Identify any steps the analyst omitted that usually would be followed during problem definition.

CASE 3-2

The Hardwood Furniture Manufacturing Company had been receiving numerous inquiries and complaints about late deliveries to retail showrooms. In order to resolve this situation, the company called in a consulting systems analyst. Management requested that the analyst define the problem within three weeks so that corrective action could be taken as soon as possible. As a first move, the analyst took steps to carefully clarify the subject, scope, and objectives of the problem definition study.

The analyst held preliminary discussions with line management and determined that completion and shipment of dinette chairs was substantially behind schedule. The analyst visited the two departments involved in making and shipping the chairs to observe work flow. The analyst noted that various materials being used to assemble the finished chairs were kept only in limited quantities due to a lack of space. In addition, she noted that the machine used to produce the chair legs had broken down several times during her observations. The analyst concluded that more space was needed to speed up the assembly process and that either new equipment was needed or the existing equipment should be thoroughly overhauled. The analyst prepared a written problem definition report and presented it to management.

QUESTIONS

1. Why should the analyst have reviewed the written procedures in the departments she visited?
2. What other step could the analyst have followed in conducting her study?
3. What could the analyst have done as an aid in task planning?

CHAPTER 4

PREPARE AN OUTLINE OF THE SYSTEMS STUDY

The OUTLINE, in some format, is used in almost every discipline practiced by people. In this chapter we adapt this familiar device to the specific needs of the systems analyst.

PLAN FOR THE STUDY

Before a full systems study can be carried out, a detailed plan of action must be prepared. This organized approach is necessary to managing all the forms, notes, statistics, and so on which will be gathered during the systems study. An effective and useful method of planning is the outline approach. The outline, along with the charts discussed in Chapter 13, serves to establish project control checkpoints. It aids the analyst in estimating both time and resources.

At this point in time the systems analyst should have a clear and concise problem definition that is understood and agreed upon by all key persons involved in the project. The analyst may also have a Type A feasibility study to help guide in the preparation of the outline. (Although, after having seen the problem, management may have bypassed the feasibility study and ordered a full systems study.)

The outline can be arranged in any way the analyst chooses. It may be arranged in order of target dates with each phase delineated by time. Or the analyst may prefer a subject approach with each type of task being a phase.

ORGANIZE THE STUDY INTO MAJOR AREAS

The primary purpose of the systems study outline is to aid the analyst in organizing the study into its major sections. The sections to be used are determined from the subject, scope, and objectives set forth in the problem definition phase.

The eleven areas listed below are representative of areas the analyst might consider (depending on the scope of the systems study) when preparing the outline.

1. *Organization structure.* Identify the current formal organization structure and, if possible, the informal organization structure. Determine how the various parts of the current system relate to the different areas and levels of the organization structure.

2. *Products.* Which products, if any, are going to be studied? Is the firm a profit maximizer, a market share maximizer, or growth oriented? Take into account the products of each of the departments that will be studied. A department's product may be a completed form, a manual or computer generated report, a transaction file of data entered on-line, or some other type of output.

3. *Marketing.* The firm must adhere to its own marketing goals. How does the system relate to the consumer's needs and to the need for internal control within the firm? Does the system provide sufficient information on consumer/customer activity to facilitate market planning?

4. *Communication.* Poor communication between interdependent areas is a major cause of systems failure. Try to determine the patterns of communi-

cation within the area under study and the reasons for any apparent communication failures. Organizational conflict is very often indicated by poor communication channels. This is obviously a subject of great importance to the systems study. Determine to what extent formal communication channels are identified in documented policies and procedures.

5. *Space or layout.* Evaluate present space allocations and layout relative to personnel, equipment, and workflow.

6. *Personnel.* Evaluate the personnel involved relative to the possible requirements of a new system. Interviewing can be a good method, if properly done. (Interviewing techniques are discussed in Chapter 6.) Consider sources of additional or replacement manpower, if necessary. Determine if the requirements of a new system would necessitate changes in job assignments, responsibilities, and the development of new job-related skills.

7. *Physical facilities.* Consider equipment, buildings, telephones, electrical outlets, bathrooms, furniture, maintenance, computer-related items, and so forth.

8. *Procedures.* Review current procedures, that is, the *what, who, when,* and *how* of the system. Determine to what extent the current procedures are documented and identify key control points and supervisory responsibilities.

9. *Policies.* Review the management policies. Include basic policies (long-range objectives), general policies (short-range objectives), and local policies (everyday departmental objectives).

10. *Records.* Examine the records in the area under study. Note the various reasons for filing and the storage methods. Determine to what extent organizational and legislative records retention requirements are applicable.

11. *Data processing.* Interface with the data processing department from the very start of the study. Are you designing around an existing data base or are you planning a new one?

The above eleven areas are some of the more important items that might be considered when formulating a systems study outline. The outline itself should be organized by the sequence in which the needed information will be gathered. It will, therefore, actually be a chronological sequence of events which can be followed in carrying out the study.

The systems study outline is nothing more than a detailed To Do list. In reality, it will be the developmental documentation of the study. Keeping within the predetermined scope, it can be divided into one or more of the five levels under study (Chapter 1). For example, if the study involves several departments within a firm (Level Three), it might also involve a functional area (Level Four) within each department and possibly a specific problem (Level Five) within that functional area. An outline of a simple systems study is shown in Figure 4-1.

Outline of a Systems Study

I. Department X, which involves a subfunction (Level IV) and a specific problem (Level V).
   A. Subfunction within department X.
      1. Study any written procedures and charts.
      2. Interview personnel involved.
      3. Observe current system.
      4. Gather hard data, for example, facts and figures.
   B. A specific problem within the subfunction.
      1. Study any written documentation.
      2. Interview personnel involved.
      3. Observe problem.
      4. Gather hard data, for example, facts and figures.

II. Department Y, which involves the entire department (Level III).
   A. Study any written documentation.
   B. Interview personnel involved.
   C. Observe current system.
   D. Gather hard data, for example, facts and figures.

III. Learn the general background and interactions between departments X and Y.

IV. Evaluate the following from departments X and Y.
   A. Understand the current system in departments X and Y.
   B. Define each department's systems requirements.
   C. Design a system compatible to both.
   D. Develop economic cost comparisons.

V. Sell the system and implement it.

FIGURE 4-1 Sample outline of a simple systems study involving Levels III–V.

It is important to develop as detailed a plan of action as possible. Include a timetable that specifies both calendar time and chargeable time. Calendar time is the overall time from start to finish, from the first day to the last day. Chargeable time is the actual number of hours to be spent in developing the new system. Since an analyst may be working on several systems projects at one time, chargeable time per week may amount to only 10 hours. It is obvious that if an analyst is supposed to be working 15 hours per week each on four different projects, the analyst is either working many hours overtime or slippage will occur on target dates.

Each activity on the outline should fit into a timetable. Gantt charts are good for estimating calendar time (see Chapter 13), whereas chargeable time can be shown in one figure (the number of hours required to complete the project). PERT also may be used for more accurate figures (Chapter 13).

A more specific example of an outline for a systems study might be similar to the following. Suppose an analyst is designing a new order entry system for a company. Also suppose that this study will encompass the following areas: Sales Branches, Order Entry Department, Credit Control Department, and the Production Control Department. The analyst might structure the outline as follows.

I. Study the Order Entry Department.
   A. Interview key personnel in order to obtain general information on the area under study and interactions between this area and the other areas within the firm.
   B. Study written procedures and observe the current system in operation.
   C. Search through any records and perform any estimating or sampling that must be done.
   D. Understand the existing system as it pertains to this department.

II. Study the Credit Department.
   A. Interview key personnel in order to obtain general information on the area under study and interactions between this area and the other areas within the firm.
   B. Study written procedures and observe the current system in operation.
   C. Search through any records and perform any estimating or sampling that must be done.
   D. Understand the existing system as it pertains to this department.

III. Study the Production Control Department.
   A. Interview key personnel in order to obtain general information on the area under study and interactions between this area and other areas within the firm.
   B. Study written procedures and observe the current system in operation.
   C. Search through any records and perform any estimating or sampling that must be done.
   D. Understand the existing system as it pertains to this department.

IV. Study various sized Sales Branches from different geographical areas.
   A. Interview key personnel in order to obtain general information on the area under study and interactions between this area and the other areas within the firm.

        B. Study written procedures and observe the current system in operation.

        C. Search through any records and perform any estimating or sampling.

        D. Understand the existing system as it pertains to this department.

    V. Tie together the general background, the interactions, and the understanding of each individual department's system in order to develop a complete understanding of the entire system as a whole.

    VI. Define the new systems requirements.

        A. Study long-range plans.

        B. Define the specific requirements of the new system.

            1. Define the outputs, the inputs, the operations, and the resources.

            2. Take into account the current requirements, the future requirements, and the management-imposed requirements.

            3. Document the new system requirements.

        C. Develop evaluation criteria for the new system.

    VII. Design the new system.

    VIII. Develop economic cost comparisons, write the final systems report, and implement the new system if management so desires.

PLAN THE ATTACK TO SUCCEED . . . TECHNIQUES

The most important factor in your favor is a positive mental attitude. Think positively and do not become a cynic since any systems study tends to take on the analyst's personality. Establish the outline in the sequence that the information is needed and seek the required data in a well-organized, efficient manner.

Aim at achievable results. Decide what the constraints are and develop the system within those constraints. For example, if upper management has decreed that they want only a computerized system in spite of the feasibility study recommendation for a manual system, *do not* design an elaborate *manual* system. They will not agree to a manual system. Even though you feel it unwise to proceed on a computerized system, you *must* do so—or perhaps begin looking for another job if you feel you cannot live with management's decision. If you are forced to design a system which you have recommended against, your best defense is to be so well organized that you can report to management what is going to go wrong next—before it happens!

This leads to another success consideration: *try to develop salable approaches.* If you cannot sell the new system to management, all of your effort has been wasted. "Blue-sky" plans are good for long-term conceptual planning, but if you never present down-to-earth, practical, usable *now* plans, management will question your ability. A salable approach has the following three principal features.

1. Technical feasibility.
2. Operational feasibility.
3. Economic feasibility.

It is imperative that the analyst realize the importance of selling his or her ideas on the new system *during* each of the steps when conducting a full systems study. (See Chapter 11 on Selling the System.)

Confine the limits of the study to the "area under study." Do not get carried away with including too much in your study. Successful completion of a few small to medium-sized systems projects gives an analyst self-confidence and shows management a proven success record. It is desirable to start with small studies and confine the scope to achievable results only.

Develop tentative timetables, both calendar time and chargeable time. Forecast approximate costs both to carry out the systems study and to install and operate the proposed new system. And always be prepared to give management periodic progress reports on your progress to date, cost to date, and any slippage in the time to completion. It is usually a good idea to submit regular, simple, written progress reports which contain the following four important elements.

1. A very short written description of the progress to date.
2. Approximate cost to date compared with estimated budgeted amount to date. (Are you overspending or underspending the estimated budget?)
3. Work completed to date compared with the estimated completion schedule. (Are you ahead or behind schedule?)
4. Report *all* unexpected problems as soon as they are encountered. Do not make excuses later!

Keep the progress report simple because management wants only enough information to evaluate your general progress on the project. Do not burden them with elaborate expositions of project detail, unless they specifically ask for it (Chapter 18 covers progress reports including a sample).

SELECTIONS FOR FURTHER STUDY

The library can be a useful source of information to amplify your study of the eleven principal considerations listed in this chapter. The information you will need depends of course upon the particular environment of your systems study.

Perhaps from your personal standpoint, the library can be of benefit in helping your plan to succeed. The works of Maltz, Drucker, and DeBono are all well known. For a lighter treatment, *The Peter Principle: Why Things Always Go Wrong* dramatizes the negative viewpoint. Since learning how to control a project is always important, Baumgartner, Gido, or Delp might be consulted.

1. BAUMGARTNER, JOHN S. *Project Management.* Homewood, Ill.: Richard D. Irwin, Inc., 1964.

2. BEHLING, JOHN H. *Guidelines for Preparing the Research Proposal.* Washington, D.C.: University Press of America, 1978.

3. BELL, MARTIN L. *Marketing: Concepts and Strategy, 2nd edition.* Boston: Houghton Mifflin Company, 1972.

4. DEGREENE, KENYON B. *Systems Psychology.* New York: McGraw-Hill Book Company, 1970.

5. DELP, PETER. *Systems Tools for Project Planning.* Bloomington, Ind.: International Development Institute, 1976.

6. DRUCKER, PETER F. *The Effective Executive.* New York: Harper & Row, Publishers, Inc., 1967.

7. GIDO, JACK. *An Introduction to Project Planning.* Schenectady, N.Y.: General Electric Company, Training and Education Programs, 1974.

8. HAIMANN, THEO, WILLIAM G. SCOTT, and PATRICK E. CONNOR. *Managing the Modern Organization, 3rd edition.* Boston: Houghton Mifflin Company, 1978.

9. HARPOOL, JACK D. *Business Data Systems: A Practical Guide.* Dubuque, Iowa: William C. Brown Company, 1978.

10. *IBM Data Processing Techniques: IBM Study Organization Plan: The Approach.* White Plains, N.Y.: International Business Machines Corporation, 1963 (F20-8135-0).

11. KAUFMAN, ROGER. *Identifying and Solving Problems: A Systems Approach.* La Jolla, Calif.: University Associates, 1976.

12. LEEDY, P.D. *Practical Research: Planning and Design.* Riverside, N.J.: Macmillan Publishing Company, Inc., 1974.

13. MALTZ, MAXWELL. *Psycho-Cybernetics: The New Way to a Successful Life.* Englewood Cliffs, N.J.: Prentice-Hall, Inc., 1960.

14. OPTNER, STANFORD L. *Systems Analysis for Business and Industrial Problem Solving, 3rd edition.* Englewood Cliffs, N.J.: Prentice-Hall, Inc., 1974.

15. PESCOW, JEROME K. *Handbook of Successful Data Processing Applications.* Englewood Cliffs, N.J.: Prentice-Hall, Inc., 1973.

16. PETER, LAURENCE J., and RAYMOND HULL. *The Peter Principle: Why Things Always Go Wrong.* New York: William Morrow and Company, Inc., 1969.

17. *Project Manager's Workplan.* Woodland Hills, Calif.: Eckman Center, 1975.

18. ROYAL, ROBERT F., and STEVEN R. SCHUTT. *The Gentle Art of Interviewing and Interrogation: A Professional Manual and Guide.* Englewood Cliffs, N.J.: Prentice-Hall, Inc., 1976.

19. SANDERS, DONALD H. *Computers in Business.* New York: McGraw-Hill Publishing Company, 1975.

20. TULLOCK, GORDON. *Organization of Inquiry.* Durham, N.C.: Duke University Press, 1966.

21. UYTERHOEVEN, HUGO E.R., ROBERT W. ACKERMAN, and JOHN W. ROSENBLUM. *Strategy and Organization: Text and Cases in General Management.* Homewood, Ill.: Richard D. Irwin, Inc., 1973.

22. WILLOUGHBY, THEODORE C., and JAMES A. SENN. *Business Systems.* Cleveland, Ohio: Association for Systems Management, 1975.

QUESTIONS

1. What is the primary purpose of the system study outline?
2. What aspects of the organization structure should the analyst consider when preparing the system study outline?
3. What aspects of communication should the analyst consider when preparing the system study outline?
4. What is the difference between calendar time and chargeable time in developing a new system?
5. What are the three principal features of a salable approach?

SITUATION CASES

CASE 4-1

The top management of a large brewery corporation requested that an analyst be assigned to assist them in correcting a problem that had been identified in the Shipping Department. Over the past three months, a backlog of incoming orders had developed and some difficulties were being experienced in setting up the delivery schedules.

Since management was preparing to launch a major sales campaign in four months, they decided to have a system analysis performed to resolve the problem before it became too serious.

After management reviewed the problem definition report, it was decided to bypass a Type A feasibility study and go directly to a full system study. The analyst's primary task was to prepare an outline of the system study. In preparing the outline the analyst considered the following areas: (1) organization structure, (2) communication, (3) personnel, (4) procedures, and (5) records.

When the analyst completed the outline, he turned it over to the systems manager with an estimate that the study would take 40 hours of his time. Six weeks later, the systems manager called in the analyst and asked what had happened. The analyst reported that he could find no way of correcting the problem unless use of a computer-based on-line delivery scheduling system was considered. Unfortunately the analyst felt there was not enough time to install such a system unless a suitable package could be found that was compatible with the corporation's computer. When asked about the delay in reporting his findings, the analyst responded that he had only been devoting 25% of his time to this project.

QUESTIONS

1. What other areas could the analyst have considered in the outline?
2. How could the delay have been avoided?

CASE 4-2

The manager of the Order Processing Department requested that an analyst be assigned to work on an urgent project in her area. The manager reported that a problem had been noted when a growing number of orders were being rejected after passing through the Data Entry Unit of the Data Processing Department.

The additional time required to correct the orders was causing delays and customer complaints were beginning to grow. As a result, the analyst was asked to complete the study within three weeks.

In order to facilitate conducting the study, the analyst developed the following outline.

I. The Order Processing Department.
   A. Study any written procedures regarding orders and their handling including manual costing of orders.
   B. Interview the mail order processing clerks.

      C. Observe the procedures for handling orders received by mail.

      D. Interview the telephone order processing clerks.

      E. Observe the procedures for handling orders received by telephone.

      F. Analyze the format of the order entry form.

II. The Data Processing Department (Data Entry Unit).

      A. Review all written procedures relating to handling order entry forms.

      B. Interview the data entry supervisor.

      C. Interview the data entry operators.

III. Identify and observe the interactions between the Order Processing Department and the Data Entry Unit of the Data Processing Department.

The analyst reviewed the outline with the manager of the Systems Department. The manager expressed some disappointment and indicated that some improvement was needed.

QUESTIONS

1. Of the eleven areas listed in this chapter, which could the analyst have included in his outline to improve it?

2. What other timing factors could the analyst have included to provide adequate assurance that the project deadline would be met?

CHAPTER 5

GENERAL INFORMATION ON THE AREA UNDER STUDY AND ITS INTERACTIONS

While human organizations all have their similarities, they all have their unique characteristics as well. These characteristics are important to the systems analyst. They are the keys to the objectives that will satisfy the users of the system. No department or function within an organization should be an island unto itself. Rather, it must interact with and serve other parts of the organization. An understanding of these interactions will guide the analyst in designing appropriate system dynamics. This chapter tells the analyst what to look for when gathering general background information on the area under study and studying its interactions with the rest of the organization.

WHY GENERAL INFORMATION?

If the analyst is not familiar with the unique characteristics of the environment in which the system must operate, the design phase presents a serious roadblock. The approach to systems design varies from one industry or organization to the next, depending upon the objectives, methods, and atmosphere peculiar to the area. For example, the payroll system for a public school district would differ in many important respects from the payroll system for a fast food service franchise. However, there would also be some similarities such as the provisions for federal and state income taxes and deductions. The differences between the two systems would be the key elements the analyst must be able to identify to ensure the new system is successful. The school district payroll system must provide the capability to process multiple jobs held by a single person (at different pay rates) and also multiple sources of funds to pay for one job (such as partial funding through private and federal grants). In contrast, the payroll system for a fast food service franchise must provide for the more traditional wage rate and time card driven capabilities. The hours worked may vary per pay period for each employee, and the wage rates may vary according to the shifts worked or by job assignment.

Beginning with this broad perspective on the environment of the study, the analyst should attempt to obtain the information needed to become familiar with each general level of the organization's operation, down to the specific area that is the detailed focus of the study.

This approach enables the analyst to both speak and understand the language of the "natives," and hence do a far better job on their system.

IDENTIFY THE AREA UNDER STUDY

As explained in Chapter 1, the area under study must be clearly defined. One should attempt to determine the organizational levels affected by the study in order to properly set the objectives and scope. There are situations, however, in which a good set of initial objectives and scope may be worked out with top management before it is known into what levels of the company the study will lead. In fact, the pertinent organizational levels involved may not be clearly known until the interviewing has been started. For a review of the five levels which may be involved in a study, the chart from Chapter 1 is reproduced in Figure 5-1 of this chapter.

Level I: A study that involves the entire firm—all its divisions and locations.

Level II: A study that involves one division of the firm.

Level III: A study that involves departmental interaction within the firm or division.

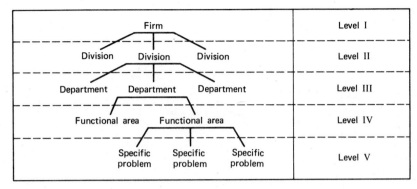

FIGURE 5-1 Area under study.

Level IV: A study that involves the functions within one department.

Level V: A study that involves one problem within a function of a department.

BACKGROUND OF INDUSTRY, COMPANY, AND AREA UNDER STUDY

Depending on the particular study, you may or may not be interested in the background of the *industry* itself. The industry background places the firm in perspective within its environment. It may also explain why the firm does something in a certain way. For example, if the industry has been using a certain accounting procedure for 50 years, that should warn you that any change may be vigorously resisted because people are used to and feel comfortable with the long-standing procedures. In addition, you will probably find some very good reasons why the system has weathered the tests of time. Nevertheless, firms that wish to survive in today's uncertain environment must prepare for survival through the use of contingency planning. This concept implies that management expects the unexpected and is ready to shift directions when the external environment, contingencies that may or may not be under management's control, changes. Contingency planning may be used at any of the above five levels.

Within the industry, some of the factors that might be considered are

1. Products and services of the industry.
2. Growth or decline in the industry.
3. Technological trends.
4. Industry-wide sales volumes and profit margins.
5. Type of industry, such as oligopoly, monopoly, or nearly perfect competition.

77

6. Effects of foreign and domestic competition.

7. Effects of unions.

8. Effects of legislative restrictions and reporting requirements.

9. Influence of industry associations, standards groups, and so forth.

10. Industry-wide subsidies, or tax advantages.

11. The size and strength of the firm within the industry.

12. Human needs the firm can satisfy.

13. General business climate.

14. Government fiscal policies.

15. Market demand for current products.

16. Market demand for prospective products.

The background of the *company* gives one a feeling for the characteristics of the firm itself. Ideas, attitudes, and opinions of key management personnel are very important to the analyst. Knowledge of the company's background will make its goals and style more understandable. Knowing why certain employees are where they are and how the company has grown over the years will help in understanding the present direction of the business. Within the firm some of the things that might be looked for are

1. Management and employee attitudes: how do they harmonize in the area under study?

2. Pattern of growth over the years and expectations for the future.

3. The products and services important to the firm's future.

4. Sales volume and profit margin trends.

5. Expansion or curtailment of any segment of the business over the years.

6. Involvement in mergers, spinoffs, or purchases of or by other firms.

7. Effects of foreign competition, domestic competition, government intervention, and unions.

8. Type of market: oligopoly, monopoly, nearly perfect competition, and so forth.

9. Effect of technology on the firm.

10. Past and present goals and objectives.

11. Long-range plans.

Obtaining the background of the *area under study* is a mandatory step. Industry and company background are not required in many systems studies, but the background of the area under study is required for *all* studies. If one does not know the background of the specific area about to be studied, one cannot appreciate its guiding policies, atmosphere, or the day-to-day problems faced by the personnel in that area. Within the specific area under study, some of the things to look for are

1. Attitudes of all involved personnel, toward the system and toward management.
2. Past and present objectives.
3. Past and present policies and procedures.
4. Increasing or decreasing budgets.
5. Growth or decline in quantity of work in the area.
6. Importance of the area to the rest of the business enterprise.
7. Apparent morale problems.
8. Power struggles between this area and other areas.
9. Respect for the manager of this area by both peers and subordinates.
10. Attitudes of personnel outside the area under study toward the area under study.

LEGAL REQUIREMENTS

Governments at the local, state, and federal level have many laws which affect company operations. In the search for general information on the area under study, remember that businesses need licenses, they must pay taxes, and they must observe the laws of the nation.

Many types of business are controlled in some way by one level of government or another. If the study is involved in public utilities, insurance, aviation, stocks and bonds, railroads, television or radio, banking, unions, or pollution, one must understand the legal requirements that govern the area (if any). For example, government regulations play an extremely large role in the aerospace industry. Nearly every function in an aerospace firm is subject to direct government regulation. The same is true in the meat packing industry, and several others.

The investigation of legal requirements should answer three basic questions. First, which government regulations help the company? The following may help the company: direct subsidies, import quotas or tariffs, protection as a monopoly, special tax incentives, franchises, labor laws, and civil rights laws.

The second question is, which government regulations restrict the company? These may include excise taxes, export limitations such as with military hardware, utility regulations, safety laws, antitrust laws, pollution laws, restrictions on the sale of stocks and bonds, privacy legislation, restrictions on corporate dealings in foreign countries, and laws that restrict a company from selling a product or freely changing the price of its product.

The third question is, which government regulations affect the recordkeeping and reporting practices of the company? The following may increase recordkeeping for the company: withholding taxes of employees, social security, workmen's compensation, corporate taxes, Federal Communications Commission reports, Civil Aeronautics Board reports, Securities and Exchange Commission reports, privacy legislation, restriction on corporate dealings in foreign countries, pollution control laws, regulated utility reports, civil rights laws, affirmative action programs and court rulings, safety records, and cost data to defend a possible contract renegotiation by the government.

THE ORGANIZATION ITSELF

There are two types of organization within a company which the analyst should consider. One is the *formal* organization and the other is the *informal* organization. These two organizations are more or less independent of each other and both must be considered when designing a new system.

The formal organization is built around the company's goals, upper management's policy statements, and the written procedures that carry out these policies. In order to see the formal organization, all the analyst has to do is look at the company organization chart. The organization chart is a who's who of each department or functional area. The normal flow of information can be written or verbal, but it is usually passed up and down the chain of command in the organization. Formal information usually includes items about the company, its operations, products, policies, procedures, and specific situations that arise such as a change in management or pay raises.

The informal organization is built around the job at hand. Whenever the system does not work or the written procedures are outdated or a manager loses control of a department, the informal organization takes over. In this case, the employees perform the job at hand in the best fashion as they see it. Written procedures are not followed or are only partially followed. The work usually gets done but not necessarily in the most efficient manner. Whenever the informal organization takes over, consistency is lost; and consistency is extremely important. For example, cost estimate reports are not perfect; but if all the cost estimates are about 10% high, management learns to compensate through its decisions. If the cost estimates suddenly vary from 10% high to 20% low (no consistency), management cannot rely on the reports, and a problem is born. Informal information usually involves verbal information exclusively. The employees exchange gripes, ideas, and problem solutions. In order to penetrate the informal organization, one should observe carefully how the work is being performed. The trust of the involved employees also must be gained.

If one concentrates on the formal organization only, certain information may never be communicated, such as employee gripes or how the work is actually being performed. Although the approval of the formal organization is needed for the system to be officially adopted, the most successful systems also get approved by the *informal* organization. Informal approval comes from letting the people who will use the new system participate in its design. Participation in the design does not mean that you ask a few questions and then return sometime later to install the new system. It means that after asking the questions, you come back with a rough draft and request ideas for more input. You may return many times for more input, and whenever possible discuss the final recommendations with the potential users of the system prior to release of the final report. This is what gets the informal organization to tacitly approve the new system.

COMPANY POLITICS

In business there is a taboo against formally recognizing that politics may be a factor in success. Political maneuvering takes place within the informal organization. Although no one advertises their personal strategy to achieve success, merit alone does not always guarantee success. Being able to understand and recognize a political situation is essential to the analyst who is attempting to understand the area under study. Politicking tends to be light in rapidly growing companies and heavier in older, well entrenched companies. Of course, there are situations with new, highly profitable ventures where everybody wants to get in "on the ground floor." Vicious in-fighting often occurs in such situations, to an extent that a nonpolitical person is either left behind or is simply replaced.

Whether you as the systems analyst will want to participate in this game of company politics will depend to a great extent on your own value system. It is advisable, however, to keep in mind that if you become too politically involved, your impartiality, and hence your effectiveness, will be lost.

We are including below some suggestions for "getting ahead" frequently made by company-politics experts. They are included not to help you become a company politician but *to aid you in recognizing others* who may be politically motivated. Some might be considered "good," others might be considered "bad," while others are simply common sense.

Be visible to the people above you. Meetings are an excellent opportunity to make yourself visible since you can ask questions or put forth new ideas. Taking up an extracurricular activity is another way of attracting notice.

Always look good. Learn your superior's value system. Join in with the group and ask advice. Endear yourself to a star in the company who is on the way up.

Perform your work thoroughly and efficiently. Never complain about work either verbally or in writing. Accept even the less desirable assignments cheerfully. Do not tell your troubles (either professional or private) to your co-workers.

Be loyal. Always ask *first* what you can do for the company and *second* what the company can do for you.

Show respect. Call superiors by their title until they indicate a first-name basis is appropriate. Do not usurp authority; it usually creates antagonism.

Think about what you say *before* saying it. Reveal only what is necessary and what should be said. Hold your beliefs until the right time.

Keep a positive outlook since nothing new will ever be attempted if all possible objections must first be overcome.

Always do things to better yourself, but do not brag about it to others. Let it be revealed in subtle ways.

If you discover the prevailing weakness of any colleague, whether intentionally or by accident, remember never to trust him where that weakness is concerned. For example, if you know a fellow-worker is a talkative drinker, do not tell that person anything you do not want to become public knowledge.

You never get promoted if no one else knows how to do your current job. The best basis for being advanced is to organize yourself out of every job into which you are put. Another is to convince superiors that you can do the job you want better than anyone else.

When angry, do not write a letter condemning someone. If you must do something about it, do it verbally. Aristotle put it this way:

Anybody can become angry—that is easy; but to be angry with the right person, to the right degree, and at the right time, and for the right purpose, and in the right way—that is not within everybody's power and is not easy.

Most managers will not openly condemn another manager, and neither should the analyst. Open condemnations can be very dangerous, especially if in writing. Criticism, if given at all, should be constructive and directed at things or procedures, not people. The analyst who thinks that criticizing people will get the system installed may win the battle for that system but may lose the war called survival. If criticism is direct and unconstructive, people will think the analyst is out to get them and will be afraid to have further dealings with the analyst. Thus the ability to obtain the needed cooperation will be lost and the analyst will not be able to function effectively.

Perhaps politicking is not your style. That is fine. But remember not to knock the political strategist too hard, because this is your competition; and the chances are good that this same person may have authority over you some day. Whether we like it or not, events occur that way sometimes and it is realistic to accept these situations for what they are.

STRUCTURE OF THE ORGANIZATION

The business process may be viewed as an integrated system which begins with inputs from the external environment on which operations are performed in order to transform these inputs into outputs. Such inputs might be paperwork for raw materials that will become a new consumer product, or a customer order, cancellation, or complaint.

There is a continuous flow of inputs, operations, and outputs, such as

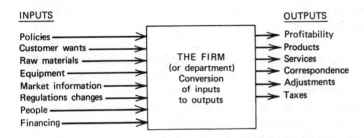

INPUTS

Policies
Customer wants
Raw materials
Equipment
Market information
Regulations changes
People
Financing

THE FIRM
(or department)
Conversion
of inputs
to outputs

OUTPUTS

Profitability
Products
Services
Correspondence
Adjustments
Taxes

This is a general picture of any organization which yields a product or service to its users. The inputs and outputs may vary, but the principle is the same for any organization: it exists to produce products or render services to its members and customers. It accomplishes its work through dynamic interaction of its members to convert inputs to outputs.

The analyst's purpose in this scheme is to calculate the members' interactions via systems which are appropriate to the objectives and goals of the organization.

After obtaining a general familiarity with the background and environment of the area under study, the analyst can proceed to examine the area's interactions with other parts of the organization.

The internal environment of the typical business firm is composed of many departments, each with its own purposes and its own problems. Each of these internal departments accepts inputs, processes them, and passes on its outputs to other departments. There is an identifiable chain of events within the firm. The analyst must look up and down that chain of events to make sure any contemplated changes will mesh with or improve the flow of business operations. A perfect system for one department might require another department to add personnel or revise its routines, perhaps at excessive cost, thus negating any savings or other expected benefits.

Since the analyst's purpose is to facilitate interaction and efficiency, a system must be designed which ensures that the benefits for a specific area will be achieved while also harmonizing with other areas and systems in the chain of operations. It must be remembered that the purpose of any one area is to render service to the other areas, all of which interact to accomplish the objectives of the organization as a whole.

The analyst should be able to detect and correct, or prevent, the functional isolation of any member or segment. If part of an organization is not serving the rest in its efforts toward attainment of overall organizational objectives, it is in need of renovation or elimination. The experienced analyst can expect to find many such fragmented areas in the organization being studied. The existence of such ineffective segments is often the source of the problems the analyst is seeking to resolve. For example, it may be discovered that the cause of delays in manufacturing sophisticated electronic devices for the military is caused by a labor intensive testing process and timing problems in scheduling components for assembly. This may be further traced to a testing process that requires tedious manual computations and a lack of timely information on the expected availability of related components for final assembly. Management often is unaware of such imbalances in the work-in-process environment until the systems analyst discovers the symptoms and identifies the related problems.

The basic interactions that should be studied are those between employees, departments, management personnel, or any combination of these elements. The analyst should start with the firm's current organization charts to view the formal relationships between the area under study and other areas. Flowcharts sometimes exist that show the work flow pattern through and between departments. These should be examined. Interviews with management also tell the analyst a lot about interactions. Job descriptions and observation of work flow will reveal more of these interactions. Finally, the informal organization, the employees who carry on the area's daily activities, can pinpoint the actual day-to-day interactions: who does what, when, how, and *why*.

In a large system study, the interactions for two or three departments up and down this chain of events must be studied. The analyst may have to go all the way to the *external* environment to understand the way some source or original input is received, or to get approval to change such an input or an output.

INTERACTIONS BETWEEN OUTPUTS, INPUTS, AND RESOURCES

Outputs can consist of completed paperwork, processed computer files, reports, or semifinished or finished products. One department's output may be another department's input, except when the paperwork or the product goes out of the firm and into the external environment.

The analyst should examine broadly the outputs from a general viewpoint in order to determine how these outputs relate to other inputs and outputs up and down the chain. The methods used by the area under study to produce its outputs must be understood and related to all departments within the overall area under study. Using a public utility as an example, the following interactions could apply.

1. Customer requests electrical service through main office.
2. Customer completes service request.
3. A copy of the request is forwarded to the responsible field unit to schedule installation.
4. An additional copy of the request goes to the billing department for processing.
5. The last copy of the request goes to credit department for approval.
6. The credit department notifies all departments of final approval.
7. The field unit initiates service and notifies the billing department of the date.
8. The billing department begins sending regular customer monthly statements.

The analyst should learn the "product line" of the area under study, that is, what service they render to the firm, and how that service fits into the general scheme of operations up and down the chain. Finally, it should be determined where all the outputs go and how they are used.

Inputs consist of raw data, raw materials, paperwork, processed computer files, reports, and semifinished products. One department's input is usually another's output.

As with outputs, the analyst should obtain a broad view of the inputs to the area under study, and should relate these inputs to other inputs up and down the chain of events. The source of each input and the reason it comes into the area under study should be determined. However, at this stage of the study only the major inputs should be observed, and trivial inputs should be avoided. The analyst should learn the characteristics of the inputs, such as manual or computer generated, approximate quantities, availability (time factor), and cyclic needs. This knowledge will be of help in determining the interactions between the area under study and the rest of the chain.

Resources are those items that are used in the day-to-day operations to convert inputs to outputs. A firm will view its resources as assets because it has an investment in them. There are four categories of resources which may require analysis in order to get a clear picture of the interactions between the areas being studied and the rest of the organization. The four kinds of resources are financial, personnel, inventory, and facilities.

The *financial* resources of the area under study consist of the budget for that area and the area manager's ability to get financial backing for new projects and systems. Is the area under study financially stable or is it a declining segment of the firm? This is a key question for the analyst to answer because it will aid in understanding the weights and priorities appropriate to the area, and to the system which will be designed.

The *personnel* resources consist of the key managers and other skilled and able personnel in the area under study. The analyst should study the personnel by skill, location within the area under study, and current function. The personnel resources may limit or enhance a system and the analysis of personalities and talents may reveal much about why the area has its current status, and why it interacts with other areas as it does.

The *inventory* resources can be looked at from two viewpoints. First, the "stock-in-trade," such as raw materials, parts, supplies, semifinished products, by-product utilization, and finished products; and second, the "files of information" or data that have been collected over the years. While studying interactions, the analyst might check the files of information throughout the area under study for the following.

1. Completeness of the file; extensiveness; why it is kept.

2. Sources of the filed documents.

3. Duplication in other departments.

4. Flow of documents within the area under study and between other areas.

5. Age, that is, how old is the oldest item in the file, and how long until new items get into the file?

6. Is the file actually used? Does it contain obsolete information?

7. How often is the file accessed? How is it indexed?

The *facilities* resources of the area under study may consist of land, buildings, data processing equipment, or other capital equipment. Carefully note which area manages the computer (if there is one) and the interactions between that area and the other areas. Also note any excesses or shortages of floor space, desks, supplies, and so forth. The analyst needs to determine physical capacity.

DEPARTMENTAL INTERACTIONS

Interactions between employees and supervisors or between departments are very important interactions for the analyst to observe.

Within a department the analyst will want to observe how the manager interacts with subordinates. What management "style" is employed? Perhaps the manager has already decided "whether it is better to be loved-than-feared or feared-than-loved." It is important to learn whether a manager tends to extremes, that is, whether "fear" or "nice-guy" techniques are used. If ruled by fear, the personnel involved in the day-to-day activities may not accept the analyst and may not cooperate unless directly commanded to do so by their manager.

It is also important for the analyst to carefully observe and understand the management styles of the managers, and the resulting differences between the area under study and other areas. Such knowledge will help the analyst understand how these managers interact with one another and how each might react to systems design features. For example, the manager who is oriented more toward Theory "X"[1] will tend to desire more stringent employee controls in a system that operates in his department. The Theory "Y"[2] manager, on the other hand, may favor fewer employee controls and a more flexible system.

The interaction between managers who view management style from these opposite viewpoints is the important focus of our discussion. The analyst must study these interactions in preparation for the role of objective intermediary. During the system design phase, the opinions of two or more managers with divergent theories of human nature will have to be considered since a balance between these views needs to be achieved. The analyst should be alert especially to a manager's style when working on a system that involves only that manager's department or personnel.

The analyst should bear in mind, as operational personnel are evaluated, that individual workers also vary widely in their reactions to management style. Some employees will agree that heavy-handed, close controls are necessary, while others will prefer an atmosphere of permissiveness to enhance creative thinking and new ideas. In short, some employees believe in Theory "X" and some in Theory "Y"—and, of course, there are those in between. In other words some employees *do* want to be closely controlled while others prefer considerable freedom.

Between departments the analyst should determine, for example, if any department is in the "empire-building" category, since an empire-building manager may not accept a system that curtails the department's power in some way, such as reducing its span of authority and control. Empire builders may be seeking personal power, prestige, and recognition, or they may be pushing causes which they genuinely believe are vital to the firm. The systems analyst is advised to avoid taking sides when a project leads to such a power struggle. The analyst may become a pawn in the battle and end up sacrificed prior to the checkmate! Systems designed in the midst of such struggles tend to be very compromised where objectivity and quality are concerned.

The systems analyst also should stay alert to the effect any individual's personality may have on the interactions between other operations personnel, managers, and departments. For example, if a particular employee is known to be demanding and unpleasant in requesting assistance from other departments, that person should not be put in a position requiring constant contact with other departments. Some people are not suited to particular types of jobs, and the analyst needs to recognize this in designing a new system.

There is one final, subtle point that may be indicative of in-fighting between the various departments and the area under study. In areas in which fending for power is common, the managers tend to take their vacations a few days at a time so they can always defend their past, present, and future actions.

In summary, to understand the area under study, the analyst must be able to identify not only the formal characteristics of the area but also, perhaps even more importantly, the informal characteristics. By understanding both, the analyst is more likely to design a workable and acceptable system.

SELECTIONS FOR FURTHER STUDY

General background information on the area under study can include just about any subject: persons, places, things, methods, legal requirements, and so forth. It can be external or internal.

External sources may be most helpful for information on the industry itself. Trade periodicals often carry articles on new products, developments in technology, and trends in sales, growth, and profit. Most of the major industries' trade publications are indexed by *Business Periodicals Index, Abstracted Business Information* (a computerized bibliographic index, see Chapter 19), or *Funk & Scott Index* (available in both paper and computerized formats). In many cases, you need only to look under the name of the industry to locate citations to relevant articles. Since the major business periodicals (*Fortune, Dun's, Forbes,* etc.) are indexed by the indexes, special industry reports often can be located quickly and easily. In addition, you may find the various financial services helpful. These would include Moody's, Standard and Poor's, and the Value Line series.

An IBM manual, on the Study Organization Plan, is important at this point. It discusses how to conduct the first phase of a systems study, or understanding the present business. It explains the reclassification of the existing business applications into goal-related activities for study and analysis.

Reports put out by a multitude of government agencies may also prove useful. (They are covered in Chapter 19.) There also are numerous business organizations that publish all types of information. These include the Business Equipment Manufacturing Association (BEMA), Conference Board (formerly National Industrial Conference Board), Electronic Industries Association (EIA), and many more. In addition, many of our Graduate Schools of Business have special publications, such as *MSU Business Topics,* the University of Colorado's *Series in Business,* and the University of California's Institute of Industrial Relations reprint series. Newspapers may also be of benefit, especially the *Wall Street Journal.*

Internal sources of information are surprisingly numerous. They may be long-range plans, old annual reports, the corporate charter, employee handbooks, speeches made by members of management, the company newspaper (called a house organ), past publications on the company, and the various *Who's Who* (-In the West, -In Finance and Industry, -In the Computer Field, etc.) The firm's employees may be a gold mine of information, especially those in management. It is wise to do some background reading on your own first, however, to avoid "foot in mouth disease"!

Legal requirements may be found in the contracts of the company, through the use of loose-leaf services such as the Commerce Clearing House (CCH) *Government Contracts Reporter* or through the law itself. Official legal pronouncements are of two types. A "statute" is a law passed by Congress or a state legislature. Federal statutes are listed in the *U.S. Code.* State statutes are listed in the various official publications of the states (example: *Statutes of California*). The second type is a "court decision." Federal Supreme Court decisions are noted in the *U.S. Reports,* the Circuit Court of Appeals decisions are in the *Federal Reporter,* and the U.S. District Court decisions are in the *Federal Supplement.* Again, each state also has its own court decisions (example: *California Appellate Reports*). Another category, that of "regulations," may play a significant role in a system study since few industries remain free of government regulation. These regulations have the force of law since the agencies that enforce the regulations are created by law and are given an implied legal stature. If you should find yourself deeply involved in legal publications, you should be aware that your local librarian may only be able to provide minimal assistance since law librarianship is a highly specialized field requiring special training. The local librarian, however, will most likely be able to help lead you in the most cost-effective and timely direction.

If you are a novice in company politics, you would do well to read up on the subject. If lucky, you may find a real "pro" who can give some pointers along the way. In the meantime, first read Machiavelli's classic *The Prince* and then *Management and Machiavelli.* In addition, Hegarty's *How To Succeed in Company Politics* gives practical "Do" and "Don't" advice.

1. BEVERIDGE, W.E. *Problem Solving Interviews.* Edison, N.J.: Allen & Unwin, Inc., 1969.

2. DALTON, GENE W., and PAUL R. LAWRENCE, eds. *Motivation and Control in Organizations.* Homewood, Ill.: Richard D. Irwin, Inc., 1971.

3. GORDON, R.L. *Interviewing: Strategy, Techniques & Tactics, revised edition.* Homewood, Ill.: Dorsey Press, 1975.

4. HARPOOL, JACK D. *Business Data Systems: A Practical Guide.* Dubuque, Iowa: William C. Brown Publishers, 1978.

5. HEGARTY, EDWARD J. *How To Succeed in Company Politics: The Strategy of Executive Success.* New York: McGraw-Hill Book Company, 1964.

6. *IBM Data Processing Techniques: IBM Study Organization Plan: The Method, Phase I.* White Plains, New York: International Business Machines Corporation, 1963. (F20-8136-0).

7. JAY, ANTHONY. *Management and Machiavelli: An Enquiry into the Politics of Corporate Life.* New York: Holt, Rinehart and Winston, Inc., 1968.

8. KOSSEN, STAN. *Human Side of Organizations, 2nd edition.* Scranton, Pa.: Harper & Row Publishers, Inc., 1978.

9. MACHIAVELLI, NICCOLO. *The Prince.* Written 1513, first published 1532.

10. MILLER, MARTIN R. *Climbing the Corporate Pyramid.* New York: American Management Association, Inc., 1973.

11. PACKARD, VANCE. *The Pyramid Climbers.* New York: McGraw-Hill Book Company, Inc., 1962.

12. PETTIGREW, ANDREW M. *The Politics of Organizational Decision-Making.* Scranton, Pa.: Barnes & Noble Books, 1973.

13. REWOLT, STEWART H., JAMES D. SCOTT, and MARTIN R. WARSHAW. *Introduction to Marketing Management: Text and Cases, 3rd edition.* Homewood, Ill.: Richard D. Irwin, Inc., 1977.

14. RICHARDSON, STEPHEN A., et al. *Interviewing: Its Forms and Functions.* New York: Basic Books, Inc., 1965.

15. SANDERS, DONALD H. *Computers in Business.* New York: McGraw-Hill Book Company, Inc., 1975.

16. WILLOUGHBY, THEODORE C., and JAMES A. SENN. *Business Systems.* Cleveland, Ohio: Association for Systems Management, 1975.

QUESTIONS

1. Why does the approach to systems design vary from one industry or organization to the next?
2. If the analyst is interested in the background of the industry itself, name five things that might be considered.
3. Name five things the analyst should look for within the specific area of study.
4. Name the three areas the analyst should explore with regard to legal requirements.
5. What is the difference between the formal and informal organization?
6. What is the risk of concentrating on the formal organization only?
7. Describe the business process when viewed as an integrated system.
8. What is the purpose of any one area within the organization?
9. Name four typical inputs to the firm.

10. Define resources.

11. Describe the financial resources of an area.

12. Name four things the analyst might check the files of information for throughout the area under study.

13. Why is it important for the analyst to carefully observe and understand the management styles of the managers, and the resulting differences between the area under study and other areas?

SITUATION CASES

CASE 5-1

In an effort to boost sales and increase market share during the past year, a department store began offering an in-house installment payment plan to its customers. Recently management noted some apparent problems with this plan when the monthly business activity reports began to show a substantial increase in the number of overdue accounts. Analyst A was assigned to look into this situation and has already completed the problem definition phase. After developing an outline for the systems study, analyst A received another job offer and left the company. Analyst B was assigned to complete this project.

After reviewing the work that had already been done, analyst B decided to concentrate his efforts in the Accounts Receivable/Collection and Credit Departments. Before beginning to gather general information on the areas under study, analyst B made up a list of some things to look for including (1) past and present policies and procedures, (2) growth or decline in quantity of work in the area, (3) apparent morale problems, and (4) respect for the manager of the area by both peers and subordinates. After completing the list, analyst B decided to review it with the systems manager to ensure he had not overlooked any key factors. The systems manager suggested that there were some points of information at the industry and company level that analyst B could pursue in addition to expanding the list for the specific areas under study.

QUESTIONS

1. What type of information could the analyst have looked into at the industry and company level?

2. What other things could the analyst have added to his list covering the areas under study?

CASE 5-2

XYZ Wine Imports, Inc. had been experiencing some difficulties with its principal warehouse facility. An analyst was asked to define formally the problem and develop an outline for the systems study. After completing those tasks, the analyst was expected to proceed in gathering general information on the area under study. Before beginning, however, she did some research at the industry and company levels. As a result, the analyst learned of some recent federal legislation that provides new tax advantages to organizations that maintain large warehouse facilities. In addition, she became aware of the firm's plans to dramatically increase sales by building a chain of branch distributorships. The analyst rightfully assumed that this could place heavy demands on the principal warehouse facility.

While gathering general information on the area under study, the analyst learned that some federal and local government legal requirements impacted warehouse operations. Federal law requires that all imported items entering the warehouse remain in the receiving area until a customs agent inspects the merchandise. This had caused some backlog and storage problems. Local government regulations mandate that no item may be stored within 30 inches of the main fuse box or within 20 inches of a small fuse box. Because of the recent problems, the company had been fined for numerous storage violations. In considering the organization of the warehouse staff, the analyst concentrated her efforts in the areas of (1) company goals, (2) upper management's policy statements, (3) the organization chart, and (4) the formal operating procedures.

With this information, the analyst completed the remaining steps in designing the new system.

QUESTIONS

1. Identify any pertinent aspects of the legal requirements not considered by the analyst.
2. What else could the analyst have considered while gathering information within the warehouse facility?
3. How could the analyst have attempted to deal with the informal organization? What is the risk if she is not successful?

CASE 5-3

EKK manufactures and sells calculators. The product line includes four styles of desktop calculators and has been expanded recently to include three additional engineering-type calculators. Management called in an analyst to evaluate the situation since financial results show a decrease in profit in spite of the increase in volume since starting the new line.

The problem definition study, an outline of the systems study, and a study of the necessary background information have all been made by the analyst. The analyst now could begin to investigate the interactions between the area under study and other departments in the company. The Accounting Department was the principal area under study.

The Accounting Department was having difficulty in meeting weekly deadlines. The accounting supervisor said this was caused by the Manufacturing Department sending in late and incomplete reports. The analyst felt the Accounting Department could not justify asking for more time or more detail. However, he agreed that the Manufacturing Department must have its data sheets filled out completely and returned on time. After further study of the records, the analyst discovered that materials purchased through the Purchasing Department were causing an increase in the overall product costs caused by a high percentage of defective units.

The search for reliable product cost data revealed considerable disagreement between the Manufacturing Department and the Accounting Department as to why sales had increased and profit had decreased. Upon interviewing the Manufacturing Department supervisor, the analyst learned that the chief complaint was that the Accounting Department required too much detailed data in reports. The supervisor said that even though the workers' morale in this department seemed pretty good, the addition of the new line had left them overworked and understaffed. They liked the new calculators, however, because of the variety they provided in assembly operations.

Further research for valid product cost involved the Purchasing Department. The Purchasing Department was in charge of purchasing all materials. The analyst's research of the past performance of the department showed there had been a big employee turnover and absenteeism had doubled in the past months. The analyst received good cooperation from the supervisor of this department. He seemed to be positive that his department was buying at the most competitive prices and assured the analyst that the department was running at peak efficiency. He said the workers' low morale was caused by the fact that they did not like the new calculator because of the extra workload it created.

The assistant manager was even more cooperative by taking the analyst aside and telling him that the supervisor was making outside purchases from a relative and that he was not buying at the most competitive prices. The assistant manager said that most of the purchases could be made for much less from any other company. The assistant manager added that the workers did not like working for the supervisor, which was the reason for low morale, as opposed to disliking the new calculator. Without interviewing the supervisor again, the analyst assumed the assistant manager was correct and reported his findings.

QUESTIONS

1. How could the analyst improve the study of interactions between the area under study and other areas of the organization?
2. How could the analyst have avoided becoming involved in company politics?
3. What are some of the risks if the analyst does become too involved in politics?

CASE 5-4

The comptroller of a large furniture company hired a systems analyst to study the existing system for handling invoices and "notice of receipt of orders" between the Receiving, Merchandising, and Accounts Payable Departments. The controller's main concern was the increase in the cash discounts lost. (An account had been set up to record cash discounts granted by wholesalers but which were not taken because payments were not made within the specified time; the balance in the account had reached a very high level.)

The analyst proceeded to study the current interactions and methods of communication between the related departments. He discovered that the flow of communications and method of receiving goods differed between two groups within the Merchandising Department.

It seemed all furniture division merchandise was received at the warehouse, located about four miles from the store. When merchandise was received at the warehouse, the boxes were inspected for their condition, and the quantities and styles were compared to the shipping invoice by the receiving clerk. In the warehouse office, a copy of each purchase order was compared to the shipping invoice and then the entire set of verified papers was sent to the Accounts Payable Department by daily shuttle truck at the end of the day. Merchandise received before 3 P.M. was accounted for and verified against its purchase order before the end of the day. The Accounts Payable Department was then notified the next morning for payment authorization.

All Accessory Division merchandise (gift items, lamps, and pictures) was delivered directly to the main store. The receiving clerk counted the number of boxes and checked their general condition. The merchandise was taken down to the marking room where it was unboxed, priced, counted, and compared to the invoice. When merchandise was ready for sale or stock the proper Accessory Department was notified. That department then verified the quantities and styles on the invoice against the purchase order. The buyer approved the acceptance of the merchandise by signing the invoice, after which the paperwork was sent to the Accounts Payable Department for payment. After the analyst noticed the delays in merchandise handling and paperwork approval, he decided this was the real problem.

The analyst decided to restudy the receiving and handling procedure for the accessory items and their corresponding paperwork. In the marking room, he noticed large amounts of gift items on trucks waiting to be moved. This merchandise was being held for inspection by the Gift Department buyer who was attending the furniture show in Chicago. The analyst investigated why merchandise for other departments was not held up for inspection. He learned that the other department buyers felt it was unnecessary for them to hold up their merchandise for this formality, so they had assigned this job to their assistants. Because of this, the merchandise in these departments was moved quickly and efficiently. The Gift Department merchandise and its corresponding paperwork, however, would sit in the marking room for up to two weeks awaiting the department buyer's return.

Upon conferring with the available supervisors and buyers, the analyst concluded that acceptance by the buyers was a mere formality. Because of the size and fragility of the merchandise, each box had to be opened and each item had to be inspected before the order could be accepted. During peak seasons, the backlog could hold up the flow of paperwork by up to four days. Even with this delay, the other departments did not miss their discount period, as did the Gift Department. Realizing that this delay was a necessary process and the paperwork would be held up, the analyst in his report to the controller recommended that the merchandise acceptance and paperwork be approved by the supervisor of the marking room and forwarded to the Accounts Payable Department for payment. This would eliminate the delay of paperwork in the merchandise office awaiting the buyer's approval.

The controller and the vice-president in charge of merchandising both accepted the proposal and issued a memorandum to the Accessory Department buyers, the supervisor of the marking room, and the Accounts Payable Department.

Upon returning from Chicago, the Gift Department buyer read the memorandum describing the changes. Because she was not consulted about the changes prior to implementation, the buyer felt this had been done behind her back. She felt that this limited her authority in the control of gift merchandise, and that she should personally inspect and approve all gift merchandise. Infuriated by this action, she filed a complaint with the vice-president in charge of merchandise.

QUESTIONS

1. Identify the key inputs and outputs discussed for the Accounts Payable Department.

2. What else could the analyst have done to improve his approach in studying the interactions between the Gift, Receiving, and Accounts Payable Departments?

3. What was the analyst's most serious mistake?

FOOTNOTES

1. Douglas McGregor, *The Human Side of Enterprise* (New York: McGraw-Hill, 1960). Theory "X" assumes that people naturally avoid work, are irresponsible, desire security above most other things, and that tight controls are required to keep them working properly.

2. Ibid. Theory "Y" assumes that, under proper conditions, people actually seek work and responsibility, that people are self-motivated, and have the capacity for a high degree of ingenuity and creativity.

CHAPTER 6

UNDERSTANDING
THE EXISTING SYSTEM

The fundamental problem of the systems
analyst is not whether to emphasize systems
analysis techniques, but how best to adapt its
offerings to the changing and challenging
needs of business. The analyst must thoroughly
understand the existing system before designing an improved version! This chapter identifies
what the analyst should do in order to understand the existing system.

WHY UNDERSTAND THE EXISTING SYSTEM?

The objective of this phase of the systems study is a complete understanding of the present operations of the area under study. It provides a "bench mark" for measurement: a clear picture of the present sequence of operations, processing times, work volume being handled, and existing costs.

A bench mark is a surveyor's mark that is used as a reference point for comparison during subsequent measurements. The analyst must obtain an understanding of the existing system for use as a bench mark or reference point in determining how much improvement can be made with a new system. Many people understand specific procedures within the system, but few understand the entire system. It is the analyst's task to learn the entire system before making any changes. Later, in order to sell a new system to management, a comparison must be made between the existing system and the proposed new one. A comparison of benefits cannot be made unless all aspects of the old system can be accurately compared with all aspects of the new one. Then management can estimate the benefits of converting.

The method of comparison may be from sys-view to prob-view. For example, a sys-view comparison is one that compares the subsystems or elements within the existing system with the subsystems or elements within the proposed new system. A prob-view comparison is one in which the overall costs or the overall numbers of people involved in the existing system are compared with the same overall things in the new proposed system. The sys-view approach is a detailed element-by-element approach, whereas the prob-view approach is an overall approach.

The analyst should first learn how the existing system came into being. How was the system designed and installed? By whom? Did it just evolve? Who, if anyone, has a stake in it? Is it someone's pet, or folly? Will anyone try to justify or defend the existing system, and thus slow up the change to a new system? What were some of the conditions that influenced the design of the system? Are they relevant now? These are questions that the analyst should try to answer in preparation for understanding the reasons behind the old system's detailed operations. In addition, special attention must be paid to identification of control strengths and weaknesses within the existing system so the control requirements of the new system can be defined carefully.

DEVELOP YOUR OWN APPROACH

As an analyst, develop an approach that best fits *you*. You can use a "formal" approach. This approach is very businesslike and follows a step-by-step logical plan with no idle talk or undue friendship. Just business! The next might be the "legalistic" approach. This is a rules approach. For example, if you work for a company you have a legal obligation to help improve it. The "political" approach is next. It is based on doing favors for people, performing only the jobs that make you look good, and acting in a prudent, shrewd, and diplomatic fashion. The last approach is the "informal" or play-it-by-ear approach. This approach follows a plan, but the analyst is able to deviate as required without losing control of the situation.

The best approach is one in which the analyst is able to combine the above four approaches. If the analyst can be "formal" when necessary, "legalistic" when challenged, play "politics" with deftness when necessary, and be "informal" in unknown situations, there is a built-in flexibility which cannot help but contribute to the analyst's success. You might remember the acronym FLIP for formal, legalistic, informal and political.

INTERVIEWING

The analyst's most important tool for gathering data is the personal interview. Systems interviewing simply is talking with people in order to find out how things operate now, and how they might like to see things operate in the future. Interviewing should be used throughout the systems study because it usually produces the most up-to-date information of all the sources available to the analyst. Organization charts and written procedures are often six months behind the times, a condition which can cause the analyst many headaches if they are used in place of interviews. During an interview the analyst has a chance to observe not only the current details of how the system works, but also the personality and attitudes of the employee. The latter can be the more important of the two in many situations. For example, the analyst may discover that a certain file clerk will be the one who will offer the strongest opposition to change. The actual influence the clerk will be able to exert over others must then be assessed, along with various methods of dealing with the situation.

Of course, the analyst should realize that one's own personality and approach will have much to do with the extent of cooperation received from those interviewed. A sloppy appearance, argumentative disposition, or an attitude of superiority can wreck the analyst's chances of getting the help of the employees. Word spreads fast if the analyst appears to be incompetent, disorganized, or untrustworthy. Operations personnel will retain this first impression and will withhold their cooperation as a group, making it nearly impossible for the analyst to ever successfully design and install a system for them.

The wise analyst will study the techniques of interviewing and will always have a planned approach to each interview. Above all, the analyst should bear in mind at all times that the function of an analyst is to *help* the people, not to enhance his own status at their expense. The analyst should have an honest respect for people at *all* levels in the organization, and they should be able to sense this through the analyst's approach.

As we implied, the interview is the best way to get the informal view. Interviews may be of a formal nature with a list of specific questions or they may be informal. The informal interview is one in which the analyst has a list of ideas to explore. The analyst talks informally with the manager or operating personnel in such a way as to get insight into these ideas, as opposed to getting answers to specific questions.

There are many interviewing techniques which the analyst can use to be effective in getting the needed information without wasting time, interfering with the firm's operations, or making a bad impression. At one extreme is the focused interview, in which the aspects of a situation have been preanalyzed and the analyst has arrived at certain conclusions. The analyst develops an interview guide as an outline to the major areas of inquiry. The interview is then focused on the subjective experiences of the interviewee. Responses of the interviewee enable the analyst to test the validity of the conclusions previously mentioned.

At the other extreme is the nondirected interview where an unstructured approach is taken. The interviewee is encouraged to discuss relevant experiences, define what is significant to him, and express any opinions as he sees fit. Whatever the approach, there are certain aspects to consider for a successful interview.

First, have a plan. Prepare ahead of time some of the guidelines that you will want to follow. Perhaps this will be in the form of an outline, or it may be some specific questions. Define the purpose of the interview and explain why the interviewee's opinions are needed. In most cases it is desirable for the interviewee to read the problem definition report (Chapter 3) ahead of time. It should be noted, however, that there are some situations in which it is undesirable for the interviewee to read this report; for example, if confidential information is included.

Always make an appointment; do not just drop in! Schedule your interview for a time when you can get the full attention of whomever you are interviewing.

It is usually desirable to interview in this order: first, the highest level of management that is involved in the area under study; then, the middle and supervisory managers; and finally, the operational employees in the area under study. In other words, try to work from the top down. It is good to have the higher level personnel introduce you to their subordinates, although sometimes there is no choice but to work upward. Try to interview people who may have differing perceptions of the same situation so you will have the benefit of contrasting points of view.

Obtain facts in advance. Whenever possible, gather background information both on the person to be interviewed and on the problems to be discussed. Try to determine what policies or personalities might be involved in the subject matter of the interview. How does the person whom you are going to interview fit into this picture? Can you determine any probable biases?

Two or three shorter interviews are better than a marathon since people get bored and lose interest during a long interview. Scheduling shorter interviews at intervals also allows both the analyst and the person being interviewed time to think between interviews. Each can go over his or her questions and answers, and perhaps decide to modify some of them. Such modification of ideas brings the analyst closer to precise understanding of the system.

A *second* aspect of successful interviewing is knowing how to conduct yourself during the interview.

Follow the plan or outline, but try not to stifle productive conversation with excessive rigidity. Try to get the needed information in a tactful way, without sounding as though you are grilling the person. Never allow the interview to sound like an investigation of something suspicious. (With the current rising trends in white collar crime, however, it is always possible that you may stumble upon some dishonest or illegal activities during a study.) Since you will probably want this person's help in the future, try not to give the impression that you are checking up on the person by comparing the person's opinions with the opinions of others. You do need to dig, however, to get to the problems and answers, so you might employ the technique of suggestion to see what reaction you get. It also is wise not to quote people directly by name since it is almost certain to get back to them. Instead, frame a hypothetical question or use some other indirect approach.

Relax and be flexible. Deviate from the outline if you think it will be more profitable. During the interview, be aware of body language on the part of the interviewee. There are many subtle involuntary reactions which can aid the interviewer. Among these are a slight change in facial expression, a change in the pitch or tone of voice, hesitancy in answering a question, or the tensing of a muscle.

There are two extremes to the personality of the area under study. In one, the people are relatively happy with their work, their surroundings, their manager, and their co-workers. For the analyst this is an ideal situation since happier people tend to be more cooperative—they want to right a wrong in their system.

At the other extreme is the group of people who are unhappy with themselves, their surroundings, their manager, and their co-workers. This attitude will be reflected in their acceptance of the analyst. They will either tend toward complete rejection, or they may open up to the analyst and pour out their troubles. Should the latter occur, try not to become involved in the group's day-to-day problems unless they pertain to the study. However, you will not want to turn them off entirely. Listen, for as short a time as possible. Then explain that you are sorry you cannot help them now on those points; but perhaps if they will complete a Problem Report Form (see Chapter 3), then someone can begin taking action. Or if those points are completely out of your control to remedy, suggest another approach, if you can; and then get down to the business at hand.

Ask for opinions and hunches, as well as for facts. All are equally important in learning about the existing system. Subjective evaluations are often closer to the person's true assessment of present and future realities.

Keep the interview moving, but do not fill periods of silence with idle chatter.

Always allow the person lots of time to think! And while doing so, use the time to phrase the next question, or to rephrase a previous one if it did not get the results you wanted. Phrase your questions clearly. If it appears that the person has sidestepped an important question, try to think of another source for the answer. If none exists and you do need an answer, explain why it is so important to have an answer, or explain some of the background or problems that led to the question. Perhaps you will get your answer, or perhaps you will learn there is something wrong with your approach to the subject.

Listen to answers clearly; do not anticipate answers by helping the person choose words. Allow people to answer in their own words at their own speed. If the answer is unclear to you, ask for clarification. If the person is one who obviously has difficulty expressing ideas, you may assist during this clarification stage, but do not put words in the person's mouth during the first answer to a question. Be alert to hidden meanings in answers.

Documentation of interviews is an important feature which should not be neglected. After all the data gathering has been completed, a "weeding out" process takes place during which the documentation is sifted for what is relevant and what is not.

Documentation of an interview takes many forms. Ask the people being interviewed if they object to a tape recorder. If there are no objections, a tape can then be transcribed for more accurate documentation. Notes may be handwritten or they may take the form of drawings or flowcharts. The written document then becomes a permanent record for this phase "Understanding the Existing System."

A *third* aspect of successful interviewing is summation. At the close of the interview verbally summarize the points covered, and verify any agreements you think you have reached with the person on important or controversial points. The object of stressing points of agreement is to verify that they do in fact exist and to end the interview in an "agreeable" mood.

Immediately after concluding the interview, summarize the interview notes in your office. Evaluate yourself! Were you really successful? Write down all the mistakes you made and decide how to avoid these mistakes in the future. Do not be discouraged. Interviewing takes practice. And even the experts run into people with whom they cannot deal in interviews. The wise analyst responds by finding other ways to obtain the needed information and cooperation without succumbing to discouragement.

WHAT TO ASK DURING THE INTERVIEW

The following general format is adaptable to most systems-type interviews. It is, of course, only indicative of the type of questions that should be asked. To ask the right questions in any study, the analyst must develop them from the study's stated objectives.

For the *operations* in the existing system the analyst should

1. Determine what is done, who does it, when it is done, how it gets done, where it gets started, and why.
2. Determine how much time it takes: per day, per hour, or per unit of work (average figures or minimum/maximum times should be obtained).
3. Gather ideas, opinions, and intuitive feelings on the systems objectives from the most experienced personnel.
4. Learn the customs and decision rules followed in the area under study, both the formal and informal views.
5. Identify the controls that ensure accuracy and completeness of processing operations.

For the *inputs* of the existing system in the area under study, the analyst should determine the following.

1. When and how is input received, and in what format?
2. From where and from whom is it received?
3. Is the input used as is, or is it checked, reworked, or just passed on?
4. What is the ultimate use, retention time, filing procedures, and disposition of all copies?
5. What are the controls that ensure accuracy and completeness of the inputs?

For the *outputs* of the existing system from the area under study, the analyst should determine the following.

1. What information is transmitted and what is the destination of that output?
2. What is the purpose of each copy and the routing of each copy of a form?
3. How is the output compiled, filled out, sorted, reproduced, checked, and what are the various times required for each?
4. What errors and omissions are referred back to the area under study, and what are the corrective action procedures for these errors?
5. How is the output controlled to ensure that it is accurate, complete, timely, and only delivered to appropriate personnel?

For the *computer* or related equipment of the existing system in the area under study, the analyst should determine the following.

1. Identification and description of the computer and its related equipment.
2. A list of the applications being run on the computer that pertain to the area under study.
3. Workload: amount of work, and the time required to do it.
4. Capability: unused capacity, programming, and systems capability.
5. What controls are in place to ensure a secure computer system?

SEARCHING RECORDS

Searching the business records is the easiest way of obtaining formal information. Informal methods are not usually documented. Therefore the analyst should remember that searching records is for learning formal information, whereas interviewing can give both formal and informal information. (See Chapter 19 for additional information on this area.)

The first, and most important, records that the analyst should look at are the written policies and procedures for the area under study. This will tell the analyst how the area's work probably should be done, if the procedures are reasonably up-to-date. However, discussions with the operating personnel will usually reveal deviations for reasons both good and bad. (Remember, the wise analyst should not criticize people directly.)

We can refer to files maintained by the area under study as "internal" files to differentiate them from files maintained in other areas of the firm or outside the firm.

The analyst should obtain a list of the internal files while interviewing key personnel. Then it should be determined which of these files are used in the existing system.

Such files will normally contain the up-to-date forms and data used by the area in the existing system. The analyst should study the documents carefully to note the various differences between transactions.

For example, it may be discovered that a form has been completed in several different ways at different times. It may be noted that on one form all spaces were filled in, and on another only some of them were used. Or there may be variations in routing, or special instructions written in, and so on. This usually indicates that the routine in which the form is involved is subject to several variations and exceptions.

The analyst might go through such a file to list such variations and exceptions, and the reasons for them (as explained by area personnel). These variations and exceptions will reveal a great deal about the existing system, and will alert the analyst to the various conditions which the new system must improve or deal with. Since this is admittedly a very time-consuming process, it is acceptable in most cases to use a random sample approach (see the next section for a discussion on sampling).

Internal records or files that the analyst might examine are operating files such as purchase orders, invoices, sales orders, shipping forms, and so forth, memoranda files, computer files, and files maintained by individuals for protection or improved efficiency. Usually the files that are the most valuable to the analyst are the ones which are in active use in the operation of the existing system. (See Appendix I for an in-depth list of considerations on how to improve file operations.) Of course, any files that are not used also are subject to question. The analyst should review the area's records retention policy. (See Chapter 15.)

"External" records or files are those files that are maintained outside of the area under study or outside of the firm. External files often affect the area's systems in various ways. Files maintained by customers, vendors, governments, and so forth, can be considered "active" to the extent that they might be used for, with, or against the firm. The situation varies widely from one industry to another. The analyst should attempt to compile a list of such files and study their potential bearing on the existing system and the proposed new system.

ESTIMATING AND SAMPLING

Estimating and sampling can be used to predict costs, quantities, time periods, and other parameters that are relevant to the existing or proposed system. These techniques are appropriate when internal and/or external records are not sufficiently complete or accurate. Estimating is the art of predicting, and therefore is uncertain to some degree. Any estimates made by the analyst should be checked against some known or reasonable overall total.

There are three kinds of estimating procedures which are generally useful in systems work. These are *conglomerate estimating, comparison estimating,* and *detailed estimating.* In *conglomerate estimating,* the representatives from each functional area within the area under study get together and confer to develop estimates based on past experience. In *comparison estimating,* the analyst meets individually with anyone, inside or outside the firm, who has a similar system. The analyst then evaluates comparable operations and comes up with estimates. In *detailed estimating,* the analyst makes a detailed study of the costs, times required, and any other pertinent factors for each step of each procedure within the system. Whenever extreme accuracy is required, perhaps because of the importance of the decisions involved, detailed estimating should be used.

Sampling is the collection of a limited quantity of data. In sampling, one collects only a fraction of the total existing data, for the purpose of studying that fraction to infer things about the total. The fractional data is the sample. The size of the sample required for such inferences varies according to the desired accuracy and certainty.

There are several approaches to determining the sample size. The following is a simple and reliable method for most applications.

$$\text{necessary sample size} = 0.25 \left(\frac{\text{desired certainty}}{\text{acceptable error}} \right)^2$$

DESIRED CERTAINTY	
For 97% certainty	Use 2.170
95% certainty	1.960
90% certainty	1.645
80% certainty	1.281

If the systems analyst wants to be 95% certain that sample data will not vary (acceptable error) by more than 5% (0.05) from the actual data, the sample size should be 384.

$$\text{necessary sample size} = 0.25 \left(\frac{1.96}{0.05} \right)^2 = 384$$

The 0.25 is a constant and is always used. The desired certainty is chosen by the analyst and is given above. The acceptable error in the above case was 5% (0.05), but the analyst can choose any percentage desired such as 2% (0.02) or 10% (0.10), and so forth.

For example, after determining the necessary sample size, the analyst can *randomly* collect a sample of 384 purchase orders to analyze. If there are an average of 2.5 mistakes per purchase order, the analyst can be 95% sure that the average of 2.5 mistakes per purchase order does not vary by more than 5% from the average for *all* the purchase orders. Obviously sampling allows the analyst to learn things about the existing system with only a fraction of the effort that 100% evaluation would require.

The systems analyst usually is not required to perform time studies, motion studies, or methods-time measurement studies. These are functions of the industrial engineer. However, the systems analyst may wish to perform a *work sampling study* in order to understand the existing system. A work sampling study consists of a large number of observations taken at random intervals. At each observation the analyst notes what the employee is doing and records it into one of a number of predefined categories. Work sampling may be carried out in the following fashion.

1. Sell the idea of work *sampling* rather than continuous observation of the job. Convince management that sampling will enable you to get the desired results in a fraction of the time required for continuous observation.

2. Define the operations of the job. (These are the "predefined categories" for recording.) Make an Observation Recording Form (Figure 6-1) for recording observations that will be made during the study.

3. For nonrepetitive jobs it is better to make many observations. If you need 2401 observations for the desired accuracy,[1] divide them by the number of days allotted to the study to get the number of observations required per day.

4. Make sure the observations are made at random times. Do not observe the people or events at the same time each day since you may not get a reliable sampling of their complete routine.

5. After making 2401 or more random observations on each item, person, or event being observed, you can compute the percentages for each predefined category, as illustrated for item 1 below (data are from Figure 6-1).

	Item being observed	
Predefined categories	1	Percentages
On telephone	241	241/2401 = 10%
Reading	501	501/2401 = 21%
Writing	362	362/2401 = 15%
Conference	831	831/2401 = 35%
Walking	345	345/2401 = 14%
Elevator	30	30/2401 = 1%
Idle	91	91/2401 = 4%
Total observations	2401	100%

6. Calculate the percentages as in step 5 above for the other items (2 through 7) from the observation recording form (Figure 6-1) and you have the percentage of time spent on each predefined category of work for each person or event (items 1 through 7).

Once the analyst has computed and studied the percentage of time spent in each category of work, there is an excellent basis for evaluating the existing system, and a bench mark for comparison with the new system design. The information gained from

OBSERVATION RECORDING FORM

Predefined categories	Items being observed: people or events						
	1	2	3	4	5	6	7
On telephone	(291) ///	†††		//			/
Reading	(801) /	///	†††	////	†††	/	///
Writing	(362) †††	///	///	//	†††	//	
Conference	// (831)			†††			
Walking	(345) ///	//	////	†††	///	/	/
Elevator	/ (30)	/				†††	
Idle	††† (91)	//					/
Total observations	2401	2401	2401	2401	2401	2401	2401

FIGURE 6-1 Observation recording form.

a work sampling may be used to realign job duties among employees, to set work standards for office employees, or to eliminate unnecessary procedures.

GAINING THE CONFIDENCE OF THE AREA UNDER STUDY

People, as a general rule, do not accept change readily. Since they may view the systems analyst as a threat to their security, communication is the key to gaining their confidence. If the analyst is able to communicate sincere willingness to help operations personnel, they will be more apt to give the needed cooperation. Every effort should be made to inform the workers of the project objectives, plan, and progress.

The methods for doing this are only as limited as your imagination. Look for skeptics, expose them, involve them, loosen up the ones with set ideas, and try to understand their problems. Circulate rough drafts of proposals and request written comments. You may not understand clearly all the ramifications of something until it is in writing, and it is best to learn of these things early! If possible, have some simulated exercises which can quickly point out any deficiencies. Ask the users to identify these deficiencies.

The users of the system should feel that they have an active part in its development. The analyst should involve them as much as possible through department meetings or any other possible way. The analyst should entertain any reasonable suggestions and attempt to explain why others do not fit the study objectives. Since the analyst is responsible for the final decision on what design elements will eventually be presented to management, the analyst's role is that of a leader. It is the analyst, in the final analysis, who hears the many viewpoints and combines them all into a feasible system.

The users of the system know the existing system and can contribute many valuable suggestions for its renovation. If the analyst heeds their suggestions and performs in a reasonable manner, the operations personnel will accept a new system more readily. They will, in fact, be obligated to make it work.

DEVELOP A METHOD TO DOCUMENT THE DATA

As the analyst goes through the process of understanding the existing system, a method of documenting the collected data must be developed. As the analyst interviews, searches records, makes estimations or samples, the data collected must be recorded, summarized, and evaluated so they will make sense. The job is to understand the existing system; it must be documented so it can be compared with the new proposed system, especially for economic cost comparisons. One possible example of developmental system documentation follows. But the analyst should be able to modify this example to fit the specific situation. Since it is used to define and analyze the existing system, another method altogether may be required for the application faced by the analyst. A data documentation example follows.

The first piece of documentation is the Area Cost Sheet. It fits each area under study into its larger context. The purpose of the area cost sheet is to document the formal organization and economic data pertaining to the area under study. Specifically the area cost sheet does the following.

1. Shows the formal organization.
2. Permits a rapid analysis of the overall structure into which the area under study fits.
3. Shows the cost impact of the area under study with respect to the overall cost structure.

The area cost sheet is divided into three sections (see Figure 6-2).

1. *The formal organization chart.* This organization chart must be broad enough to cover the entire area under study. Each box shows the number of full-time equivalent employees (F.T.E.'s) in that area, except that the lowest box gives the total F.T.E.'s from that level down.
2. *The budget breakdown section.* These are the annual or monthly costs for each organization in the lowest level of the organization chart. The upper level costs are prorated into this tabulation. In this example, personnel, equipment, supplies, miscellaneous, and training were used as categories. The analyst should use whatever categories fit the study. Departmental budgets or historical costs from accounting are the best sources for these costs.
3. *The area under study section.* Assume that the area under study is Collection Costs, that is, the cost of collecting money from customers who charge purchases. Using cost data and the judgment of the supervisory personnel involved, isolate only those costs that pertain to the area under study, which in this example is collection costs.

The analyst now has a complete cost picture of the area under study. Include the organizational structure that lies above the specific area being studied plus the overall costs and specific costs of the area under study. For example, the new accounts department has a total annual cost of $27,500, but only $5,100 of that $27,500 pertains to collection costs.

Another general method of documenting the existing system is to use a Documented Flowchart. The documented flowchart (Chapter 13 covers flowcharting) traces the flow of a single activity through its sequence of operations. It begins with

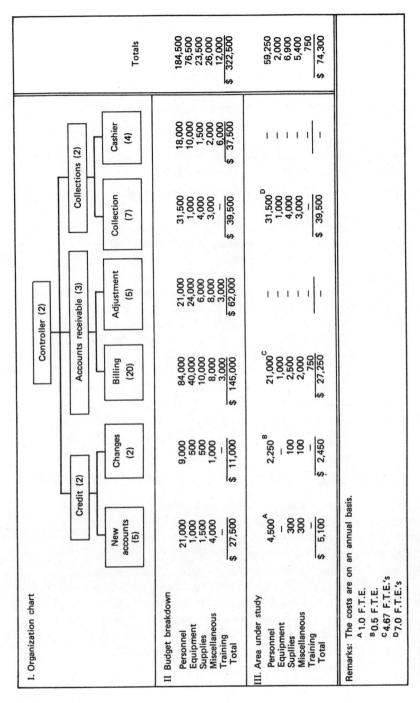

I. Organization chart

Controller (2) → Credit (2); Accounts receivable (3); Collections (2)

Credit (2) → New accounts (5); Changes (2)

Accounts receivable (3) → Billing (20); Adjustment (5)

Collections (2) → Collection (7); Cashier (4)

II. Budget breakdown

	New accounts (5)	Changes (2)	Billing (20)	Adjustment (5)	Collection (7)	Cashier (4)	Totals
Personnel	21,000	9,000	84,000	21,000	31,500	18,000	184,500
Equipment	1,000	500	40,000	24,000	1,000	10,000	76,500
Supplies	1,500	500	10,000	6,000	4,000	1,500	23,500
Miscellaneous	4,000	1,000	8,000	8,000	3,000	2,000	26,000
Training	–	–	3,000	3,000	–	6,000	12,000
Total	$ 27,500	$ 11,000	$ 145,000	$ 62,000	$ 39,500	$ 37,500	$ 322,500

III. Area under study

	New accounts (5)	Changes (2)	Billing (20)	Adjustment (5)	Collection (7)	Cashier (4)	Totals
Personnel	4,500[A]	2,250[B]	21,000[C]	–	31,500[D]	–	59,250
Equipment	–	–	1,000	–	1,000	–	2,000
Supplies	300	100	2,500	–	4,000	–	6,900
Miscellaneous	300	100	2,000	–	3,000	–	5,400
Training	–	–	750	–	–	–	750
Total	$ 5,100	$ 2,450	$ 27,250	–	$ 39,500	–	$ 74,300

Remarks: The costs are on an annual basis.
[A] 1.0 F.T.E.
[B] 0.5 F.T.E.
[C] 4.67 F.T.E.'s
[D] 7.0 F.T.E.'s

FIGURE 6-2　Area cost sheet.

whatever started the sequence of operations and goes on to its completion. The documented flowchart itself (see Figure 6-3) is just a typical systems flowchart, but there are numbers in each box. These numbers are used to cross reference each flowchart box back to the Documentation Section (Figure 6-4).

In the documentation section of the documented flowchart the analyst records such items as

1. What is done, with what resources (financial, personnel, inventory, or facility), under what conditions, how often, and to produce what results.
2. Elapsed time, frequencies, volumes, and decision points. It should be as detailed as necessary: prob-view, feas-view, or sys-view level of detail.
3. Collect any forms used in the system. Using a standard instruction sheet (if one exists) for each form, fill in the form with sample data. Attach the forms to the documentation section to which it fits (Figure 6-4).

The documentation section is almost a narrative of the flowchart and is cross referenced into the flowchart by the individual box numbers that are in the flowchart itself. The narrative and numerical values collected are recorded in Figure 6-4 (documentation section) and attached to the flowchart. The set then becomes the documented flowchart. One of these documentation sections should be used for each box in the flowchart.

Both the area cost sheet and the documented flowchart are excellent methods of documenting the existing system and/or the new proposed system. More documentation ideas are offered in Chapter 7.

PREPARE A SUMMARY ON THE EXISTING SYSTEM

Although the developmental documentation already prepared is an excellent summary of the existing system, a summary of findings still needs to be prepared. This summary should include everything of importance learned during this phase of the study. It is, in fact, the analyst's written understanding of the existing system.

The summary should include any design ideas, notes on whether currently used forms are adequate or inadequate, who was helpful, who hindered progress, and overall impressions. Notes on all aspects of the study should be included whether gained from interviews, meetings, records, flowcharts, or work sampling.

In general, what is needed here is something which can be referred to later for data when defining the systems requirements and designing the new system. It is the "bench mark" to be used for later comparisons.

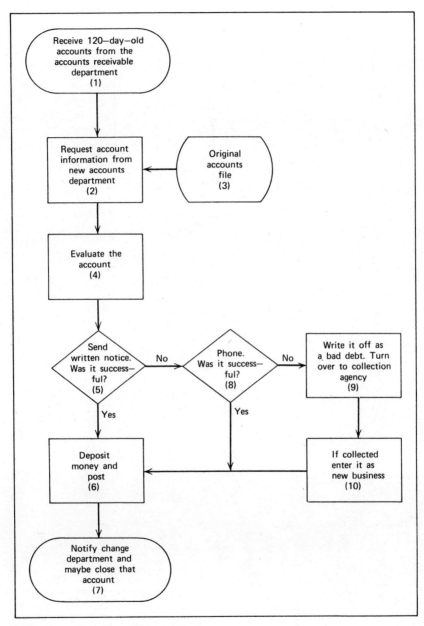

FIGURE 6-3 Documented flowchart.

DOCUMENTATION SECTION			
Flow chart box number	Average number of items	Peak number of items	Average time per item
Analyst's name			
Narrative:			

FIGURE 6-4 Documentation section of the documented flowchart.

A sample outline for the summary of the existing system may look like this.

I. Description of the existing system.
   A. Inputs.
   B. Operations.
   C. Outputs.
   D. Resources.
      1. Personnel.
      2. Inventory.
      3. Financial.
      4. Facility.
   E. Operational and accounting controls.

II. Documentation.
   A. Interviews.
   B. Records.
   C. Flowcharts.
   D. Work sampling.
   E. Cost analysis.

III. Positive benefits of existing system.

IV. Weaknesses in existing system capabilities and controls.

V. Any other items relevant to the existing system.

SELECTIONS FOR FURTHER STUDY

You must understand the current system in order to set goals for the new system. At this point you may wish to read *Goal Setting: Key to Individual and Organizational Effectiveness.*

Since interviewing is of primary importance at this stage, it is essential that you learn how to go about it. To learn how, observe what you like and dislike during television interviews. Does the interviewer listen to the interviewee's answers or is the interviewer too busy thinking of a cute comment or the next question? Observe the very successful interviewers and compare them with some of the newcomers.

Social researchers have been covering the techniques of interviewing for years, so any good text on methods of sociology would be helpful. In addition, there are a number of books specifically on interviewing. It is important that you know to whom you are speaking, particularly when dealing with upper management personnel. Consult your library's resources to determine ahead of time your interviewee's expertise on the subject, political orientation, or any other factor that might have a bearing on the interviewee's personality. Some firms maintain a file of biographical information on all management personnel. Others maintain files of reprints by authors within the company, while still others maintain newspaper clipping files relevant to their personnel. These files are maintained for just this type of use. With the newly available computerized bibliographic indexes, it also is possible to have your librarian perform a quick search on your interviewee as an author.

Searching internal records is a fairly straightforward way of obtaining information. Files maintained by employees to protect themselves or to function more efficiently are very useful. For example, a person working with classified government documents may feel the necessity of keeping extra files on orders processed to avoid being turned down twice for the same document request. At first notice it appears to be a terrible waste of time, but to the engineer working on a short-term contract where all ordering takes valuable time, this record is a blessing and its cost is minimal where large contracts are at stake.

Quite a bit has been written on both estimating and sampling. The systems analyst will find Maynard's *Industrial Engineering Handbook* to be good basic information on the subject. For a more comprehensive treatment consult Cochran's *Sampling Techniques*. Work sampling and time study also are covered in Maynard's *Industrial Engineering Handbook*, but you may wish to consult a volume that deals exclusively with that subject, such as Barnes' *Motion and Time Study*. The four phases of work measurement are outlined with ample use of tables in *Guidelines for Organizing a Work Measurement Program*.

Since communication is the key to gaining the confidence of the area under study, you would do well to examine some of the numerous books available in the area of communications.

Documentation has been an essential step in scientific method for many years. It is equally essential for the systems analyst. An IBM manual describes in detail an approach to documenting the study, analysis, and design of business systems. Your sources for the budget breakdown section are departmental budgets, updated historical costs from accounting, and interviews with and estimates from management personnel.

1. BARNES, RALPH M. *Motion and Time Study, 7th edition.* New York: John Wiley & Sons, Inc., 1980.

2. BASSETT, GLENN A. *Practical Interviewing.* New York: American Management Association, 1965.

3. CARIN. *Creative Questioning and Sensitivity: Listening Techniques, 2nd edition.* Columbus, Ohio: Charles E. Merrill Publishing Company, 1978.

4. COCHRAN, W.G. *Sampling Techniques, 3rd edition.* New York: John Wiley & Sons, Inc., 1977.

5. DONALDSON, HAMISH. *A Guide to the Successful Management of Computer Projects.* New York: Halsted Press, 1978.

6. GORDON, R.L. *Interviewing: Strategy and Techniques.* Homewood, Ill.: Dorsey Press, Inc., 1969.

7. GEETING, BAXTER, and CORINNE GEETING. *How To Listen Assertively.* New York: Monarch Press, 1978.

8. *Guidelines for Organizing a Work Measurement Program.* Cleveland, Ohio: Association for Systems Management, 1971.

9. HAMILL, B.J. *Work Measurement in the Office: An MTM Systems Workbook.* New York: International Publications Service, 1974.

10. HARPOOL, JACK D. *Business Data Systems: A Practical Guide.* Dubuque, Iowa: William C. Brown Publishers, 1978.

11. HODNETT, EDWARD. *Art of Working with People.* New York: Harper & Row Publishers, Inc., 1959.

12. HUGHES, CHARLES L. *Goal Setting: Key to Individual and Organizational Effectiveness.* New York: American Management Association, 1965.

13. *IBM Data Processing Techniques: IBM Study Organization Plan: Documentation Techniques.* White Plains, N.Y.: International Business Machines Corporation, 1961 (C20-8075-0).

14. JOHNSON, IDA MAE. *Developing Listening Skills.* Hayward, Calif.: Activity Resources Company, Inc., 1974.

15. KOCHHAR, A.K. *Development of Computer-Based Production Systems.* New York: Halsted Press, 1979.

16. LUNDBERG, GEORGE A. *Social Research: A Study in Methods of Gathering Data, 2nd edition.* Westport, Conn.: Greenwood Press, Inc., 1968.

17. MAYNARD, HAROLD B. *Industrial Engineering Handbook, 3rd edition.* New York: McGraw-Hill Book Company, 1971.

18. MUMFORD, ENID, and DON HENSHALL. *Participative Approach to Computer Systems Design.* New York: Halsted Press, 1979.

19. QUICK, JOSEPH H., et al. *Work-Factor Time Standards: Measurement of Manual and Mental Work.* New York: McGraw-Hill Book Company, 1962.

20. SANDERS, DONALD H. *Computers in Business.* New York: McGraw-Hill Book Company, 1975.

21. SMITH, GEORGE L. *Work Measurement: A Systems Approach.* Columbus, Ohio: Grid Publishing, Inc., 1978.

22. VAN FLEET, JAMES. *How To Put Yourself Across to People.* Englewood Cliffs, N.J.: Prentice-Hall, Inc., 1971.

23. WILLOUGHBY, THEODORE C., and JAMES A. SENN. *Business Systems.* Cleveland, Ohio: Association for Systems Management, 1975.

QUESTIONS

1. What is the fundamental problem of the systems analyst?
2. With regard to understanding the existing system, what is a "bench mark"?
3. Name the four approaches the analyst could use in gaining an understanding of the existing system.
4. Name three aspects to consider for a successful interview.
5. Name five things the analyst should determine during an interview for the inputs of the existing system in the area under study.
6. What is the easiest way to obtain formal information when seeking an understanding of the existing system?
7. Describe conglomerate estimating.
8. Describe detailed estimating.
9. What is the purpose of the area cost sheet?
10. What does the documented flowchart (see Figure 6-3) accomplish?
11. Why are control strengths and weaknesses important for the analyst to examine for understanding the existing system?

SITUATION CASES

CASE 6-1

The president of the ABC Company has noticed that sales for the past six months have declined steadily. In order to get a more complete answer as to why the sales have declined, the president called on the company's senior analyst to help solve the problem.

During the analyst's study, she identified the problem as purchase orders that were filled too slowly, thus causing many customer complaints. Further study showed the cause of the problem to be a combination of overwork and lack of organization in the company's procedure for completing the purchase orders and delivering the goods.

After defining the problem, outlining the study, obtaining general information, and studying the relative departmental interactions, the analyst began her project to gain a complete understanding of the existing system.

First, the analyst learned how the existing system came into being by finding out who designed the system and how it was designed. She discovered that the system was designed by a former analyst who was no longer employed by ABC.

The analyst studied the company's organization charts and written procedures but found that these documents had not been kept up to date. In order to get a better idea of how the existing system works and to observe the personality and attitudes of the employees, the analyst decided to conduct a series of interviews.

The analyst determined that informal interviews would yield the best information relative to what she was seeking. She carefully planned her approach to the interviews and wrote an outline of all the items that she wanted to cover during the interviews.

Before beginning the interviews, the analyst obtained all the facts that she could find concerning the area under study. With the help of the personnel office, the analyst also gathered background information on each individual whom she was going to interview.

The analyst began her interviews with the workers who filled the incoming orders and worked her way up to the head of the department. She felt that this order of interviewing would give her a better feel for how the system was working and how it could be improved. So that she would have sufficient time to gather useful information, the analyst planned to conduct each interview over a two-hour period.

After completing all of the interviews, the analyst summed up her data and correctly finished the remaining stages in understanding the existing system. She was now ready to define the requirements.

QUESTIONS

1. What error did the analyst make in setting up her schedule of interviews?
2. What important feature of interviewing did the analyst overlook? Why is it important?

CASE 6-2

ACE Steel Company is a small steel manufacturer that has an electric arc furnace, a continuous casting machine, and a rolling mill. The organization structure has a typical line and staff arrangement. (See Figure 6-5.)

119

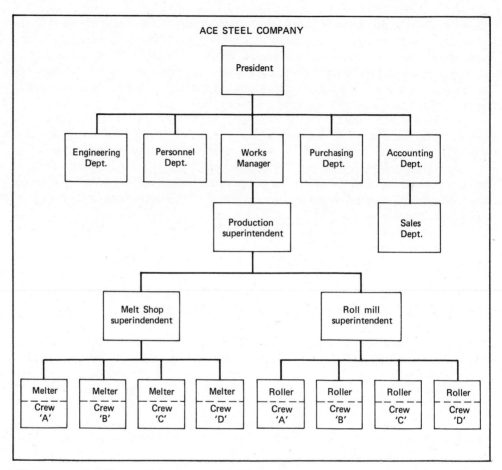

FIGURE 6-5   Ace Steel Company organization chart.

The president of ACE Steel Company had two major problems in the Melt Shop area. They were: (1) a need for exact control of costs, and (2) the need for improved product quality and performance. He asked the chief engineer to make a thorough study of the Melt Shop area to determine how the existing system keeps track of costs and how the product might be improved.

The chief engineer then asked his system analyst to gain an understanding of the entire current Melt Shop area system. After this, the analyst was to design an improved version which would keep exact cost records and improve the product quality and performance.

The analyst felt that his most important tool for gathering data of this type would be the interview. With this tool, he would have the opportunity to observe the current system at work, and the personality and attitudes of the Melt Shop personnel.

In order to gain the confidence of the Melt Shop personnel, the analyst decided to adopt the "informal" approach to interviewing. He planned to outline his plan and follow it; however, he would deviate as the situation required as long as it was profitable and relevant.

The analyst learned that the current Melt Shop system was designed and installed by the present production superintendent. The analyst felt that this superintendent might try to justify or defend the existing system. With this thought in mind, he decided that he would begin his interviews with the Melt Shop superintendent.

Before beginning these interviews, the analyst sat down and developed an informal outline of the things that he wanted to accomplish. In keeping with his "informal" approach to interviewing, he dropped in on the Melt Shop superintendent at a time that he considered to be convenient. He was in luck; the superintendent was happy to give him a little of his time. The analyst explained his reason for the meeting. The analyst was very tactful and informal in asking questions. He did not give an impression of checking on anyone nor did he compare the superintendent's opinions with other opinions. The interview moved along in a relaxed mood. One of the melters came in to confer with the Melt Shop superintendent about some current operational problem. The analyst used this time to phrase his next questions. When the Melt Shop superintendent finished with the melter, they continued the interview. Some of the questions that the analyst asked were: What is done? Who does it? When? How? Where is it started? Why? How much time or what unit of work does it take?

He obtained average time figures, and gathered ideas, opinions, facts, and intuitive feeling from the Melt Shop superintendent. In addition to this information, he also learned the customs and key decision criteria that they followed in the Melt Shop area. He learned about the inputs: when and how they were received, what their format was, where and from whom they were received, whether the input was used as is, checked, reworked, or passed on, their ultimate use, retention time, filing procedures, and the disposition of all copies. Of outputs, he learned what information was transmitted and where it was going, what the purpose and routing of each copy was, and how the output was compiled, filled out, sorted, reproduced, or checked. He also determined what types of errors are returned to the Melt Shop area. When the interview was over, the analyst asked the Melt Shop superintendent to introduce each of the melters. The superintendent did, and the melters' interviews proceeded in the same manner as the interview with the Melt Shop superintendent. After concluding the interview with the last melter, he felt ready to interview the man who was responsible for the design and installation of the existing system.

The analyst dropped by the production superintendent's office. The production superintendent's secretary kept the analyst waiting for some time in the outer office. When the production superintendent finally saw him, it seemed to the analyst that his manner was somewhat restrained. The analyst explained his reason for the meeting. The analyst began the interview but was interrupted several times by the secretary. He concluded the interview as quickly as possible because he felt that this man was no longer objective as far as this interview was concerned. Although the analyst had a vague feeling of having left something unfinished he felt ready to design the new system.

The analyst designed a new improved system which took into account all those points that he had learned during his interviews. He then wrote a report about his new system and turned it over to the chief engineer.

The chief engineer reviewed the report the analyst had written and recommended to the president that they implement the new system as soon as possible. The president agreed and it was implemented with these results: the new system began to have trouble right away. The Melt Shop superintendent and melters complained that the analyst had misunderstood key points of the old system. They also said that when he performed the study, he interfered with their production and operation. The production superintendent allowed the new system to operate for three months; then he asked for, and received, permission from the president to return to the old system.

QUESTIONS

1. What important step did the analyst leave out of the interviewing technique and how did this affect his system?

2. In what way did the analyst misuse his most important tool (the interview)?

3. What other activities could the analyst have performed before designing the new system?

FOOTNOTE

1. $N = 0.25 \left(\dfrac{1.96}{0.02}\right)^2 = 2401$ (95% certain that the sample data will not vary from the true data by more than 2%)

CHAPTER 7

DEFINE THE
NEW SYSTEMS
REQUIREMENTS

**The conversion of ideas to realities requires
the use of real methods and resources. Thus the
systems analyst must define the real methods
and resources that will be required to move the
system from the drawing board to actual
operation. To accomplish this, the broad performance objectives must be established.**

WHY DEFINE REQUIREMENTS?

The object of defining the *new* system requirements is to assemble an overall picture of the inputs, outputs, operations, and resources required by the system to meet the present and future needs of the organization. Another activity of this phase is to outline the evaluation criteria which will be used to evaluate the new system's performance. In this chapter we will discuss the broad performance criteria which are required to design the new system. We do not include here the technical aspects of the system which will be included in Chapter 8 "Designing the New System." In other words, we are not yet designing the new system; we are *preparing* to design it.

At this point in the systems study, there is usually a problem definition, perhaps a feasibility study report, an outline, some general information on the area under study, and an understanding of the existing system. Now all of this information should be pulled together. The first objective is to define what the new system must be able to do. The second objective is to determine methods that can be used to evaluate the performance of the new system.

In this chapter we will use as an example a new system being designed for a department that has the responsibility for collecting overdue accounts.

Let us assume that the problem has been defined as not enough overdue bills being collected by this department. Let us further assume that the symptoms of this problem have been singled out as being overwork and disorganization in the Collections Department, and we have concluded that a new system is needed to straighten things out. We have the authorization to begin developing the new system's requirements.

LONG-RANGE PLANS

If the analyst has not done so already, the long-range plans should be reviewed at this time. Decisions must be made concerning which work operations and activities will be retained, modified, or eliminated. This requires an investigation of any long-range plans directly affecting the area under study. An understanding of any such plans is necessary before the new systems requirements can be properly defined. This approach enables the analyst to define system requirements that will not be made obsolete by later implementation of long-range plans which were available at the time of system design. For example, if the firm intends to extend more lenient credit terms in the future, the system must be capable of processing the attendant increase in overdue accounts, data, and paperwork flow. Otherwise, the new system will be obsolete soon after the new credit terms are put into effect. Other examples of long-range considerations are

1. Changes in company goals and policies.
2. Development plans for new products or new services.

3. Projections of changing sales, manpower, or cash flow.

4. Research and development projects.

5. Major capital expenditures: new plants, computer, and so forth.

6. Possible changes in product mix.

7. A new level of file security.

The term "long-range plan" may have a different meaning to different people. In practice, some people view long-range plans as making long-run forecasts and then adjusting to them, defining what the company should be, or developing a long-range program for the entire company. An analyst who makes decisions based simply on reading a long-range plan is not totally fulfilling his or her commitment to the company. Elements that distinguish a good analyst in this respect are: (1) Instead of defining the new system's requirements on the assumption that present conditions will continue, the analyst should try to read the future, challenge the status quo, and apply imagination in evaluating the systems requirements for the years to come. (2) The analyst should attempt to fit the implications of the requirements of the system into some kind of consistent pattern that fits the company's long-range plans.

The long-range plan of a company is its "road map" of the future. The long-range plan points in the direction that the company will be moving in the future. As we have already stated, the analyst looks at the long-range plan in order to make sure the new systems requirements fit in with the direction that the company will be moving. Another possibility is that the long-range plan itself might be affected by the results of a major systems study. The analyst should always be alert to the fact that a systems study evaluating something as big as the acquisition of another company may affect the long-range plans of the parent company. Therefore, usually the long-range plans affect the requirements of the new system, but in some cases the systems study itself may affect the long-range plans of the company.

It may seem absurd that a system could actually be designed and implemented, only to become useless to the firm shortly thereafter because of the implementation of previously made plans. The fact is that it happens *often* in moderate to large organizations. The usual reasons are complexity of the organization, communication problems, lack of an organized knowledgeable approach, or the analyst's difficulties in learning of all the activities and information that may affect an area under study. In our example, the analyst may receive approval to an area under study. In our example, the analyst may receive approval to locate the customer record files in an area strategically related to other components of the system, and proceed to design the new system and office layout around the location of the file bank. Then, to everyone's dismay, it is learned that the person who gave authorization did not know about the new blueprint files that were already approved for eventual placement in that location. Or the analyst might design exception routines into the system to handle certain special requirements, only to learn later that the firm will not be dealing with that type of situation by the time the system gets into operation.

For example, in our new collection system the analyst might design a routine to process the overdue accounts of an owned subsidiary company. This could easily require the inclusion of numerous inconvenient steps in what might otherwise be a smooth system. That is, if we had only our own accounts to process, things would be much simpler—but with the subsidiary's bills to do, too, we have to design "exception" routines to handle their differences. This could require a redesign of many forms, new files, and dozens of other problems. But without asking any questions about future plans for the subsidiary, we design the necessary exception procedures, at the cost of considerable time and effort. If only we had asked the right person the right question we could have learned that the subsidiary company will have been sold by the time our new system goes into operation!

The analyst's step of checking for existing plans or anticipated changes is a mandatory one! And above all, the information received regarding such plans and contingencies must be reliable information. The analyst should investigate the possible existence of conflicting plans or contingencies for every major aspect of the system being designed.

DEFINE SPECIFIC REQUIREMENTS OF THE NEW SYSTEM

In order to define what the new system must be able to do, all the data collected during the study of the existing system should be reviewed. The objectives set during the problem definition phase also should be analyzed and refined. This information, and any other relevant information such as long-range plans affecting the area under study, should be used as a basis for developing the new system requirements.

The analyst must be very careful when defining the new system requirements because these requirements are the analyst's "road map" that will be used to design the new system. These requirements must be broad enough to cover each and every detail of the system. They must be flexible so the analyst can mold them or modify them during the design phase (Chapter 8). The system requirements are the "heart" of the new system that the analyst will be designing. The analyst will be designing a new system to carry out and achieve the system requirements that were defined during this phase of the systems study.

While the new system requirements are being determined, they should be summarized continually in a write-up that outlines the objectives of the new system, and spells out the requirements for each activity or job procedure involved in the new system. The analyst should determine the requirements of the new system in terms of

1. Outputs it must produce.
2. Inputs it needs to produce the outputs.

3. Operations it must perform to produce the outputs.

4. Resources it must use to produce the outputs.

5. Operational and accounting controls.

Define outputs first. Then define *inputs.* Next define *operations* and *resources.* Last, define *controls.* By working from the anticipated end product (outputs) BACKWARD into the system, the analyst can be assured of a "clean" design unencumbered by existing system routines. Put another way, the analyst can be innovative in the use of new approaches since the only concern with the existing system will be at those points which interface with other systems.

When defining outputs, inputs, operations, and resources for the new system, four major questions should be considered.

1. What are the *current requirements* of the new system?

2. What are the *future requirements* of the new system?

3. What are the *management-imposed requirements,* such as time limits, or restrictions on money expenditures?

4. What are the key operational and accounting *control points?*

Once the analyst has satisfactorily answered these questions, a system requirements model can be developed. Figure 7-1 depicts one of these models in general terms. Note that it is concerned only with the new system, not the existing system.

DOCUMENT THE REQUIREMENTS FOR THE NEW SYSTEM

Figure 7-1 shows the System Requirements Model as applied to our hypothetical collection system. The level of detailed description will vary from one item to another, but the system requirements model ensures that *all* requirements (current, future, management-imposed, and controls) of the new system are taken into account. It provides a checklist which ensures that the analyst has not forgotten any of the outputs, inputs, operations, controls, or resources that will be necessary in operating the new system. It is an excellent device also for communicating the picture of a system to others. The model should always be checked for completeness by the personnel in the area under study. Figure 7-2 shows the detail of Figure 7-1.

An Input/Output (I/O) Sheet (Figure 7-3) should be used to document the *inputs* and *outputs* that will be different from those used in the existing system (usually new forms or changes in existing forms). At this stage the I/O sheets probably will only be used in general terms since actual design of the system has not yet taken place. In the analyst's mind, however, the two phases of defining the new system requirements and

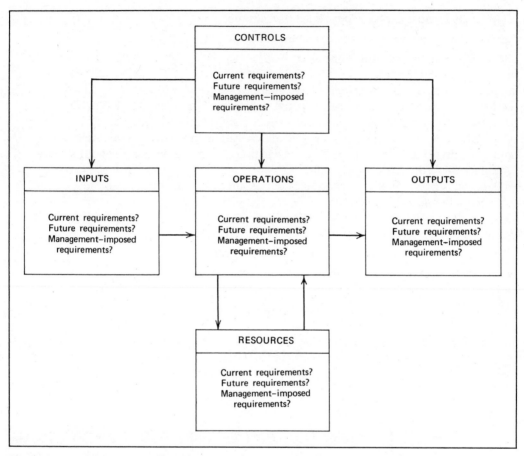

FIGURE 7-1 System requirements model.

design of the new system often are so closely related that the two phases cannot be separated. Obviously, if a very good idea occurs to the analyst, it should be documented immediately lest it be entirely forgotten. The I/O sheets will not be completed, of course, until the new system is completely designed.

Perhaps some of the input and output formats of the old system will be used "as is" in the new system. Others, however, will require changes, and still others must be completely redesigned. The analyst's documentation package should be composed of samples of the I/O formats which will be used "as is," in addition to I/O sheets for the changed or new formats.

The I/O sheet should bear a *functional description* of the input or output and

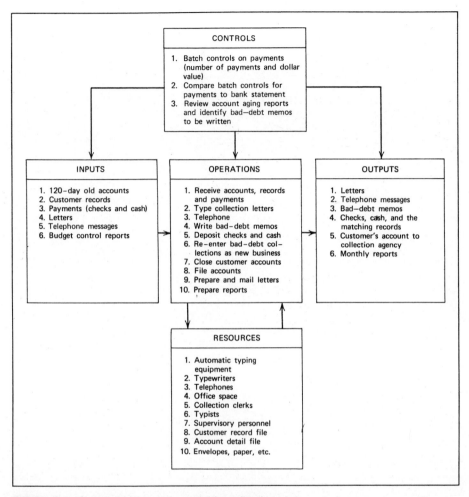

FIGURE 7-2 System requirements model for collection costs.

describe its purpose and use, and whenever practical, a *sketch* of the input or output. For example, in our new collection system we may need more data on the input *Customer records* (see Figure 7-2) so we can supply some extra details on our monthly reports to management. Possibly we will need to revise or completely redesign the form on which these data are received. For example, we may decide that we need data on the customer's financial standing, such as the firm's present D & B rating, current ratio, and whether discounts are being taken with other creditors. We need these data to compute the probability that we will be paid the overdue amount owed us. (We wish to include this probability figure in our monthly report to management.)

INPUT/OUTPUT SHEET	
NAME OF THE INPUT OR OUTPUT	SYSTEMS ANALYST
FUNCTIONAL DESCRIPTION OF THE INPUT OR OUTPUT (WHAT IT IS FOR, AND HOW IT WORKS)	
SKETCH OF THE INPUT OR OUTPUT, INCLUDING NUMBER OF CHARACTERS WHICH MUST BE ALLOWED FOR IN EACH FIELD	

FIGURE 7-3   Input/Output (I/O) sheet.

The existing form does not supply the data to us, so we design some changes, or a new form, or some means of securing the needed input.

We must be specific about our new requirements.

- What specific data items are needed?
- How are they to be expressed—in words? numbers?
- If the data will be processed with tabulating or electronic equipment, the maximum number of characters must be specified for each data item. Also, we must specify whether the item is numeric (all numbers), alphabetic (all letters), or alphanumeric (mixture of numbers, letters, and possibly other special characters).

We would describe the purpose and function of the data items in the upper portion of the I/O sheet. Of course, as the title indicates, the input/output sheet is used to record either input or output data.

In defining the input and output requirements for a new system, the analyst must be able to identify potential applications where a video display terminal or Cathode Ray Tube (CRT) can be used in place of the traditional input documents and printed output reports. In the past, CRT's have been used primarily as an input device for cardless data entry. Today, however, CRT's are being used by management and staff much more frequently as an on-line inquiry device. For example, a bank teller can enter a customer's account number and display the balance in the savings and checking accounts. As an output device the CRT can display standard computer reports, respond to specific inquiries, and, when needed, function as a sophisticated graphics terminal capable of displaying charts and diagrams designed to facilitate management decision making.

When the requirements of a new system warrant the use of CRT's as a medium for input and output, the analyst must explore thoroughly the pertinent hardware considerations and define the CRT terminal requirements. The computer hardware must be able to support data communication and provide the user response times needed. The term "response time" usually is defined as the elapsed time between the last operator entry and the initial display of the response. The analyst must be able to identify into which of the three categories of data communication configurations the company's computer fits.

The first category is the stand-alone computer which is designed to handle a specific set of communication facilities and terminals. The circuity to handle this is built directly into the computer so that it can interact effectively in a real time mode.

The second category is the general purpose computer which is not normally designed with built-in communication interface hardware. This type of computer can handle a small data communication network, but as the number of terminals increases there is a risk of severe degradation of overall performance and response time.

The third category is the front-end configuration where a general purpose computer is used for both batch and on-line processing. In this configuration there is a distinct division of labor between the front-end module and the general purpose computer. Once the analyst has become familiar with the type of computer configuration to be used, he or she is in a better position to evaluate the response time requirements. If the response time must be minimal, the analyst should study the impact this requirement could have on certain aspects of the hardware and software design, such as

- File organization.
- Telecommunication network.
- Type of terminal.
- Control program.
- Paging and scheduling algorithms.
- File addressing scheme.

In order to define the CRT terminal requirements the analyst should examine several factors. First, the intended use of the terminal must be identified as discussed earlier (input or output, data entry, inquiry or graphics display). Second, the analyst must determine whether the terminal is to be used by a full-time or occasional operator. Full-time operators can be trained thoroughly for the job and will have plenty of time to practice. The occasional operator, such as the manager, will not be highly trained in terminal use. In this case special care must be taken to ensure that instructions and responses are simple and clearly understood. As a final step in defining the terminal requirements, the analyst should determine which of the following characteristics will be important in using the new system.

- Typewriter keyboard.
- Numeric pad (3 × 4 matrix).
- Specially labeled keys.
- Good cursor controls.
- End of line bell or indicator.
- Data can be keyed into a buffer and modified prior to transmission.
- Skip and tab keys.
- Erase or backspace keys.
- Page or scroll keys.
- Screen that has an acceptable angle for viewing ease.
- Screen displays with enough characters.

- Characters that are large enough and easy to read.
- Screen that does not flicker.
- Screen color that does not tire operator's eyes.
- The image is bright enough for operator to see clearly and is protected from external glare.
- Diffused light intensity to reduce screen glare.
- Handles graphics.
- Individual fields can be highlighted.
- Upper and lower case characters.
- Selectable horizontal tabs and other formatting capabilities.
- Unique terminal identification by the computer.
- Lockable keyboard.
- Nonprinting password when keying it in.

Once the analyst has identified how the CRT is to be used, explored the hardware implications, and determined the terminal requirements, the foundations have been laid for the input and output design phase. Chapter 8, Designing The New System, includes a section on designing CRT terminal input and output screens.

Document the required *operations* of the new system with Flowchart Segments, Decision Tables, or Narrative Descriptions. A description of each follows.

Use of a Flowchart Segment is shown in Figure 7-4. The flowchart segment enables the analyst, and others, to get a comprehensive view of the operation, and provides a format which is easy to use when working out the design for the new system. Flowcharting templates with standardized symbols are available at most office or drafting supply companies. In Figure 7-4 the two diamond-shaped figures contain decision criteria which lead to one or the other actions specified in the boxes. (Conventions and methods of flowcharting are covered in Chapter 13.)

Use of a Decision Table in our new collection system is illustrated in Figure 7-5. This example parallels the flowchart segment example (Figure 7-4). Note that the decision criteria and actions are the same in both examples. The logic is also the same. Decision tables are very useful for listing all the possible decision criteria that might be involved in an operation. As such a list is compiled and entered into the table, the analyst (with the help of management and operations personnel) can specify the action to be taken in each situation. Thus the completed decision table is an excellent communication and coordination device. It can be used to verify inclusion of all possible situations that will require decisions in the system. More important, it can also be used to compel managers to decide exactly what their policy will be in each situation. (Decision tables are covered in more detail in Chapter 13.)

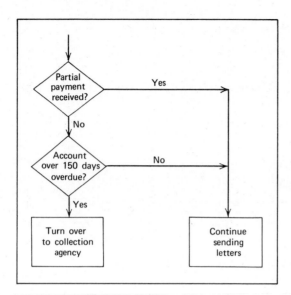

FIGURE 7-4 Flowchart segment.

A Narrative Description of the operations is a description written in story form. It is a verbal model in which the analyst details the sequence of steps involved in the necessary operations. The narrative should begin with whatever starts an operation, and follow through each step to the completion of the operation.

A narrative description of the key operational and accounting control points also should be prepared. The analyst must identify the techniques to be employed in establishing controls in the areas of (see Chapter 9)

PARTIAL COLLECTION PROCEDURE			
Partial payment received?	Yes	No	
Account over 150 days overdue?		Yes	No
Continue sending letters	X		X
Turn over to collection agency		X	

FIGURE 7-5 Decision table.

1. Completeness of input data.

2. Accuracy of input data.

3. Completeness of processing and related file updates.

4. Accuracy of processing and related file updates.

5. Supervisory/managerial authorization for both input and processing activities.

6. Movement controls over input documents, reports, and files.

7. File maintenance/run-to-run controls.

Document the required *resources* of the new system with Equipment Sheets, Personnel Sheets, or File Sheets. A typical Equipment Sheet appears in Figure 7-6. Equipment sheets are used to describe needed equipment such as typewriters, copy machines, desks, telephones, computers, and so on. Usually, only relatively major items are listed.

A typical Personnel Sheet appears in Figure 7-7. The analyst should specify the job description, job title, number of positions available, and approximate pay range for each needed position in the system. The personnel department can be consulted regarding pay scales for the various classifications involved.

A typical File Sheet appears in Figure 7-8. The file sheet describes a collection of information, that is, a file. Each item on the file sheet is described below.

1. File name: unique name of this file.

2. Location: physical location of the file.

3. Storage medium: tub file, computer disk drive, magnetic tape, file cabinet, three-ring binder, and so forth.

4. Age of file: age of the oldest record in the file or the retention time before items are to be removed from this file and sent to a records retention area.

5. How current: age of the record when it is first entered into the file.

6. Sequenced by: sequencing item of the file, for example, customer name, part number.

7. Format: a character-by-character layout of any record to be processed by tabulating or electronic equipment.

8. Characters per record: an estimate of the average number of characters per record and of the maximum number possible.

9. Records per file: an estimate of the average number of records per file and of the maximum number.

EQUIPMENT SHEET			
NAME AND DESCRIPTION	LOCATION	QUANTITY	APPROX. VALUE

FIGURE 7-6 Equipment sheet.

PERSONNEL SHEET			
JOB TITLE	NUMBER OF POSITIONS	JOB DESCRIPTION	APPROX. PAY RANGE

FIGURE 7-7 Personnel sheet.

FILE SHEET			
FILE NAME (1)	LOCATION (2)		
STORAGE MEDIUM (3)	AGE OF FILE (4)		HOW CURRENT (5)
SEQUENCED BY (6)			
COMPLETE BELOW FOR COMPUTERIZED FILES			
FORMAT (7) (8)			(9)
CHARACTERS PER RECORD		RECORDS PER FILE	
AVERAGE	PEAK	AVERAGE	PEAK

FIGURE 7-8   File sheet.

The analyst may use all or part of the devices described for documenting the new system requirements, or entirely new methods may be devised. In any case it is necessary to document clearly the requirements of the new system in some manner.

System requirements vary widely from one situation or firm to another. The analyst is encouraged to be creative in defining the system requirements and documentation. And it is important to keep not only the present requirements in mind, but the future requirements as well. New systems can be very expensive to develop; thus a prime objective should be long life.

PREPARING A SUMMARY OF THE NEW SYSTEM REQUIREMENTS

The analyst should summarize the notes and documentation package as a memory aid for the future. The summary should include the important points from both the *existing system* documentation (Chapter 6) and the *new system requirements* documentation (Chapter 7). The summary should be structured in such a way that it will remind the analyst of all key design considerations that must be observed during the detailed design phase. Thus the summary should be written as a comprehensive overview of all work done by the analyst thus far on the project, and not as a report to others. It is an exercise that compels the analyst to make many necessary decisions on the procedures to be followed during the detailed designing of the new system.

DEVELOP EVALUATION CRITERIA FOR THE NEW SYSTEM

While the analyst is defining and documenting the new systems requirements, a determination of evaluation criteria should be made. The object is to provide management with methods of valid measurement to evaluate the new system's performance. Evaluation of the new system is very important. A new system should never be installed without a valid set of evaluation criteria. This means that performance standards must be devised for key characteristics of the system. The performance standards should be geared to the objectives set for the system. For example, in our collections system, one of the objectives should be to pay for the cost of the new system with increased collections. That is, we wish to add more to revenue than to cost. Therefore, one criterion of evaluation will be the net revenue of the new system compared with the net revenue of the old system. This will require that we carefully analyze (and document) the costs and collections of each.

Perhaps the disorganization and confusion that characterized the old system caused frequent errors in mailing the collection letters. Perhaps good customers were often mailed strongly worded letters by mistake. We should set an objective to reduce these errors, and should define to management how we will measure our success in doing so.

The analyst will always be ahead if valid, reasonably accurate, conservative standards and measurement criteria for the system can be devised. Once the analyst is convinced of the validity of the standards and measurement criteria, concurrence from management must be obtained so that everyone will judge the system's performance in the same terms.

Usually, if the analyst does not provide such evaluation criteria, management will adopt its own. This can have very unfavorable implications for the analyst's future with the firm. Managers are busy people and may lack sufficient time to analyze the validity of their adopted criteria of the system's merits. They have not had the advantage of the analyst's comprehensive overview of the system, nor do they have the analyst's technical skills. The analyst owes it both to management and him or herself to provide the necessary system evaluation criteria.

We do not mean to imply that a firm's management is unqualified to decide the merits of a system. Quite the contrary, theirs should be the final word. Our point is that no matter how good the new system is, the analyst must explain and justify its *relative* merits so management can agree or disagree with a minimum of effort. If the new system costs more than the old one (which is usually the case), the analyst must point out the system features that appear to justify the higher cost. There are many system characteristics that must be evaluated when assessing the overall merit of a system. Some of these evaluation criteria are

1. Time: processing, elapsed, response, or operations.
2. Cost: annual, per unit, or maintenance.
3. Quality: better product or less rework.
4. Capacity: average or peak loads.
5. Efficiency: increased productivity.
6. Accuracy: fewer errors.
7. Reliability: fewer breakdowns.
8. Flexibility: many possible operations.
9. Acceptance: employee or management.
10. Controls: fewer operational, accounting control breakdowns and increased security.
11. Documentation: written/pictorial descriptions.
12. Training: how to operate the system.

SELECTIONS FOR FURTHER STUDY

You will use essentially the same tools in defining the new system requirements as you did in understanding the existing system. The major difference will be in your utilization of the company's long-range plans. If the firm has no long-range plans in writing, you will have to try to determine, through interviews, whether any plans or contingencies exist which may affect the area under study. If the study is especially large, you may wish to go beyond your coverage of the firm's long-range plans by studying the periodical literature to determine the industry's long-range outlook.

1. ACKOFF, RUSSELL L. *Concept of Corporate Planning.* New York: John Wiley & Sons, Inc., 1970.

2. ARGENTI, JOHN. *Systematic Corporate Planning.* New York: Halsted Press, 1977.

3. HARPOOL, JACK D. *Business Data Systems: A Practical Guide.* Dubuque, Iowa: William C. Brown Company, 1978.

4. *How To Set Objectives.* New York: Preston Publishing Company.

5. *IBM Data Processing Techniques: IBM Study Organization Plan: The Method Phase II.* White Plains, New York: International Business Machines Corporation, 1963 (F20-8137).

6. JENSEN, RANDALL W., and CHARLES C. TONIES. *Software Engineering.* Englewood Cliffs, N.J.: Prentice-Hall, Inc., 1979.

7. KELLY, JOSEPH F. *Computerized Management Information Systems.* New York: Macmillan Publishing Company, Inc., 1970.

8. LUCAS, HENRY C., JR. *The Analysis, Design, and Implementation of Information Systems.* New York: McGraw-Hill Book Company, 1976.

9. O'CONNOR, ROCHELLE. *Planning Under Uncertainty: Multiple Scenarios and Contingency Planning.* New York: Conference Board, Inc., 1978.

10. SANDERS, DONALD H. *Computers in Business.* New York: McGraw-Hill Book Company, 1975.

11. WILLOUGHBY, THEODORE C., and JAMES A. SENN. *Business Systems.* Cleveland, Ohio: Association for Systems Management, 1975.

12. YOURDON, EDWARD. *Design of On-Line Computer Systems.* Englewood Cliffs, N.J.: Prentice-Hall, Inc., 1972.

QUESTIONS

1. What are the two objectives of pulling all of the information together that has been collected up to this point in the system study?

2. Why should long-range plans be reviewed at this time?

3. Name four long-range considerations.

4. In what terms should the analyst determine the requirements of the new system?

5. What does the flowchart segment allow the analyst to do?

6. Name four areas where the analyst must identify the techniques to be employed in establishing controls.

7. Name six criteria that could be used in evaluating the overall merit of a system.

8. Name the three types of data communication configurations that a company's computer could be classified as.

9. Name two factors an analyst should examine in order to define CRT terminal requirements.

SITUATION CASES

CASE 7-1

The Coleman Manufacturing Company is a small firm that manufactures and distributes replacement automobile parts. Within the last three years, sales have increased 300% and management has decided that in order to meet this demand more effectively it may need to change from a manual system to a computer system.

Mr. Thomas, a senior system analyst, was in charge of the changeover. After six months of working on the project he had completed the feasibility study, the problem definition study, preparation of an outline of the study, gathered general information, and he had gained a good understanding of the interactions of the areas involved in the existing system. The next step would have been to define the requirements of the new system. Before Mr. Thomas could begin, however, he was offered a higher position with another firm and left Coleman Manufacturing.

His assistant, Ann Jones, a junior system analyst, was called upon to finish the project. The analyst had just graduated from a local university which was known for its fine Data Processing Department. Since this was Ms. Jones first job, she wanted to make a good impression.

The first thing Ms. Jones did was research Coleman's existing long-range plans to identify any potential impact on the new system. By interviewing key personnel, she learned that the company planned to open a new branch downtown within a year if business continued to grow at the present rate.

Ms. Jones then set to work defining the specific requirements of the new computer system for processing accounts receivable. She began by identifying the outputs that

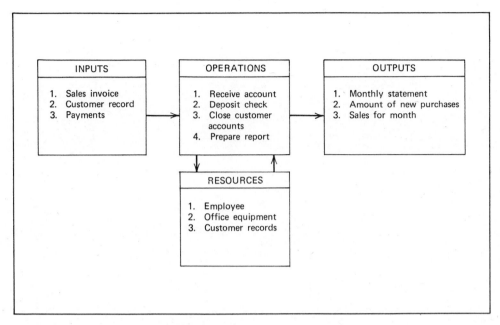

FIGURE 7-9 Systems requirements model.

must be produced. Management wanted each customer to receive a montly statement reflecting their balance and the amount of any new purchases for that month, if any. It also wanted a report of all new sales by month, and an indication of the percentage increase or decrease in sales from the previous month. Ms. Jones identified the inputs needed to produce these outputs, the operations that must be performed, and the resources needed. After she documented the requirements using a system requirements model and a system flowchart (see Figures 7-9 and 7-10) she prepared a summary of the system. It included the important points from both the existing system and the new system requirements. Ms. Jones was then requested to meet with top management to present a summary of the progress of the accounts receivable system. She was told to be prepared to answer any questions management might ask.

After Ms. Jones had completed her presentation, management asked a few questions.

"What will the processing capability of the new system be compared to the old one?"

The analyst replied, "I don't know exactly, but I guarantee it will be an improvement over the old one."

"This new system will cost us a lot more than the old one. Why should the company implement the new one?"

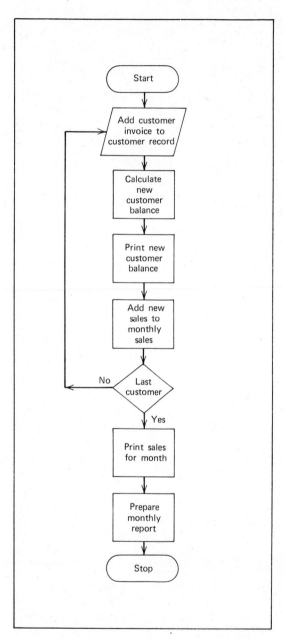

FIGURE 7-10 Flowchart of accounts receivable new invoice processing.

Ms. Jones answered, "Every other large company is using a computer. If we don't implement one we will be left out in the cold."

"In the past some of our good customers have been getting unnecessary collection letters. What provision have you made to guarantee this doesn't happen in the future?"

"Oh, I wasn't aware of that," the analyst said. "but I'll look into the problem and see what I can do."

QUESTIONS

1. In what two major areas was the analyst's work deficient?
2. What changes should be made in Ms. Jones' work?

CASE 7-2

The Water-Grow Sprinkler Manufacturing Corporation produces lawn and irrigation sprinklers. The manufacturing system employs automated machinery for pulling together subassemblies and uses conveyor belts where employees manually assemble the subassemblies. The demand for sprinklers has been booming and back orders have been building up. An analyst was called in to determine how to alleviate the bottleneck. After much effort, he pinpointed the problem as being "the conveyor belt operations cannot put together subassemblies fast enough to keep up with the increased production requirements." He noted that in order to try to meet demand, the conveyor personnel were very overworked. As a result, he concluded that a new system was needed to increase production. He was authorized, by the plant manager, to begin defining the new system's requirements.

The analyst reviewed the data collected while studying and understanding the existing system. He analyzed and refined the objectives set during the problem definition phase and then proceeded to develop the proposed new system's requirements.

The analyst first viewed the new system's requirements in terms of the outputs it must provide. He noted that the current output requirements were 18,000 sprinklers daily and that next year's requirements were expected to be more than 20,000 sprinklers daily. He also knew that management would not allow poorly assembled sprinklers to leave the plant until they had been disassembled, recycled through the conveyor operations, and then inspected by the quality control team. The analyst determined what the necessary inputs would be to meet the production requirements. He calculated the total number of subassemblies that would be needed and the necessary storage areas required for work-in-process. He noted that the automated machinery was capable of meeting and surpassing the new system's requirements.

The analyst listed the operations needed to assemble the sprinklers on the conveyor belt and the time involved per sprinkler for each of the operations. They were:

nozzling, 10 seconds; pinning, 10 seconds; spring insertion, 10 seconds; nozzle adjusting, 20 seconds; gauging, 10 seconds; and stamping, 10 seconds. The analyst also noted that the conveyor belt workplaces were filled, leaving no room to add more employees along the conveyor belt to help speed up assembly.

The analyst listed the existing resources and the resources needed in the future. The existing conveyor belt had workplaces for fifteen workers and there were fifteen employees filling these places. The new system would require seventeen employees in the first year and nineteen in the second year. The analyst noted that management had no objections to increasing the work force, as long as the additional employees supported increased production.

After collecting the data, the analyst documented it using an input/output sheet to describe the new and different inputs and outputs. He used a process flowchart to document the necessary operations and then developed a system requirements model, using both an equipment sheet and a personnel sheet to document the resources. The analyst also drafted a layout of the conveyor area showing storage and queuing areas.

The analyst developed two feasible alternatives. These were: (1) running the conveyor belt two shifts per day with nine workers per shift; and (2) investment in a 10-foot extension for the conveyor and the hiring of the needed additional help. A cost comparison of the two alternatives showed that keeping the entire plant open an additional shift was more expensive than investing in a conveyor extension and hiring additional help.

The analyst presented his recommendation for a conveyor extension to the plant manager who thought it was a good idea. He agreed that it would meet the needed capacity requirement, that quality of assembly operation would remain satisfactory, and that line flexibility would not be hampered. He further indicated that there had never been any problems with employee acceptance of a lengthened conveyor, which had been done once previously.

The manager approved of the idea; but, because it involved an expense that was not previously allotted for in his budget, he took the recommendation to his superior, Mrs. Green, the head of manufacturing.

Mrs. Green shook her head as she replied no to the plant manager. A project had been initiated earlier in the year that would result in a machine that would automatically perform nozzle inserting and aligning in one operation and at a rate that was much faster than existing methods. The new machine would cause the availability of four new workplaces on the conveyor belt where these operations were presently done. These four new workplaces would be used to double up on other operations on the belt and the result would be increased productivity. She then pointed out that sales forecasts indicated a steep continual rise for the next five years and that an extended conveyor would probably be unable to meet production requirements within the next three years. Mrs. Green stated that the sprinkler industry was depending more and more on technology and automation and that, if Water-Grow Corporation wanted to remain competitive, it too would have to pursue these goals.

QUESTIONS

1. What did the analyst fail to consider when he recommended that lengthening the conveyor would solve the problem? Why was this an unsatisfactory solution?

2. What physical aid did the analyst lack when he made the proposal to the plant manager?

CHAPTER 8

DESIGNING
THE NEW SYSTEM

To improve things we usually must have clear
concepts of where things stand at present and
what the objectives and requirements are for
the future. We can then pull all of our ideas together into a design that will accomplish our
purposes. After the analyst has performed
these preparatory steps, the actual technical
design can take place.

PROBLEM SOLVING VERSUS DECISION MAKING

A problem is defined as a question proposed for solution or consideration.[1] The process by which one finds the solution to the problem is called problem solving. Technically, problem solving is the process of recognizing the symptoms of a problem, identifying exactly what the problem is, determining what is causing the problem, and providing a solution to the problem.

Decision-making, on the other hand, is *choosing* the best solution to the problem based upon the alternatives identified. In current business practice, a systems analyst is primarily a problem solver and a manager is primarily a decision maker.

This is not meant to imply that the analyst *never* makes a decision. To the contrary, the analyst constructs a new system by considering the alternative solutions and then evaluates the solutions according to established criteria. Decision-making is inherent in this process. The difference is more in the interaction between the two processes. That is, the analyst provides solutions to the problem and recommends that solution which appears to be the most favorable. Management then makes a decision based on the considered judgment of the analyst.

DESIGN BACKGROUND

By the time the analyst is ready to begin designing the new system, certain items already should be established. There should be a problem definition, general background information on the area under study, a feeling for the interactions within the area under study and with other areas, a good understanding of the current system, and a set of requirements for the new system. It should be noted at this point that few companies today have a formalized systems development process that encompasses the above phases.

Systems design is a highly creative process which can be greatly facilitated by the following.

1. Strong problem definition.
2. Pictorial description of the existing system.
3. Set of requirements of the new system.

By definition, *design* means to map out, to plan, or to arrange the parts into a whole which satisfies the objectives involved. Systems design is mainly concerned with the coordination of activities, job procedures, and equipment utilization in order to achieve organizational objectives. The systems designer is faced with choices. There is a "choice-set" which is the set of all available alternatives. Each alternative is a different system or a slightly modified version of another system alternative. The analyst in the design phase

1. Determines the choice-set, that is, all possible systems.
2. Divides the choice-set into the attainable and the unattainable sets. The at-tainable set contains only those alternatives that have a reasonable chance of acceptance by management.
3. Places in order the alternatives in the attainable set, from the most favored to the least favored.
4. Presents the most highly favored alternatives to management for review and, hopefully, approval of one of the alternatives.

The designer, at this point, should know already whether the new proposed system is going to be maximizing something, optimizing something, or satisficing something. To *maximize* is to get the highest possible degree of use out of the system without regard to other systems. To *optimize* is to get the most favorable degree of use out of a system taking into account all other systems; an optimal system does just the right amount of whatever it is supposed to do, which is not necessarily the maximum. To *satisfice* is to choose a particular level of performance for which to strive and for which management is willing to settle.

The designer must also be aware that the individual job procedures within the new system have three levels of dependence.

1. Random dependence: a procedure is required because of some other procedure.
2. Sequential dependence: one procedure *must* precede or follow another proce-dure.
3. Time dependence: a procedure is required at a *set time* with regard to another procedure.

The analyst should look at these procedures during the design phase in terms of their flexibility, maintainability, and expandability. Put another way, the system's proce-dures should be able to meet new needs over a period of time; they should not be so rigid that they are ignored; and they should be open-ended to allow for growth.

SYSTEMS CONCEPTS

An *empirical* system is a working system, whereas a *conceptual* system exists only in thought. During the design phase the new system slowly takes shape. The formerly independent facts are all pulled together into a single idea or concept. Later, when the system has been approved and is installed, it has evolved from the conceptual to the empirical.

An *open loop* system is one which does not have automatic self-regulating controls. A *closed loop* system, on the other hand, does have automatic self-regulating controls—like a furnace system that has a thermostat which will automatically turn on the heat when it gets cool. A closed loop system has automatic feedback and control, where an open loop system has manual control that calls for human interaction. (See Chapter 1 for a more complete discussion of open and closed systems.)

Large systems usually are comprised of many *subsystems* and each subsystem usually is composed of many procedures. There is no restriction on the number of subsystems allowed in a large overall system.

Variables are those elements of an activity or job procedure that are subject to change or variation. In systems design the analyst may experiment on paper with various alternatives. Each alternative may have several constants which remain unchanged from one alternative to another, such as management-imposed conditions. Other factors, the variables, may change with each alternative. It is the variables that provide system flexibility.

Parameters are the elements of an activity or job procedure that are almost always unchanging. They vary so little that they are considered to be constant. Parameters remain uniform through time and are the opposite of variables.

Components are the moving parts of a system. They are the parts that compose the whole and they may be personnel, facilities, paperwork, or computers. A component can be either a variable or a parameter, depending on its constancy.

For example, using the System Requirements Model for Collection Costs (Chapter 7, Figure 7-2), the parameters might be the automatic typing equipment (assuming there is no available funding to purchase more equipment) and the available office space. The variables might be the monthly reports requested by management and the alternative methods available to the analyst to achieve better collections. The components include all the inputs, operations, resources, controls, and outputs whether they are variables or parameters (constants).

Scope, or range, refers to the extent of activity covered by a system. It is sometimes referred to as a boundary or a constraint placed on a system. The system parameters may influence the scope. In our collection costs model the scope would include the inputs from other departments, the Collection Department itself, and departments receiving outputs—for example, the manager receiving the monthly reports. (Scope was discussed in Chapter 3.)

The inexperienced analyst should become familiar with these terms since they tend to be standard in most literature dealing with systems information. The analyst also should realize that design ability improves rapidly with experience. Things will almost certainly get off to a good start if the analyst has a strong problem definition, a flowchart of the existing system, and a clear list of the new system's requirements. With this knowledge the analyst is in a position to design a system that fits the present and future needs of the organization.

PATTERNS OF SYSTEMS DESIGN . . . A STEP-BY-STEP APPROACH

The pattern of systems design follows an iterative technique. Systems design is a creative process in which the analyst iterates through various activities or job procedures, one at a time, mentally tracing through the entire system. You, as the analyst, should keep these two very important points in mind.

1. Solve one problem at a time. Do not let your mind get clogged with too many problems at the same time.
2. Your new system must conform with the overall objectives and goals of the area under study and the firm itself.

The *first logical step* in designing a system is to define the problem accurately (this should be done already). Many problems never get solved because they never are defined correctly. Keep an open mind since the definition of the problem may need refinement as the steps toward solution are completed.

The *second step* is to assemble the facts which seem to pertain to the problem. Examine your feasibility study, problem definition, outline, general information, interaction between areas, understanding of the existing system, and the new system requirements. The object is to have *all* of these facts in your mind so you can put them together in a better design than the original system.

The *third step* is to THINK! The designer can use vertical thinking or lateral thinking.[2] Vertical thinking starts with the most promising method of approaching the problem and proceeds from that point to a solution. Lateral thinking explores all of the different ways of looking at the system.

For example, suppose you are to play a game of chance with a person of questionable honesty. You are to pick a marble from a hat and if the marble is red, you lose; if it is blue, you win. (The hat is supposed to contain one blue marble and one red one.) The problem is your opponent's dishonest character. Your objective is to win a fair game. But your opponent is to be the one who puts the two marbles into the hat.

What are the vertical and lateral approaches for verifying your opponent's honesty? The vertical approach is to make sure that there is one red and one blue marble in the hat *before* choosing. This is a perfectly acceptable solution! But lateral thinking might further suggest that the first marble drawn will be discarded (thrown away) and its color determined by the color of the remaining marble.

Notice that vertical thinking started at the *beginning* of the event when the marbles were first put in the hat. Lateral thinking usually starts at the *end* of an event (where the marbles were picked out of the hat). Starting with the systems outputs and working backward through the system is a good example of lateral thinking. Vertical thinking goes from inputs to the output, whereas lateral thinking goes from outputs to the input.

One of the toughest problems to overcome in either vertical or lateral thinking is the preoccupation with a dominant idea. Dominant ideas tend to block out creative thinking. The designer cannot fruitfully search out all of the alternatives. Some suggestions for dealing with this problem include: stop thinking about that idea for 24 hours; or carry the idea to its extremes in order to make it appear foolish; or write down the idea so your mind does not have to hold it "in memory."

There are many methods available to the analyst to help stimulate the creative aspects of design work. Just talking with other people either to use them as a "sounding board" or to "pick their brains" can be beneficial, although the analyst has to learn to recognize the plateau that occurs when nothing new is being said. Then it's time for the talking to taper off and the designing to pick up!

Group brainstorming can be very effective as long as people are stimulated. If the group is too large and little cliques break off, it can evolve into a gab session.

Another method is to prepare lists of all the ideas and then try to rank them according to objectives, inputs, outputs, or any other criteria.

The manipulation of data collected to discover any recurring patterns can be especially effective if there is a computer available. The computer can manipulate the data in such a way that one factor can be balanced against another. This technique is called *factor analysis*. And, of course, the computer can manipulate the data thousands of times faster than a human.

For extremely large systems where many people are working on the design, each alternative can be given to a different team for analysis.

The important thing is to try to think in pictures rather than words. Visualize the activities and job procedures as well as the various ideas of design. Begin to see pictures instead of words. No matter how ridiculous an idea, try to visualize it. Do not immediately discard an idea because it seems wrong or foolish. The desire to be perfectly correct in each step of the design process is a great barrier to creative thinking. Let your mind roam. You may discover a whole new approach to a section of the design—one which may provide new benefits. One such benefit will be the establishment or reinforcement of your image as a creative thinker in the minds of your evaluators, that is, your manager! One idea is to listen to yourself on tape; the bad or good ideas may stand out!

Always review the unresolved problems, in the design, each day to stimulate the subconscious mind's ability to solve them. The subconscious mind has a way of putting things together on its own while your conscious mind is busy with other problems. It will announce its progress by "striking" you with bright ideas.

The *fourth step* is to evaluate the ideas or systems that have occurred to you. Without being too rigid, or as they say, "without setting your ideas in cement," do the following.

Decide what the system will do, working from what inputs to what outputs (vertical) or from what outputs to what inputs (lateral), through what operations, using what resources. Go back and talk to the management in the area under study in order to get a tentative approval on the alternatives or on the design itself.

In summary, then, the points to follow when designing a new system are

1. Examine all of the available data.
2. Think hard, and think creatively!
3. Devise various input, output, operation, control, and resource techniques.
4. Evaluate the most important procedures first.
5. Examine various alternatives.
6. Do not spring the final result on management since surprise endings usually are rejected.

Another consideration in the design phase is the amount of control to be exercised from within the system. Some controls will be determined by various system parameters such as the applications and inputs. Certain quality controls may need to be designed into the system. For instance, all inputs should be prepared in a consistent manner to maintain reliability, a built-in check and balance system may be desirable, or a special error section may be developed.

THE STRUCTURED DESIGN PROCESS

The design of a successful computer-based system is dependent largely upon the intelligence and capabilities of the analyst responsible for its design. Many analysts have found that the structured design process facilitates describing the kind of reasoning and logical progression of analysis that must be followed in order to arrive at a sound design. As a result, the use of this technique is growing rapidly.

The basic purpose of the design process is to formulate a suitable framework for the subsequent development of software. The structured design process is based on the principle that the most logical design is one that begins with a high-level picture of the system identifying the major component parts, and ends with each component being broken down to a sufficiently detailed level to allow efficient and effective software engineering. The structured design methodology is based upon some concepts originated by Larry L. Constantine and later published by him while working with Ed Yourdon, and by Glenford J. Myers.

The structured design process utilizes a number of concepts, measures, analysis techniques, guidelines, rules-of-thumb, notation, and terminology. The process consists of six steps, the first three of which are iterative. The order of iteration, however, is not rigid. A data flow diagram is used in the first step to identify and illustrate the expected flow of information through the software. In the second step, a structure chart is used to determine the most effective organization of the software. And in the third step, the data table/structure is used to summarize data definitions and to show relationships between pieces of data.

The purpose of the data flow diagram is to define the changes (transformations) that data undergoes during its flow through the system. The diagram does not necessarily define a time sequence of actions and does not contain any control information. The data flow diagram describes data along directed lines and shows transformations proceeding from left to right. When data is shown entering a circle it must, by definition, undergo some change (transformation) to result in the data output from the circle. Figure 8-1 shows a basic data flow diagram.

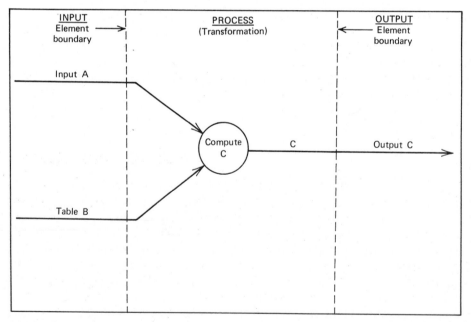

FIGURE 8-1 Example of data flow diagram.

The activity described could be found as a component of almost any system. The diagram describes a situation where there is one direct input item A, and a table item B being referenced for additional information. Next, in reading the diagram from left to right, the processing is done to compute the result C by acting on the two items A and B. Finally the result C is passed as output. The input element boundary line shown on the diagram represents the point at which the input data stream is most processed but it still is considered an input. The output element boundary line is the point at which the output data stream is least processed but it is still able to be considered as an output.

As the data flow diagrams are drawn, redrawn, and combined, the placement of these boundary lines will impact the way the resulting structure charts are laid out. A good data flow diagram possesses certain characteristics that can aid in evaluating the state of the design. Inadequacies in the diagram can aid in discovering and resolving problems in the thinking process. When the data flow diagram is reviewed, the following factors should be considered.

1. All input to the system should be identified, starting at the left.
2. All output is identified, ending at the right.
3. The data must be named clearly and its exact contents identified to prevent confusion with other data in the system.
4. The change (transformation) data undergoes must be in clear, precise terms.
5. It is incorrect to show a single input producing a single output because it implies that some change was achieved without a catalyst.
6. It is incorrect to show multiple inputs producing multiple outputs because it implies that some change was achieved without a catalyst.

The purpose of the structure chart is to define and illustrate the organization of the system on a hierarchical basis in terms of modules and submodules. In the chart, the relationship between modules and their respective interfaces are defined along with the method of control. Thus, the structure chart includes a complete description of the system with respect to data, modules, their interrelationship, and controls.

A basic structure chart consists of three major branches: (1) the input branch that collects and formats data into a state ready for processing, (2) the transform or process branch that performs the basic function of the system, and (3) the output branch that formats and disperses the output. Figure 8-2 illustrates a basic structure chart with one input A being processed and transformed into one output B. The structure chart normally is created from the data flow diagram. The transformations (data changes shown as circles) on the data flow diagram become the modules or nodes on the initial structure chart. As a rule, the "best" structure chart results when the data flow diagram is set up so the sum of the conceptual data streams cut by the input and

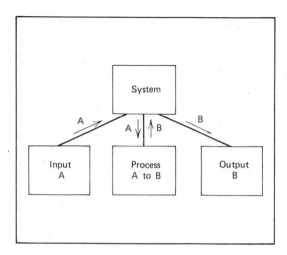

FIGURE 8-2 Basic structure of system.

output element boundary lines is minimized. This structure chart is then further broken down into submodules which are identified at a level below their respective modules. The data being sent from a module to a submodule, or the reverse, is shown along the connecting line with an arrow indicating the direction of flow. The data passed between modules may be used either as information to modules or as a control flag to the module.

Figure 8-3 illustrates how a structure chart would be constructed from the data flow diagram shown in Figure 8-1. The input item A is shown being obtained from a device (represented by a parallelogram). Item A is then shown as being passed by the input module to the system along the connecting line. The arrow and label indicate the direction of data movement. The system routes item A to the process module. The process module also obtains data item B by referring to a table (represented by a hexagon). Once the process module has all the information it needs, the resulting item C is computed and passed to the system. The system passes item C to the output module. Again an arrow and label are used to indicate direction of data movement. Finally item C is passed to a device for final disposition.

Figures 8-4 through 8-7 illustrate how the placement of the input and output element boundary lines on the data flow diagram can reduce the number of conceptual data streams cut. The resulting structure chart becomes better defined and clearly shows the terminal modes doing the processing and the higher level nodes performing a control function. Figure 8-4 shows a data flow diagram describing the computation of net pay. There are three input items, that is, (1) Regular Hours (RH), (2) Overtime Hours (OH), and (3) Payroll Deductions (PD). In addition two tables are referenced, that

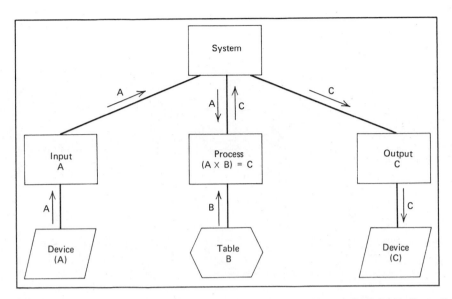

FIGURE 8-3 Basic structure chart drawn from the data flow diagram illustrated in Figure 8-1.

is, (1) the Regular Rate table (RR), and (2) the Overtime Rate table (OR). The process section of the diagram contains four transforms or subprocesses, that is, (1) calculation of Regular Pay (RP), (2) calculation of Overtime Pay (OP), (3) computation of Gross Pay (GP), and (4) computation of Net Pay (NP). In the first transform the regular hours data element is input and the regular rate table is accessed. These two data elements are acted upon to produce regular pay. In the second transform the overtime hours data element is input and the overtime rate table accessed. The two data elements are acted upon to produce overtime pay. The third transform receives the regular pay and overtime pay data elements calculated in the first two transforms and treats this information as inputs. The gross pay computation is performed and the result is passed to the fourth transform. The fourth transform also receives a direct input of payroll deduction information. The net pay computation is performed and the result NP is passed to the output section of the data flow diagram. Note that in this diagram the input element boundary line cuts across five conceptual data streams. Figure 8-5 is the structure chart constructed from the data flow diagram shown in Figure 8-4.

Figure 8-6 illustrates the same data flow diagram as Figure 8-4. Note, however, that the input element boundary line has been moved to the right side of the first and second transforms. This reduces the number of conceptual input data streams cut by the boundary line from five to three. According to the rules this should improve the structure chart drawn from the diagram. Figure 8-7 illustrates the resulting structure chart drawn from Figure 8-6. The "get transformed data" node in Figure 8-7 is now

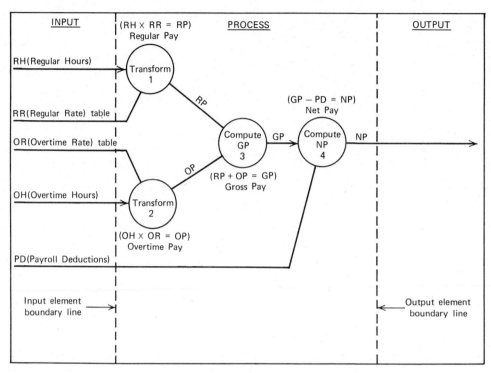

FIGURE 8-4 Data flow diagram illustrating computation of net pay.

acting as a higher level control over all input data and the results of preliminary transformations. The "compute result" node in Figure 8-7 is now acting as a higher level control over the computations of gross and net pay.

As the analyst reviews a structure chart the following factors should be considered.

1. Terminal nodes do the processing while all higher level nodes perform a control function.
2. Only terminal nodes will interface with data tables.
3. A node may not interface with a table and another node because this implies that a dual function is being performed by the node.
4. All data communicated between nodes must be identified along with the direction of flow.
5. The control constructs must be identified.

During each structured design iteration, the structure chart is analyzed to identify the logical data tables. Each table should describe a grouping or collection of data that

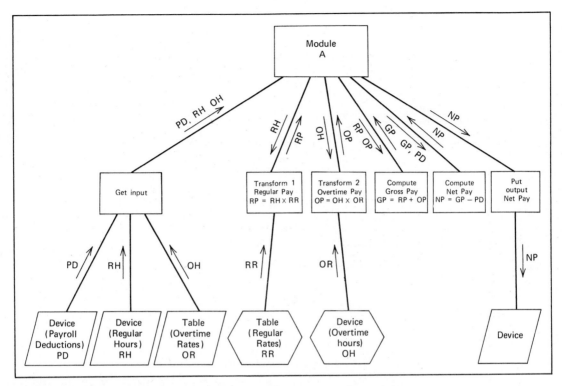

FIGURE 8-5   Structure chart drawn from data flow diagram shown in Figure 8-4.

is organized as an entity within the system. The regular pay rate and overtime pay rate tables shown in Figure 8-4 are a good example. Initially data is defined at the conceptual level, but as the system processing is broken down, the contents of the data table/structures are identified in more detail from the data flowing into and out of them.

The data flow diagrams, structure charts, and data table/structures are used interactively to establish a framework for structuring the system as the design proceeds from initial to intermediate stages, and to final detailed structuring. The data flow diagram is used initially to discover the overall information flow within the system. From this initial overview, the structure chart is used to identify the top-level structure and functional relationships of the system. The data table/structures are used to define the conceptual data relationships. This completes the first iteration of data flow diagrams and structure charts. The process, as illustrated in Figure 8-8, now repeats itself beginning with the breaking down of the initial top-level structure chart. The process ends when all of the branches (modules) of the structure chart have been broken down to their logical conclusion.

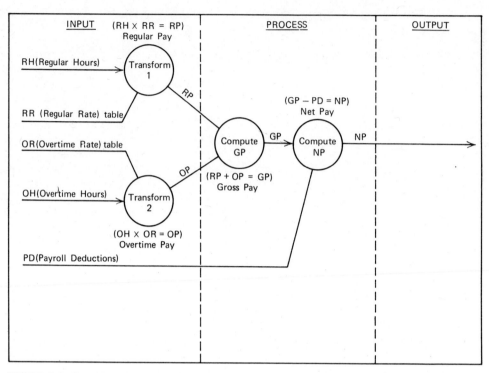

FIGURE 8.6 Data flow diagram (same as Figure 8-4) showing shift of the input element boundary line.

The following rules apply when the structure chart is to be broken down functionally into more detail.

1. Make the structure chart a hierarchical tree structure with each node being a function and controller to all functions subordinate to it.
2. Every function must influence its subfunctions in some way.
3. Every subfunction must be influenced by its parent in some way.
4. The priority of a function must be greater than any of its subordinate functions.
5. Ensure that the decomposition of a function is complete, that is, the immediate lower-level subfunctions express the entire parent function.
6. Use at least two functions at every level.
7. Make each functional node of the tree unique.

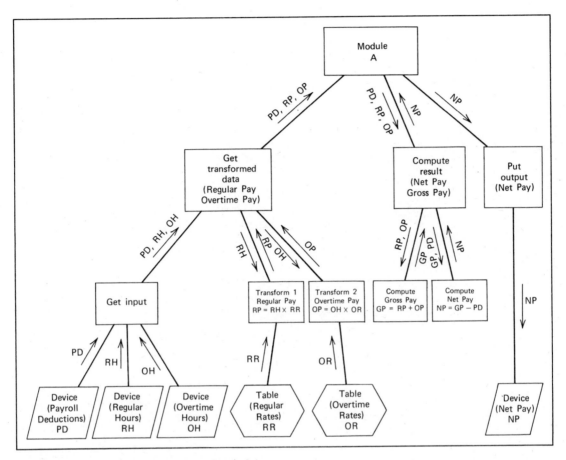

FIGURE 8-7 Structure chart drawn from Figure 8-6.

8. Break down the system by one of the following methods.

a. Functional composition: A module is broken down into a series of modules, all of which are executed in sequence to accomplish the function.

b. Set partition: A module is broken down into a series of modules, only one of which will be executed. Once a subfunction is selected on a given level, other subfunctions on that level are no longer required.

c. Class partition: A module is broken down into a series of functions; each subfunction acts upon a subset of the input variables to produce a subset of the output variables.

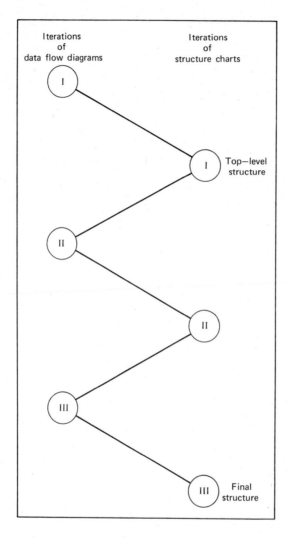

FIGURE 8-8 Iterations of data flow diagrams with structure charts.

The fourth step is to analyze the flow of data between each independent section of the first structure chart to determine data streams. The four types of data streams are: (1) input to table, (2) table to table, (3) table to output, and (4) input to output. Independent data streams are identified and a separate data flow diagram is drawn for each. When all the independent data streams have been diagrammed, they must be combined into a single (more detailed) data flow diagram for that entire section. When all the sections have been analyzed in this manner, all the diagrams are connected using the following rules.

1. Connect all data streams that end at the same major table.
2. Connect all data streams that begin at the same major table.
3. Remove all minor tables.
4. Continue connecting all data streams until a single data flow diagram results showing (a) all system inputs on the left, (b) all transformations between data table/structures, and (c) all system outputs on the right.

When all of the branches of the structure chart have been broken down to their logical conclusion, the design structure chart is complete.

The fifth step in the structured design process involves conducting a structured walk-through, which is an informal review of the design with key technical and management personnel. Any changes identified during the structured walk-through are incorporated into the design.

The sixth and final step in the structured design process calls for packaging the software design. This step involves the compression and design of modules that actually will be coded. The design process has created many modules to aid in understanding the design; however, since the interface between modules requires additional program code, it may be desirable to compress a number of modules into a single module for the sake of efficiency. The following guidelines can be used in compressing modules.

1. Modules should not be larger than fifty lines.
2. If a module has only one submodule, they should be combined.
3. If a module calls another module in a loop, it could be compressed.
4. Modules that use the same data may be compressed.
5. If compression increases complexity, do not compress.

At this point, the design structuring has been completed. The only activity that remains is to specify the internal workings of the modules so they can operate as indicated in the structure chart. The Hierarchical Input Processing Output (HIPO) is a methodology that can be used for this. HIPO is further discussed in Chapter 13.

SYSTEM DESIGN THROUGH SIMULATION

Computers can be used to simulate the firm's business environment. A "model" of the environment can be programmed into the computer, and then inputs can be fed to this model to simulate the actual operation of the firm's business systems. The model is actually a mathematical representation of the proposed system.

A system model may be developed that will enable the organization to predict and explain how well the system will work. The model may show the outputs that result from certain inputs, speed, accuracy, and so on.

For example, in the simulation of a manufacturing system the system designer develops a model that represents the system. The simulation model tells the computer the physical structure of the manufacturing plant such as the number and type of machines, the number of machine operators, machining sequence, which worker can work on each machine, an order-priority scheme, overtime rules, the orders pending, and any other pertinent facts. The computer then simulates running the orders through the plant and routing them from one machine to another. During the course of simulation the computer keeps track of how well the simulated system is performing, according to whatever evaluation criteria the designer has specified in the computer program.

A simulated system can be operated any number of times with variations in the number of orders or men or machines, and so forth. Dynamic analysis of alternative systems under a variety of conditions often uncovers difficulties that static analysis cannot find.

The subject of business modeling can become very complex. This complexity arises because each model must be expressed mathematically. What is a model? A model is any representation of reality that embodies those features of reality that are of particular interest to the user of the model. For example, a model, using a theoretical basis, portrays the realities of the business situation. For a very long time there have been two particular models of a business in widespread use: the balance sheet (a static model) of the assets and liabilities of the business, and the income statement (a dynamic model) representing the income and expense streams through the business; therefore, a model can be either static or dynamic.

In order to develop a model the analyst must first define the problem, analyze the data requirements, and determine the availability of sources of the data required. One technique for developing and using business models is simulation. Computer simulation has enabled the model designer to develop very complex models because the computer is able to keep track of all the many details and techniques that the human mind was unable to take into account. The overall model of the total business is not the place to start. The place to start is to develop many small models that can interact with each other. Each of these small models can represent a portion of the overall business, such as production control, quality control, purchasing systems, and so forth. Each of these various areas can be called subsystems and they can then be joined together in order to make up the overall model of the total business.

Subsystems of large business problems can be identified by analyzing the system in one of three ways. The first method of analysis is the *flow* approach. This method is used for a subsystem having dynamic properties that are determined by the flow of physical entities through the system. In this approach the analyst studies the flow of physical entities and builds a model that simulates the flow of these items through the business organization. The flow approach would be used to analyze a production or manufacturing system.

A second method of identifying the subsystem is the *functional* approach. This method is used when there are no observable flowing entities through the system, but when there is a relatively clear sequence of events to be performed. In this approach the analyst studies the sequence of events and builds a model that simulates that sequence of events as they happen throughout the business organization. The functional approach might be used to analyze a production control process even though a production control process might have paperwork flowing.

The third approach in identifying subsystems is known as the *state-of-change* approach. This approach is useful for a system that has a large number of interdependent relationships for which no specific sequence can be observed. This method requires that the person building the model check the state of the subsystem at various points of time in order to determine if any change has occurred within the subsystem. Various checkpoints are monitored in order to determine what changes occur as the inputs to the subsystem are varied.

If the individual subsystems cannot be identified and models developed that describe these subsystems, the model building process will be much more complex. Without these submodels the entire business model must be formulated and validated all at once. This is difficult to do because a large number of assumptions have to be built into the simulation of the overall business model since each subsystem could not be tested and validated beforehand.

In the next stage the model designer combines the subsystem models into a simulation model of the entire business situation. At this point the analyst has already developed the mathematical description of each of the subsystems and will be combining each mathematical description of each subsystem into the mathematical description of the entire business enterprise. To do this one must choose a simulation programming language.

The most popular simulation language, and for many types of problems the most useful, is General Purpose Simulation System by IBM (GPSS). GPSS is a complete language oriented toward problems in which items pass through a series of processing and/or storage functions. Another language is SIMSCRIPT. SIMSCRIPT is a complete FORTRAN-like language oriented toward event-to-event simulation in which discrete logical processes are common. Another simulation language is JOB SHOP SIMULATOR. JOB SHOP SIMULATOR is a program package that can be set up to represent a variety of job shops. GASP is a set of subroutines in FORTRAN that perform functions useful in simulations. CSMP is a complete language, oriented toward the solution of problems that are stated as nonlinear equations. And finally,

DYNAMO is a complete language oriented toward expressing microeconomic models of firms by means of difference equations.

After the program simulation language is chosen, a programmer must program the business model using this language. After the program is debugged and operational, the business model will be simulating the firm itself. The analyst can now change the various inputs and determine what the final outputs will be. One advantage in this method is that time can be compressed. The analyst can put in data that represent a six-month time period for the firm and get the results back in a matter of minutes or hours. The business planners can make better decisions because of the results from the simulation model. For example, if the outputs are not what is desired the analyst can vary the inputs in each of the individual subsystem models. By continually varying these inputs, the analyst can obtain the desired outputs and the sensitivity of the various inputs can be obtained. In this way the business organization can determine which inputs should be used.

In summary, the analyst defines the problem, determines the data requirements and sources, formulates models of the various subsystems, combines these models of the subsystems into an overall model of the total business, and decides upon a simulation programming language in which the overall model of the business system can be programmed. After programming and debugging, the model is used in order to assist in making forecasts or planning decisions.

ON-LINE INPUT AND OUTPUT DESIGN CONSIDERATIONS

As discussed in Chapter 7, CRT terminals are being used much more frequently as a medium for both input and output in the design of new systems. As an input device, the CRT terminal can be used in the traditional data entry role or as an on-line medium for data collection. As an output device, the CRT terminal can be used to: (1) display traditional reports of financial and operating significance, (2) answer simple queries, or (3) provide sophisticated graphic displays to facilitate management decision-making.

The use of CRT terminals as an integral part of the data entry (system input) function has grown rapidly over the past decade. Initial activities in this area involved the use of stand-alone data entry systems. This type of system was set up to collect the information entered and then to store it on magnetic tape. Periodically the information would then be transferred to the main computer. Numerous technological advances in computer hardware and software capabilities, however, have made it possible for the analyst to consider incorporating on-line input and output features into the design of new systems.

When an analyst is considering the use of on-line inputs in designing a new system, the possible advantages and disadvantages must be weighed carefully.

Advantages

1. The operator can immediately identify and correct manual keying errors by scanning the screen.
2. Redundant key verification procedures, typically used in punched card data entry, are not required.
3. Input data is usually available for processing much sooner. Depending upon the situation, the system could process input data immediately.
4. The data entry function can be dispersed out into the user departments where the information originates. This facilitates the timely entry of data and the immediate correction of data that fails any detailed editing and validation criteria.
5. CRT terminals may, in some cases, be less expensive than card punching equipment.

Disadvantages

1. The computer hardware and communications costs to support on-line data entry may be too high for some organizations.
2. Provisions must be made for backup equipment or manual procedures if the computer is down for extended periods of time.
3. The data entry routines must be programmed and maintained.

The input of information to a computer via on-line CRT terminals can be handled by two basic types of operators, that is, the dedicated data entry operator, and the general information collection and entry operator. The analyst must determine which type of operator approach best fits the needs of the company and the system being designed. A combination of both approaches may be necessary.

When the dedicated data entry approach is utilized, the operator is going to perform a function similar to a full-time keypunch operator. If the analyst decides to use this approach, the following points should be considered carefully.

1. The emphasis should be on facilitating operator ease and speed of entry.
2. Interaction with and evaluation of data displayed on the screen should be minimized.
3. On-line editing and validation should be limited to: (a) ensuring all mandatory data fields have been entered (a data field is that part of the input which contains specific information such as an employee name or employee identification number); (b) ensuring data entered does not exceed or overlap the maximum size of the field; (c) ensuring that all numeric fields contain only numeric data.

4. The operator should be able to visually scan items entered if a keystroke mistake is suspected.

5. Users should batch documents to be entered (by item count), and establish hash totals on significant data fields.

6. The system should accumulate automatically batch totals and compare them to previously entered manual control totals.

7. The operator should be able to recall the entire batch, if necessary, to locate and correct errors noted.

In using the dedicated data entry approach, the most common method of identifying fields of data to the system is to have them entered in a fixed sequence. As a part of the detailed system design process, the analyst must work with users to establish a fixed entry sequence of data fields for each type of document or transaction type to be entered. As an example, the system could be set up to require that the data on a payroll time card be entered in the following sequence.

1.	Employee name	John Doe
2.	Employee number	1234
3.	Date	8/12/80
4.	Regular hours worked	32.0
5.	Overtime hours worked	0.0
6.	Vacation hours taken	0.0
7.	Sick time hours taken	8.0

If the data fields to be entered must be fixed in length, then either the operator or the data entry program must be capable of ensuring that the spacing within fields is correct. This may require zero or blank filling fields where no data is to be entered, or where the data must be right or left justified. Figure 8-9 illustrates how a typical payroll time card entry would look. Note that the name field has been left justified with trailing blanks, and the numeric fields have been zero filled or right justified depending on the value.

If the system design requires that the data fields be set up as variable in length, then the operator can use a field separation character such as '/' to distinguish between the end of one field and the beginning of the next. By using a special character to separate fields, the operator can avoid entering fill characters and can identify which fields have no data by entering dual separators. Figure 8-10 illustrates how the same time card information would be entered in a system using variable length fields and a separation character '/'. Note that the name field does not need to be padded with blanks. The employee number field, the regular hours field, and the sick time field do not need to

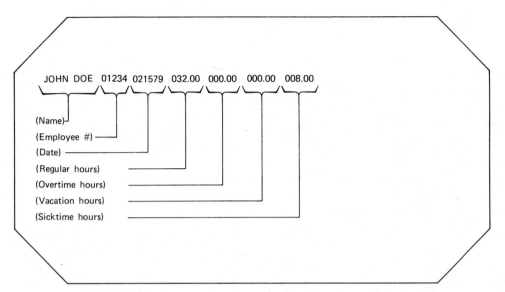

FIGURE 8-9   Illustration of how a typical payroll time card entry would be entered on-line.

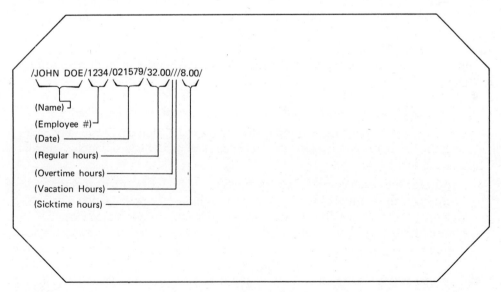

FIGURE 8-10   Illustrates how the same time card information would be entered on-line using field separators.

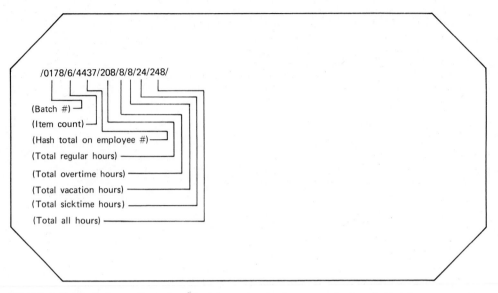

FIGURE 8-11   Illustrates on-line entry of batch number and user supplied control totals.

be zero filled. In addition, the overtime hours and vacation hours fields were not entered because there was no data. In this case, the data entry program interrupts two consecutive special characters to mean the field was not entered and default values should be assigned.

The typical sequence of events that the dedicated data entry operator goes through include

1. Entry of the batch number and user supplied control totals. (See Figure 8-11.)

2. Entry of the batched transaction data with a symbol that denotes end of batch so the balancing can be done ('*' has been used as the end of batch symbol). (See Figure 8-12.)

3. Scanning the batch control message produced by the data entry program to determine if the computed batch totals match the user supplied batch totals. (See Figure 8-13.) If the batch control totals are in balance the message example A would be displayed. If an out of balance condition is noted, message example B would be displayed.

When an operator has entered a batch of input transactions and an out of balance condition is noted, corrective action must be taken immediately. Normally the operator would review the source documents and the entries on the screen to identify

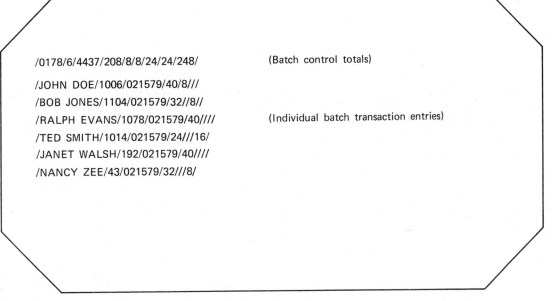

/0178/6/4437/208/8/8/24/24/248/   (Batch control totals)

/JOHN DOE/1006/021579/40/8///
/BOB JONES/1104/021579/32//8//
/RALPH EVANS/1078/021579/40////   (Individual batch transaction entries)
/TED SMITH/1014/021579/24///16/
/JANET WALSH/192/021579/40////
/NANCY ZEE/43/021579/32///8/

FIGURE 8-12   Illustrates the on-line entry of a batch of payroll time cards using field separators.

which of the fields noted as "out-of-balance" was entered incorrectly. If an erroneous entry is located, the operator would move the cursor (place marker) across the screen to the incorrect character and write the correct one over it. The final step would be to signal the data entry program to rebalance the batch, so that the operator can ensure all corrections have been made.

In recent years, the dedicated data entry approach has lost some ground to the general information collection and entry approach. Although the traditional approach will probably always be used to some extent because of its high speed, more and more systems are being designed that incorporate the use of the general approach. Under the general approach the operator will probably have such responsibilities as

1. Collection of information entered directly from customers, users, or from output reports from other systems.
2. Entry of the data and immediate resolution of all editing and validation problems.
3. Use of the output reports produced by the system.
4. Provision of immediate feedback to the customer or user after entering the data.

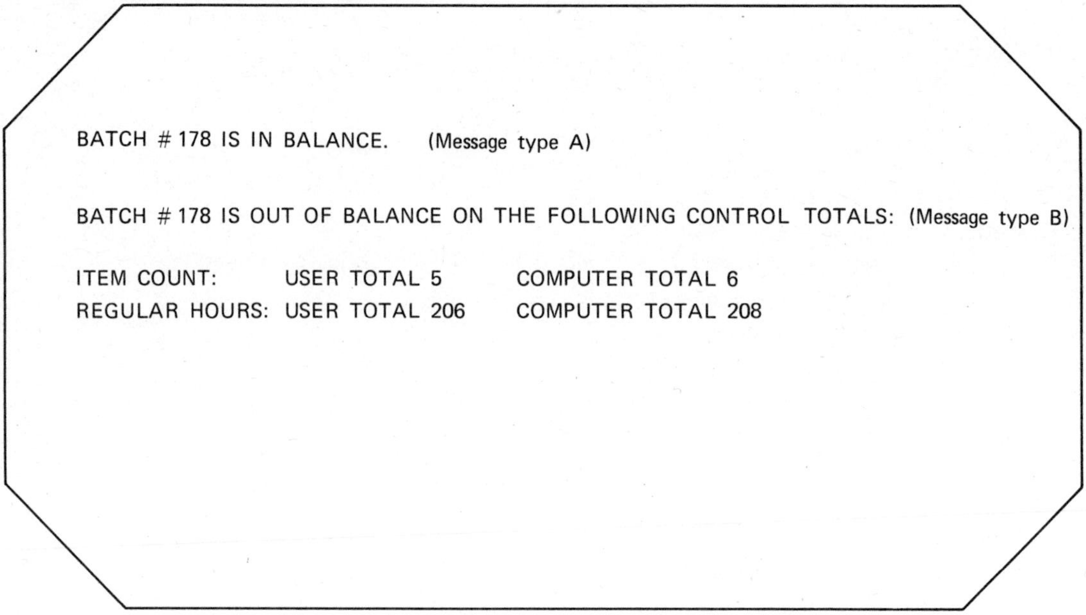

```
BATCH # 178 IS IN BALANCE.     (Message type A)


BATCH # 178 IS OUT OF BALANCE ON THE FOLLOWING CONTROL TOTALS: (Message type B)

ITEM COUNT:      USER TOTAL 5       COMPUTER TOTAL 6
REGULAR HOURS:  USER TOTAL 206      COMPUTER TOTAL 208
```

FIGURE 8-13   Illustrates two types of batch completion control messages that could be produced by the data entry program.

Since the operator in this approach has other duties, his or her level of technical data processing and data entry expertise may be limited. The type of operator used in this approach could include a bank teller, an airline reservations clerk, or even a salesman who wishes to check the status of a customer order. In the general approach, the dialogue required between the operator and the terminal/computer system assumes greater importance. The procedures to be followed must be as simple and easy to understand as possible. The most popular procedure begins by providing the operator with a menu of possible functions from which to choose (see Figure 8-14). Once the function to be performed has been selected, the analyst must decide what type of dialogue best fits the needs of the application and the users. The two most common techniques are (1) the detailed prompting technique, and (2) the simulated forms completion technique.

The detailed prompting technique is relatively slow, but it allows an unskilled operator to be walked through each step required to perform the desired function. Figure 8-15 illustrates how this technique would be applied to opening a checking account. Each field to be entered is individually prompted. If an obvious error is made, the operator can re-enter the data immediately. If editing and validation criteria are exceeded, the operator can be advised and prompted to re-enter the data. When the data

```
BANK TELLER MENU
ENTER LINE NUMBER OF FUNCTION TO BE PERFORMED
1.  OPEN CHECKING ACCOUNT
2.  OPEN SAVINGS ACCOUNT
3.  DEPOSIT IN CHECKING ACCOUNT
4.  DEPOSIT IN SAVINGS ACCOUNT
5.  WITHDRAW FROM SAVINGS
6.  QUERY BALANCE OF SAVINGS ACCOUNT
7.  QUERY BALANCE OF CHECKING ACCOUNT
8.  TRANSFER FUNDS FROM SAVINGS TO CHECKING ACCOUNT
9.  TRANSFER FUNDS FROM CHECKING TO SAVINGS ACCOUNT

"1"   (represents the operator's response in selecting function '1')
```

FIGURE 8-14   Illustrates a displayed menu of possible function a bank teller or clerk could perform.

is completely entered, the operator is given one last chance to scan the results to be sure they are correct. If everything is accurate, then a Y is entered to complete the session. If a mistake is noted, an N is entered which allows the operator to make last-minute corrections (see Figure 8-16).

The simulated forms completion technique normally is used when the operator is more familiar with the equipment and functions to be performed. In this technique, the operator is presented with a simulated form to fill in. The form identifies the fields to be entered and their relative sizes by initializing them with blanks. The user then moves the cursor to the appropriate place and fills in the blanks. Figure 8-17 illustrates a simulated form that could be used for opening a checking account.

As previously mentioned, CRT terminals are being used more frequently as a medium for system output. The approaches used include (1) the display of traditional type reports, (2) answering user queries about specific data, and (3) providing sophisticated graphic displays. When the design of a system calls for the use of CRT terminals as a medium for output, the analyst should consider the following questions carefully.

```
FUNCTION '1' SELECTED — OPEN CHECKING ACCOUNT
ENTER NAME:
'JOHN Q. DOE'
ENTER STREET ADDRESS:
'1411 E. MARKET STREET'
ENTER CITY:
'SAN FRANCISCO'
ENTER STATE & ZIP:
'CALIFORNIA,  94014'
ENTER SSN:   (xxx–xx–xxxx)
'595–14–3817'
ENTER SPOUSES NAME:
'JANE A. DOE'
```

FIGURE 8-15   Illustrates the detailed prompting technique in setting up a checking account.

```
ENTRY IS COMPLETE.
PLEASE SCAN FOR ENTRY ERRORS.
IS THE DATA CORRECT (ENTER Y OR N)
'Y'
```

FIGURE 8-16   Illustrates the dialogue the operator must respond with when an entry has been completed and a final check of the data for accuracy has been made.

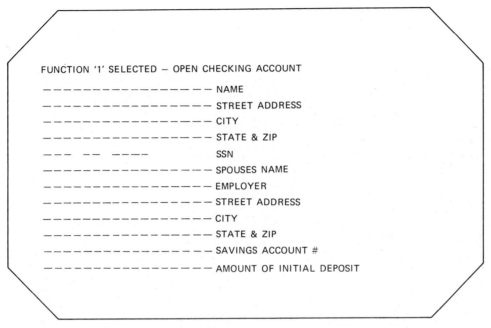

FUNCTION '1' SELECTED — OPEN CHECKING ACCOUNT

— — — — — — — — — — — — — — NAME

— — — — — — — — — — — — — — STREET ADDRESS

— — — — — — — — — — — — — — CITY

— — — — — — — — — — — — — — STATE & ZIP

— — — — — — — — — SSN

— — — — — — — — — — — — — — SPOUSES NAME

— — — — — — — — — — — — — — EMPLOYER

— — — — — — — — — — — — — — STREET ADDRESS

— — — — — — — — — — — — — — CITY

— — — — — — — — — — — — — — STATE & ZIP

— — — — — — — — — — — — — — SAVINGS ACCOUNT #

— — — — — — — — — — — — — — AMOUNT OF INITIAL DEPOSIT

FIGURE 8-17   Illustrates the simulated forms fill-in technique for setting up a checking account.

1. Which type of output approach best fits the needs of the intended user?
2. Does the size of the CRT screen place any limitations on how the output can be formatted?
3. Does the CRT terminal have sufficient local memory to store extra lines or pages of data when there is too much for one screen?
4. Is the CRT terminal capable of displaying all the required characters (uppercase and lowercase), symbols, and punctuation marks?
5. Can the CRT terminal distinguish between variable data and protected data as field labels? (Some terminals highlight the intensity of protected data, see Figure 8-18.)

If the system design calls for the display of the traditional style output reports, spacing will be an important consideration. The horizontal lines of data should be presented with adequate spacing between fields. This serves to minimize user eye strain in distinguishing field values. If insufficient horizontal space is available, the report may need to be split into two parts. Vertical spacing should be used whenever possible between report ID information, column headings, data, and report totals. Figure 8-19 illustrates how a traditional type report could be displayed.

```
ACCT NUMBER — 135792468017          ACCT STATUS — INACTIVE
NAME — SMITH*JOHN D                  SERVICE START DATE —
STREET  —  123 N LAUREL AVE          TYPE SERVICE — RESIDENCE SINGLE METER
CITY/STATE — CHICAGO, ILLINOIS  /60691   CIS ACTIVITY — NONE PENDING

* * * * * * * * * * * * * TURN—ON SERVICE ORDER * * * * * * * * * * * * * *
ENTER ALL KNOWN FIELDS — HIGH INTENSITY FIELDS ARE REQUIRED
DATE WANTED — 10/14/80               CUSTOMER NAME — RAY CARLSON_____
TELEPHONE — 312–157–2308             SPOUSE — _____
MAIL ADDR — _____        CITY/STATE — _____   ZIP — _____
PREV ADDR — 786 MAPLE RD _____   CITY/STATE — DALLAS, TEXAS   ZIP — 75080
EMPLOYER    _____        CREDIT RATING —
DEPOSIT AMOUNT — $ RN                DEPOSIT DATE — — 10/10/80
SPEC INST — _____
LIGHT APPLIANCES — X WELCOME LETTER — X LOAD CHECK —   AIR CONDITIONING —
ORDER RECEIVED FROM — _____ BY — ____ TIME — __:__ DATE — __/__/__
```

FIGURE 8-18   Illustrates how some terminals highlight the intensity of protected data.

When the inquiry technique is used, the output may be relatively small in quantity. In this case, the formatting considerations would probably be minimal. Figure 8-20 shows an example where a bank teller has inquired about the balance of a customer's checking account. When an inquiry may involve substantial amounts of data, it may be necessary to segment it so that selective portions can be displayed. Figure 8-21 illustrates an example where the output from a bank teller's inquiry against a customer's checking account has been divided up into four logical segments. When the teller selects the segment to be reviewed, the number is entered and the segment displayed. Figure 8-22 illustrates how segment 2 would look if selected for display.

The use of graphic displays is the most complex of the output approaches. The techniques employed vary widely. The most common approaches include displaying Gantt charts and bar graphs (see Figures 8-23 and 8-24). The most important factors for the analyst to consider in using this approach are

1. The information needs of the user.
2. The level of information required (detailed or summary).
3. The use to which the information is going to be put, that is, information only or timely decision-making.
4. The most appropriate method of presenting the data.

REPORT ID – CR013   CASH RECEIPTS JOURNAL   PAGE 01
                         PERIOD ENDING 6–30–81   DATE 7–9–81

CUSTOMER NAME	CUSTOMER NO.	INVOICE NO.	CHECK NO.	AMOUNT DUE.	DISCOUNT	CASH RECEIVED
JONES MFG	1347	81–21346	80–4512	2,156.47	21.57	2,134.90
A & E SUPPLY	4714	81–20312	65–213	984.23	9.80	974.43
BELMONT AUTO	4808	81–09845	201452	3,465.45	34.66	3,430.79
SF AUTO & WRECKING	5002	81–10134	1032	10,148.00	102.00	10,046.00
			TOTALS	16,754.15	168.00	16,586.12

FIGURE 8-19 Illustrates how a traditional type output report could be displayed.

'ACCOUNT NUMBER: 12345678'
'BALANCE ?'
CURRENT BALANCE OF ACCOUNT NUMBER 12345678 IS $478.53.

FIGURE 8-20 Illustrates a bank teller's inquiry as to the balance of a customer's account.

```
FUNCTION 3 SELECTED — CHECKING ACCOUNT INQUIRY
IDENTIFY SEGMENT OF DATA TO BE DISPLAYED
1 — PERSONAL DATA
2 — HISTORY OF DEPOSITS
3 — CHECKS BEING HELD FOR NEXT STATEMENT
'2' (Teller has selected segment 2 for display)
```

FIGURE 8-21   Illustrates the inquiry technique where the data available is too much for a single screen so it has been segmented into three pieces.

```
HISTORY OF DEPOSIT
SEGMENT 2 — HISTORY OF DEPOSITS
LAST 30 DEPOSITS:
LAST DEPOSIT:  8/15/80      875.38

1/15/80    742.19      5/4/80      432.00
1/31/80    836.22      5/15/80     742.19
2/11/80    123.00      5/31/80     836.22
2/15/80    742.19      6/15/80     742.19
2/28/80    836.22      6/30/80     836.22
3/15/80    742.19      7/15/80     875.38
3/31/80    836.22      7/31/80     978.56
4/15/80    742.19      8/15/80     875.38
4/30/80    836.22
```

FIGURE 8-22   Illustrates what segment 2 of data would contain for a typical customer.

PROJECT NAME	S or C	JAN 7 14 21 28	FEB 7 14 21 28	MAR 7 14 21 28	APRIL 7 14 21 28	MAY 7 14 21 28	JUNE 7 14 21 28	JULY 7 14 21 28	AUG 7 14 21 28	SEPT 7 14 21 28	OCT 7 14 21 28	NOV 7 14 21 28	DEC 7 14 21 28
A/P SYSTEM DESIGN	S					x x x x							
	C					x x x x							
A/P SYSTEM TEST	S						x x						
	C						x x						
A/P SYSTEM START	S						x x						
	C						x x						
G/L SYSTEM DESIGN	S							x x x x					
	C							x x x x					
G/L SYSTEM TEST	S								x x x x				
	C								x				
G/L SYSTEM START	S												
	C									x x x x			
A/R SYSTEM DESIGN	S								x x x x				
	C								x x x				
A/R SYSTEM TEST	S												
	C									x x x x			
A/R SYSTEM START	S												
	C										x x x x		
	S												
	C												
	S												
	C												
	S												
	C												
	S												
	C												

FIGURE 8-23 Illustrates how a Gantt chart would look graphically displayed.

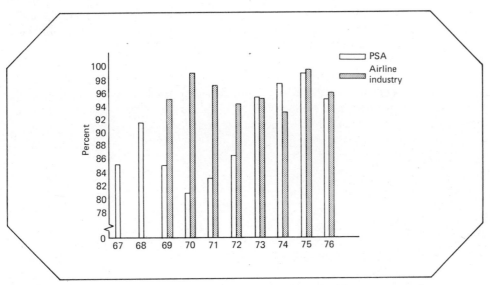

FIGURE 8-24 Illustrates how bar graphs can be displayed as output for management.

AUDIT TRAILS

Whether the analyst is defining the new system requirements or designing for a manual or for a computer-based system, audit trails must be taken into account. An audit trail provides a thread of continuity or traceability between various reports. Audit trails allow the user of the system to trace forward from the source document to the end report or to trace backward from the end report to the original source document. Whenever the analyst is defining the requirements for a system that will have control of the firm's resources, such as money, the firm's auditors will demand that there is an audit trail.

The audit trail in a manual system consists of source documents, journals, ledgers, and reports. This trail enables an auditor to trace an original transaction either forward or backward through the system. Figure 8-25 shows a traditional audit trail which should be present in any system that deals with the firm's resources.

When the analyst is designing the new system for a computer-based system, the audit trail may look different than for a manual system. The capabilities of the computer change certain key elements that relate to the audit trail. These possible changes include the fact that source documents once transcribed into machine readable format are no longer used in the data processing cycle, and may be filed in a manner that makes them hard to retrieve. In some systems, such as an on-line real-time system, source documents may be eliminated entirely by the use of direct input devices. Ledgers (a summary of the transactions) may be replaced by direct access files which do not show the amount leading up to these summarized totals. Files kept on magnetic drums, magnetic disks, or magnetic tapes cannot be read except by use of the computer. The sequence of the records and the processing activities cannot be viewed because these activities are taking place within the computer system. It is for these reasons that the analyst, when designing computer-based audit trails, should provide a means for identifying the account from which the transaction was written. The analyst should provide a means for tracing the summary amount back to the individual source document, a regular provision should be made to print out the record necessary, upon request, and there must be a means for tracing any account even though a regular provision for this trace is not made.

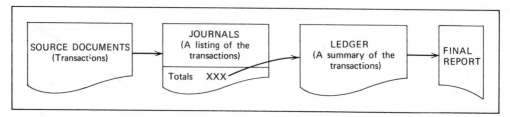

FIGURE 8-25 Traditional audit trail.

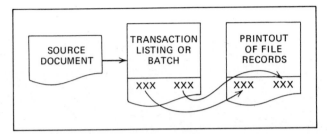

FIGURE 8-26 Computerized audit trail.

The method by which a systems analyst provides for a computerized audit trail is limited only by the analyst's ingenuity. One example might be that the data recorded on a disk file would give the current balance of that account and also would contain reference numbers to all changes that were made to that account. These reference numbers can then refer back to a batch number or a transaction listing that will lead the auditor back to the record (source document) that was originally made machine readable and read into the computer. It is this record (source document) that updated the file. This type of audit trail is shown in Figure 8-26. The source document is recorded on a transaction listing, which can be interpreted as a machine readable input, and a printout of the file record refers back to the transaction listing or batch. In summary, the computerized file record gives the current balance in the printout of the file record and it references all changes by transaction listing or batch number. The auditor need only look at the transaction listing or batch in order to determine where the source document is located.

For another type of audit trail, the printout of the file records gives the balance only. This balance is directly traceable, by its unique amount, back to the transaction listing or batch. The auditor then locates the transaction listing or batch by this unique total, checks the individual listings on that transaction listing or batch input, checks the reference number and goes directly to the source document by using that reference number. Figure 8-27 shows this type of an audit trail pictorially.

The amount of detail in an audit trail depends upon the organization's particular needs. Several reports that are generated by rearranging a given set of data from a computer file will require a more comprehensive audit trail. The analyst should design the new system for the audit trail so there is a minimum amount of file space used for the audit trail. This minimizes the cost of preparing data, it minimizes clutter in the printout of the file records, and it minimizes the chance that errors will be generated within the computer system. Obversely, a simple error may be fatal to the audit trail because it would end the trail at the point where the error was made.

The auditor uses the audit trail to test the reliability of the system and the dependability of the data it generates. Management uses the audit trail to trace items in order to

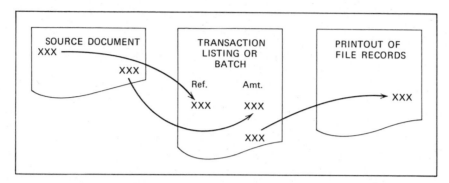

FIGURE 8-27   Computerized audit trail.

troubleshoot or find the cause of some problem. Management also uses the audit trail to verify the accuracy of processing. The systems analyst should not make audit decisions alone. The analyst can design the technical aspects of developing and using the audit trail and the related computer programs, but an auditor *must* be called in to assist the analyst in determining what test data are needed and what information is to be obtained. See Chapter 9 for more on controls.

SELECTIONS FOR FURTHER STUDY

Although the systems analyst is concerned primarily with problem solving as opposed to decision-making, his work necessarily requires that certain decisions be made along the way, for example, of all the alternatives available, which are the best to present to management. Churchman's work, *Thinking for Decisions,* is aimed at stimulating thinking that will solve problems. Other volumes have been mentioned in earlier chapters, and there are many more available. Some of these works stress operations research, a concept made famous by the British during World War II.

A system design that utilizes a computer in the design process is called simulation, modeling, or gaming. There are numerous works also on these subjects including those of Greenblat, Emshoff, or Fullerton. Books on modeling within a wide variety of subject areas (such as the social sciences) are too numerous to mention. The IBM manual on *The Method Phase III* analyzes equipment configurations, and implementation costs are compiled and added to other costs to determine the economic impact on the business.

Periodical articles on all the subjects mentioned in this chapter can be located through such indexes as *Business Periodicals Index, Applied Science and Technology Index, Engineering Index* (called *COMPENDEX* in its computerized format), and the

Computer and Control Abstracts (called *INSPEC* in its computerized format). The first two generally are easiest to use, but the latter two are international in their coverage and contain short annotations to indicate what is in the article. Your librarian should be consulted about the computerized versions of these indexes.

1. ACKOFF, RUSSELL L. *The Art of Problem Solving: Accompanied by Ackoff's Fables.* New York: John Wiley & Sons, Inc., 1978.

2. ACKOFF, RUSSELL L. *Scientific Method: Optimizing Applied Research Decisions.* New York: John Wiley & Sons, Inc., 1972.

3. *Applied Science and Technology Index* (formerly *Industrial Arts Index*). New York: H.W. Wilson Company, 1913–

4. ATWOOD, JERRY W. *The Systems Analyst: How to Design Computer-Based Systems.* Rochelle Park, N.J.: Hayden Book Company, Inc., 1977.

5. BAKER, FRANK. *Organizational Systems: General Systems Approaches to Complex Organizations.* Homewood, Ill.: Richard D. Irwin, Inc., 1973.

6. CHURCHMAN, CHARLES WEST. *Design of Inquiring Systems: Basic Concepts in Systems Analysis.* New York: Basic Books, Inc., 1972.

7. CHURCHMAN, CHARLES WEST. *Thinking for Decisions: Deductive Quantitative Methods.* Chicago: Science Research Associates, 1975.

8. CLIFTON, H.D. *Systems Analysis for Business Data Processing, revised edition.* New York: Van Nostrand Reinhold Company, 1974.

9. COCHIN, IRA. *Analysis and Design of Dynamic Systems, preliminary edition.* New York: Harper & Row Publishers, Inc., 1977.

10. *Computer and Control Abstracts.* London: The Institution of Electrical Engineers, 1969–

11. CONDON, ROBERT J. *Data Processing Systems, Analysis and Design.* Englewood Cliffs, N.J.: Reston Publishing Company, Inc., 1974.

12. DAVIS, GORDON B. *Auditing and EDP.* New York: American Institute of Certified Public Accountants, Inc., 1968.

13. DENEUFVILLE, RICHARD, and JOSEPH H. STAFFORD. *Systems Analysis for Engineers and Managers.* New York: McGraw-Hill Book Company, 1971.

14. EMSHOFF, JAMES R., and ROGER L. SISSON. *Design and Use of Computer Simulation Models.* London: The Macmillan Company, 1970.

15. *Engineering Index.* New York: Engineering Index Company, 1884–

16. FOX, KENNETH L. *Auditing Objectives.* Columbus, Ohio: Grid Publishing, Inc., 1975.

17. FULLERTON, HERBERT H., and JAMES R. PRESCOTT. *Economic Simulation Model*

for Regional Development Planning. Ann Arbor, Mich.: Ann Arbor Science Publishers, 1975.

18. GORDON, GEOFFREY. *System Simulation, 2nd edition.* Englewood Cliffs, N.J.: Prentice-Hall, Inc., 1978.

19. GREENBLAT, C., and R. DUKE, eds. *Gaming-Simulation: Rationale Designs and Applications.* New York: Halsted Press, 1975.

20. HARPOOL, JACK D. *Business Data Systems: A Practical Guide.* Dubuque, Iowa: William C. Brown Company, 1978.

21. *IBM Data Processing Techniques: IBM Study Organization Plan: The Method Phase III.* White Plains, New York: International Business Machines Corporation, 1963 (F20-8138-0).

22. JENKINS, BRIAN, and ANTHONY PINKNEY. *An Audit Approach to Computers: A New Practice Manual.* London: The Institute of Chartered Accountants in England and Wales, 1978.

23. JENSEN, RANDALL W., and CHARLES C. TONIES. *Software Engineering.* Englewood Cliffs, N.J.: Prentice-Hall, Inc. 1979.

24. KUONG, JAVIER F. *Audit and Control of Advanced On-Line Systems, Manual.* Wellesley Hills, Mass.: Management Advisory Publications, 1978.

25. MAIR, WILLIAM C., DONALD R. WOOD, and KEAGLE W. DAVIS. *Computer Control and Audit: How To Reduce Management Exposure to Traps, 2nd edition.* Altamonte Springs, Fla.: The Institute of Internal Auditors, 1976.

26. PORTER, W. THOMAS, and WILLIAM E. PERRY. *EDP Controls and Auditing, 2nd edition.* Belmont, Calif.: Wadsworth Publishing Company, Inc., 1977.

27. SANDERS, DONALD H. *Computers in Business.* New York: McGraw-Hill Book Company, 1975.

28. WARNIER, J.D. *Logical Construction of Programs.* Leiden: H.E. Stenfert Kroese B.U., 1974.

29. WILLOUGHBY, THEODORE C., and JAMES A. SENN. *Business Systems.* Cleveland, Ohio: Association for Systems Management, 1975.

30. YOURDON, E., and L.L. CONSTANTINE. *Structured Design.* New York: Yourdon, Inc., 1975.

QUESTIONS

1. Define problem solving.
2. What items should already be established by the time the analyst is ready to begin designing the new system?

3. With what is systems design mainly concerned?

4. Define maximize.

5. Define optimize.

6. Define satisfice.

7. What does random dependence mean?

8. What is the difference between an empirical system and a conceptual system?

9. What two types of thinking can the designer use?

10. Upon what principle is the structured design process based?

11. Describe the purpose of the data flow diagram.

12. What is the purpose of the structure chart?

13. Identify the four types of data streams.

14. Describe the functional composition method of decomposing a module.

15. What is the set partition method of decomposing a module?

16. What does the final step in the structured design process require?

17. Name three ways that a CRT terminal can be used as a medium for system output.

18. Identify four possible advantages to using on-line inputs in the design of a system.

19. List five points an analyst should consider if the dedicated data entry approach is to be used.

SITUATION CASES

CASE 8-1

The XYZ Corporation manufactures and distributes skateboards. The firm has been in business for five years and employs ten people. Its sales area consists of 1000 stores throughout the United States, including several major department store and toy store chains.

The sales manager indicated that market conditions were such that sales could be doubled if production was increased. The production manager, however, indicated he was having some difficulties expanding production. As a result, the president decided to expand and requested that a senior system analyst be assigned to help decide how to go about it.

The first thing the analyst did was to remind himself to only attempt to solve one problem at a time. The analyst divided his work into three major steps as a means of following a logical step-by-step approach toward finding available alternatives. The analyst's first step was to assemble the facts that pertained to his assignment by considering the feasibility study, problem definition, outline, and an understanding of the existing system. The second step was to pursue a line of lateral thinking which involves starting with system outputs and working backward through the system. The analyst's third step was to evaluate the ideas that had occurred to him without being too rigid.

In checking with the production manager, the analyst noted that the four different machines used in manufacturing skateboards were more than five years old and as an assembly line operation had reached their capacity. At this point, the analyst was faced with the problem of identifying what all the possible alternatives were, that is, the choice-set. With the help of the production manager, the analyst determined that there were four alternatives. First, the company could simply keep the existing four machines. Second, they could add one more final assembly machine and increase production marginally. Third, they could sell all the old machines and purchase new high-speed equipment. Fourth, they could extend operations to three shifts instead of the current two. The analyst's next step was to divide the identified alternatives into attainable and unattainable sets. During this process, the analyst noted that due to cash-flow problems, the company was not in a position to purchase all new equipment. As a result his attainable set included alternatives one, two, and four.

Since the company's labor force was unionized, the effort required to gain approval for running a third shift was expected to be substantial. The first alternative was considered to be the least favored alternative because of the president's desire to expand production. This was followed closely by the fourth alternative. The analyst decided the second alternative was the most favored and returned to the production manager to get his tentative approval before making a final recommendation to the president.

QUESTIONS

1. What should the analyst have done first in following a logical step-by-step approach in system design?
2. What other pertinent areas should the analyst have examined pertaining to his assignment?

CASE 8-2

A junior analyst was assigned to a system design team which was working on the design of a new payroll system. The team had been utilizing the structured design process; however, the junior analyst had no experience in this technique. The team leader and a senior analyst spent several days reviewing the structured design process with the junior analyst. In addition, the junior analyst read several articles discussing the process. In order to determine if the junior analyst fully understood the basic principles of constructing flow diagrams, she was asked to analyze a sample diagram and indicate if anything was wrong. The junior analyst reviewed the diagram and indicated that she felt the following things were wrong: (1) input D should be shown coming in from the right, and (2) H was not shown coming out as output.

QUESTIONS

1. Identify any additional problems in the diagram (Figure 8-28) that the analyst failed to note.

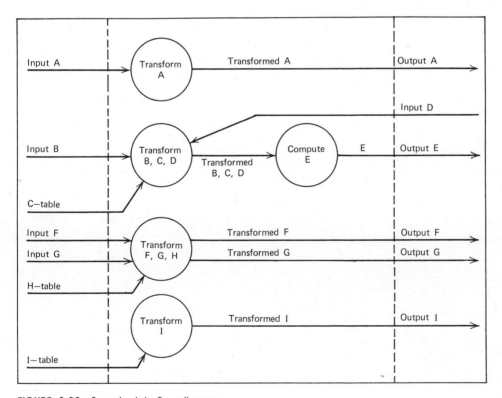

FIGURE 8-28 Sample data flow diagram.

FOOTNOTES

1. *Webster's New World Dictionary of the American Language* (Cleveland, Ohio: The World Publishing Company, 1960), p. 1161.

2. Edward de Bono, *New Think; the Use of Lateral Thinking in the Generation of New Ideas* (New York: Basic Books, Inc., 1968).

CHAPTER 9

DESIGNING NEW SYSTEM CONTROLS

A successful new system in today's sophisticated, rapidly changing business environment must be built upon a solid framework of operational and accounting controls. This chapter provides the analyst with a general background in control theory which can be applied easily in designing the application controls for today's new systems.

THE NEED FOR CONTROLS

The management of a firm is responsible for establishing and maintaining adequate internal controls. The establishment and maintenance of a system of internal controls is a significant management obligation. A number of developments have triggered considerable increased interest in the subject of internal controls. Among these developments have been

- Explosive growth in data processing technology.
- Wider use of data processing by organizations of all sizes, especially smaller organizations because of growth in the use of minicomputers and microprocessors.
- Increased dependence upon on-line distributed data processing systems and the growing recognition of the impact data processing has on every facet of today's organization in providing accounting, production, and management-related information.
- Increased utilization of data communications (teleprocessing).
- Widely publicized cases of alleged computer fraud or breaches of privacy.
- Increased liability by directors and officers of organizations and the growing concern over the shift in control responsibilities and techniques that must be utilized to maintain an adequate infrastructure of organizational controls.
- Increased attention as expressed by the ever-growing number of audit committees at the director level.
- The emphasis by public accountants on internal control systems.
- The Foreign Corrupt Practices Act of 1977.

To further emphasize the need for controls, it should be noted that in recent years, organizations increasingly have become dependent upon computer hardware, software, and data processing personnel. This commitment to computerization has changed the potential vulnerability of the organization's assets because the traditional security, audit, and control mechanisms take on a new and different form in a computer-based system.

A complex on-line data communication-oriented system consists of various combinations of hardware, software, facilities, people, and the policies and procedures that interrelate these components. The many diverse components and potential entry points into a complex on-line system make it possible for a person with sufficient technical or applications knowledge to enter the system to make unauthorized manipulations of data, programs, or operational procedures. Furthermore, control procedures for an on-line system cut across many lines of responsibility within an organization, creating a control problem in itself. For instance, several departments within an organization may share in the exercise of the control procedures and each department may be responsible for only one segment of the overall control plan. The integration of controls among the various components of such a complex on-line system is the infrastructure upon which a secure on-line data communication-oriented system must be based.

This increased reliance on computers, the consolidation of many previously manual operations onto computer systems, the shared responsibility between different departments for control procedures, and the fact that on-line systems cut across many lines of responsibility have increased management's concern about the adequacy of the present control mechanisms in use in the EDP environment.

While the use of computers is rising, it is also evident that a far greater potential exists to control errors and omissions, disastrous events, fraud, and other adverse occurrences in automated systems than in the manual systems they have replaced. Finally, management's concern over adequate controls will be negated if the data processing system designers, EDP auditors, and their managers do not have the proper training and control techniques to utilize when designing or reviewing the internal controls associated with on-line computerized systems.

The systems analyst must be able to appreciate both environmental and application controls in order to design and implement effective, efficient, and well-controlled systems. Specific application control areas are discussed in a later section in this chapter. The typical data processing system can be viewed as having up to nine major control components in its environment including (see Figure 9-1)

- General organizational controls.[1]
- Input controls.
- Data communication controls.
- Program/computer processing controls.
- Output controls.
- On-line terminal/distributed system controls.
- Physical security controls.
- Data base controls.
- System software controls.

It is the responsibility of the systems analyst to determine which of the major control components may impact the environment of the system being designed.

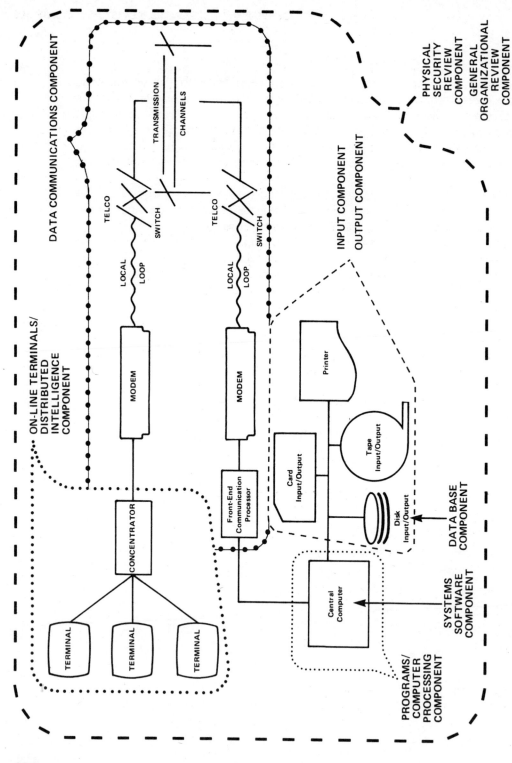

194

FIGURE 9-1 Control components of a data processing system.

©Copyright Jerry FitzGerald 1978

THE CONTROL MATRIX APPROACH

The control matrix approach facilitates organizing the review of each control component that could impact the system being designed. The first step is to list the system resources and assets on the left side of the matrix (see Figure 9-2). The second step is to list the concerns and exposures (sometimes called threats) relative to the system across the top of the matrix. The final step involves reviewing all the potential controls that could be applied to the system and deciding, for each control component, which control techniques provide some protection of the resource/asset from an exposure/concern.

The controls are listed on a separate sheet and numbered sequentially. The number of each control that applies is listed in the matrix cell that represents the intersection of the resource/asset being protected to the exposure/concern being protected against (see Figure 9-2). Appendix II contains an extract from Fitzgerald's book which provides examples of the control matrix approach applied to the following control components: (1) input controls, (2) data communication controls, (3) programming controls, and (4) output controls.

GENERAL APPLICATION CONTROL CONCEPTS

The analyst has traditionally had to deal with controls in the form of the age-old audit trail (see Chapter 8). The audit trail was designed to provide a thread of continuity or traceability between various reports. This allowed the user of a system to trace forward from the source document to the end report or to trace backward from the end report to the original source document.

The following general application control concepts are intended to provide the analyst with a basic framework within which to understand the fundamentals of application controls as they relate to the computer environment. Many of the techniques discussed, however, are equally applicable to the controls needed for manual systems. They may be broken into the following major control concepts for analysis.

1. *Input completeness:* All transactions must be converted to a machine readable media and reach the computer intact.
2. *Input accuracy:* All transactions that reach the computer must be entirely accurate.
3. *Processing completeness:* All accepted transactions must be processed and recorded on files as appropriate.
4. *Processing accuracy:* All accepted transactions must be accurately recorded on files as appropriate.

CONCERNS / EXPOSURES

RESOURCES / ASSETS	ERRORS AND OMISSIONS	MESSAGE LOSS OR CHANGE	DISASTERS AND DISRUPTIONS	PRIVACY	SECURITY/ THEFT	RELIABILITY (UP-TIME)	RECOVERY AND RESTART	ERROR HANDLING	DATA VALIDATION & CHECKING
CENTRAL SYSTEM	1-4, 39, 41-43, 47, 48	1-5, 48, … 89	1, 8, 1… 51, 5…, 58, 64, 65, …	6, 8, 24, 35, 53 … 70, 72-74, 77-80	6, 8, 24, 35, 53, 56, 60, 6…, 77-80	1, 13, 16, 29, 63-65, 68, 81, 88	50, 51, 63-65	48, 85, 89	6, 24, 47, …
SOFTWARE	1-4, 39, 41-43, 46-49, 52	1-5, 41, 48, 49, 52, … 89	1, 8, … 51, 54, 57, 58, 64, 65, …	6, 8, 24, 35, … 70, 72-74, 77-80	… 56, 60, 62 … 77-80	63-65, 68, 81, 88	50-52, 61, 63, 64, 68	48, 61, 85, 89	6, … 55
FRONT-END COMMUNICATION PROCESSOR	1-4, 34, 39, 41-44, 46-48	1-5, 7, 34, 37, 39, 41, 42, 49 (89)	1, 8, 40, 44, 46, … 51, 54, 57, 58, 65, 79, 85	6, 8, 24, 35, 37 … 70, 72-74, 78-80	35, 37, 39, 45, 60, 62, 68, 70, … 78-80	…, 30, 32, 34, 36, 40, 43, 44, 50, 51, 63-65, 81, 88	37, 50, 51, 63-65	43, 48, 85, 89	6, 24, 39, 45, 47, 48, 88
MULTIPLEXER, CONCENTRATOR SWITCH	1-4, 7, … 41, 44, 46, 47	1-5, 7, 37, 39	1, 8, 13, 16, 29, 44, … 54, 57, 58, 65, 79, 85	6, 8, 24, 35, 37, 45, 60, 62, 68, 78-80	6, 8, 24, 35, 37, 39, 45, 60, 62, 68, 70, 74, 78-80	1, 13, 16, 29, 30, 32, 34, 36, 40, 44, 50, 51	37, 50, 51, 64		45, 47, …
COMMUNICATION CIRCUITS (LINES)	28, 70, 91	28, 70, 91	10, 15, 16, 18, 44, 49, 63, 64, 66, …	25, 28, 68, 70, … 75, 76, 78-80	25, 28, 68, 70, 75, 76, 78, … 91				
LOCAL LOOP	12				… 75, …				
MODEMS	12, 18					9-11, 13-18, 21, 23, 36, 88	9-11, 14, 15, 63, 64	18-20, 22, 23	
PEOPLE	5, 39				24, 29, 71, 74, 80	81, 82, 85-87	50, 51, 86, 87	49, 86, 87, 89, 90	6, 88
TERMINALS/ DISTRIBUTED INTELLIGENCE					29, 53, 56, 62, 70	1, 40, 88	63, 64		6, 24, 45

Callout — DEFINITION OF A CONCERN/EXPOSURE: THE LOSS OF MESSAGES AS THEY ARE TRANSMITTED THROUGHOUT THE DATA COMMUNICATION SYSTEM OR THE ACCIDENTAL/INTENTIONAL CHANGING OF MESSAGES DURING TRANSMISSION.

Callout — EXAMPLE OF A SPECIFIC CONTROL FROM THE LIST GIVEN IN THIS CHAPTER: 89. REVIEW ERROR RECORDING TO REDUCE LOST MESSAGES. ALL ERRORS IN TRANSMISSION OF MESSAGES IN THE SYSTEM SHOULD BE LOGGED AND THIS LOG SHOULD INCLUDE THE TYPE OF ERROR, THE TIME AND DATE, THE TERMINAL, THE CIRCUIT, THE TERMINAL OPERATOR, AND THE NUMBER OF TIMES THE MESSAGE WAS RETRANSMITTED BEFORE IT WAS CORRECTLY RECEIVED.

Callout: ALL THE CONTROLS LISTED DOWN THIS COLUMN ARE THOSE THAT CAN BE UTILIZED TO SAFEGUARD AGAINST A "MESSAGE LOSS OR CHANGE."

Callout: ALL THE CONTROLS LISTED ACROSS THIS ROW ARE THOSE THAT CAN BE UTILIZED TO SAFEGUARD THE "FRONT-END COMMUNICATION PROCESSOR."

Callout — DEFINITION OF A RESOURCE/ASSET: A HARDWARE DEVICE THAT INTERCONNECTS ALL THE DATA COMMUNICATION CIRCUITS (LINES) TO THE CENTRAL COMPUTER OR DISTRIBUTED COMPUTERS AND PERFORMS A SUBSET OF THE FOLLOWING FUNCTIONS: CODE AND SPEED CONVERSION, PROTOCOL, ERROR DETECTION AND CORRECTION, FORMAT CHECKING, AUTHENTICATION, DATA VALIDATION, STATISTICAL DATA GATHERING, POLLING/ADDRESSING, INSERTION/DELETION OF LINE CONTROL CODES, AND THE LIKE.

SEE CHAPTER 7 FOR MOST OF THESE CONTROLS

SAMPLE MATRIX: DATA COMMUNICATIONS CONTROLS

FIGURE 9-2 Sample matrix on data communications controls.

5. *Authenticity:* All transaction data must be checked for its appropriateness and proper authorization.
6. *Maintenance:* All data on computer files must remain correct and up-to-date.

Figure 9-3 illustrates graphically most of the control requirements noted and the various techniques that commonly are found in the computer environment needed for new systems.

While the techniques are discussed in detail later, brief explanations have been provided here for use with Figure 9-3.

	One—for—one checking	Batch control totals	Computer sequence check	Computer matching	Programmed checks	Prerecorded input	Key verification
Input completeness	X	X	X	X			
Input accuracy	X	X		X	X	X	X
Processing completeness	X	X	X	X			
Processing accuracy	X	X		X	X		
Authenticity					X		
Maintenance	X	X			X		

Note: *Accuracy applies to keying activity only.

FIGURE 9-3 Common computer control requirements.

1. One-for-one checking refers to the manual checking of each individual input or source document against detailed computer reports of documents processed.

2. Batch control totaling is the process of grouping transactions on some basis such as document count, item count, dollar totals, or hash totals. This control total is then balanced manually, or by computer, with the total of the transactions processed.

3. Completeness can be established by having the computer check the sequence of numbers on input documents and report on missing or duplicate numbers for manual investigation.

4. The matching procedure uses the computer to match data on input documents with information held on master or suspense files. This control anticipates the transaction being input because there is sufficient information recorded on the file to identify the transactions that are expected to be processed. Computer matching can detect not only missing documents but, in some cases, even transactions that should have been but were not recorded.

5. Accuracy checks can be embodied in computer programs in the form of edit routines, that is, checks (e.g., reasonableness, dependency, existence, etc.) against specific data fields to provide additional assurance with respect to accuracy of input.

6. Prerecorded input is a technique involving the preprinting of certain information fields on a document (often called a "turnaround document") which is output by the computer for subsequent input.

7. Key verification is a common technique for controlling the conversion of information from input documents into machine-readable form. Although it initially was used for punched cards, it also has been used extensively in key-to-disk and key-to-tape systems. It involves a repetition of the keying process by a separate individual using a machine which checks the key strokes with what was originally punched on the card.

INPUT CONTROLS: COMPLETENESS

In designing the input controls for a new system, the analyst must consider both the completeness and accuracy of the input. Controls for completeness of input are designed to ensure that all transactions are recorded, submitted to the computer, and accepted by the computer. Controls on the accuracy of input are discussed in the next section.

The analyst has several techniques available to choose in establishing input completeness. The key is to select those techniques that best fit in with the organization and provide the most coverage. They can be considered under the following headings.

1. One-for-one checking of printouts.
2. Agreement of manually established batch totals.
3. Computer sequence check of serially numbered documents.
4. Computer matching with a file of previously accepted transactions.
5. Rejections.

One-for-one checking is a technique that calls for each individual input or source document to be compared against a detailed computer listing of documents processed or being processed. This technique is used primarily with the input of master file data which acts upon normal transactions and which could cause cumulative errors if not processed correctly and in a timely manner. This type of data is relatively low in volume and is referred to as maintenance data. An example would be a change in pay rate, which would be used in processing all subsequent time cards. One-for-one checking is used also in many on-line systems where a CRT operator first inputs a transaction and then visually checks the entry against the input document.

The *batch control total* technique involves the manual grouping of transactions prior to input and establishing a control total over the batch on some relevant basis such as document count, item count, dollar totals, or hash totals. Batch control totals normally are used in one of the following two ways.

1. The total of the batch is established manually, using one of the previously described methods, and recorded in a batch control register. The batch is input to the computer, which adds the batch based on the same variable(s) used manually and prints out the total. The total recorded in the batch control register is compared with the total printed out. The reconciliation of the total allows the input completeness to be established. There must be procedures, however, for adjusting the total in the batch control register for rejected items and to ensure that rejected items are corrected and later reprocessed.
2. The batch total is established manually and, rather than recording this total in a register, a batch header slip is prepared indicating the batch total and this is input to the computer with the batched data. The computer adds the batched data and internally compares the total to the total input from the header slip.

In order for the batch control technique to work effectively, the analyst must establish adequate procedures to ensure that all documents are batched and all batches are presented for processing.

The *sequence check* technique can establish input completeness by having the computer check the sequence of numbers on the input documents and report missing or duplicate numbers for manual investigation.

The *matching* technique uses the computer to match data on input documents with information held on master or suspense files. This control anticipates the transaction being input because there is sufficient information recorded on the file to identify the transactions that are expected to be processed. Computer matching will detect not only missing documents; but even transactions that have not been recorded (e.g., the failure of an employee to prepare a time card). To use this procedure, the computer must have data already recorded for matching with input; therefore, matching is normally associated with a chain of events rather than the initial recording of the first transaction in a sequence of events. (For example, sales orders are input to the system and then are recorded on a suspense file to await the confirmation of shipment. When the confirmation of shipment is input, it is matched by the computer with the order held on the suspense file before processing is continued.) Items not matched for any reasonable period of time usually are rejected.

The *rejection* concept is based upon the assumption that some input data will contain errors and therefore will be rejected from the processing stream. In batch processing, it is impractical to make corrections at the time erroneous data is identified and rejected. If the analyst is to be confident that data is input completely, he or she must ensure that adequate controls exist to correctly reprocess all rejections. Computer systems normally will handle input errors in one of two methods.

1. Reject the items from further processing with no record other than a printed report.
2. Reject the items from further processing but maintain within the computer an error suspense file of items awaiting correction and produce a cumulative report of all uncleared items on the error file.

In order for the rejection concept to work effectively, the analyst must prepare detailed, standardized, manual procedures to

1. Ensure that rejection reports and items held in error files are promptly investigated, corrected, and resubmitted.
2. Adjust properly any previously established control totals.

INPUT CONTROLS: ACCURACY

Once the analyst has selected the techniques to be used in establishing input completeness, the next step is to decide on a compatible approach for establishing input accuracy. In general, controls on the accuracy of input are designed to ensure that errors will be detected when transactions are initially recorded, converted to machine-readable media, and accepted by the computer. Accuracy is not concerned with documents as a whole. Instead, it is concerned with individual, predefined fields or elements. The real question to be answered here is, given that a document is processed, is all the significant information correct?

To evaluate input accuracy, the analyst must first identify the significant information such as "reference" data or "financial" data. Reference data includes such information as account numbers, dates, part numbers, name and address information, and so forth. Financial data usually includes dollar values, quantities, and prices.

The analyst has a variety of techniques available to establish input accuracy. They can be considered under the following headings.

1. Extensions of completeness techniques.
2. Programmed edit checks.
3. Review of output.
4. Prerecorded input.
5. Key verification.

The techniques used to control input completeness can, in some cases, be extended to control input accuracy. This applies to one-for-one checking, agreement of manually established batch totals, and computer matching. Sequence checking and document counts, however, will not establish accuracy since they do not establish the correctness of data fields input. Control techniques such as one-for-one checking and computer matching may be used effectively to simultaneously establish the accuracy of input for several fields within the transactions. While control totals also could be used to assure the accuracy of input for several fields, the batching process could be too time consuming.

Edit routines can be included in computer programs to provide a means of ensuring the accuracy of input. There are four basic types of programmed edit checks for the analyst to choose from, including

1. Dependency checks.
2. Current data matching checks.
3. Reasonableness checks.
4. Validity checks.

Programmed edit checks are designed to identify erroneous or suspect data and print it out on a report for manual review and followup. As a result of this process, this data may be

1. Rejected and dropped from subsequent processing. It therefore requires correction and reinput.

2. Rejected and held in suspense. It requires an amendment to the suspense file for any errors or, alternatively, a manual override of the edit check if the transaction is correct.

3. Not rejected but identified and carried through processing in the normal way. In these circumstances, corrective action is necessary only if the transaction is in fact wrong. A great deal of manual review and followup is necessary to ensure that incorrect data is identified and corrective action taken.

The *dependency check* technique of editing input data is based upon the premise that the transaction being submitted does possess a logical relationship between two or more data fields, which the program is designed to check. An example of this technique would be where a new part is input to an inventory system. The input data includes part number, part group, and unit price. If the unit price is not correct for that part group, the transaction will be reported on an exception report for subsequent manual investigation.

The *current data matching check* technique edits by requiring that the input transaction include the current values of the data fields to be changed. The current data portion of the input transaction is checked by the computer to ensure that it agrees with the data already held on the file. If the current data matches, the new data is then updated to the file. A personnel system could serve as an example of this technique. When a new person is hired into a recently vacated position, a transaction is generated. To load the new hire onto the file will require that the following data be input.

1. Name of previous person holding position, date of hire, job classification.
2. Name of new person to hold position, date of hire, job classification.

The computer would then compare the data input concerning the previous position holder with the information presently stored on the file. If it matches, the new data is written on the file, otherwise the transaction is rejected. This technique is particularly useful where maintenance data (such as pay rate) is being input which could have a cumulative error effect if updated to the wrong record. It ensures that the user is keeping maintenance data current, and it also helps to establish that the correct generation of the file is amended. This technique, however, does not ensure the accuracy of the new information itself.

The *reasonableness check* technique ensures that a data field lies within predetermined limits. For example, an hourly wage pay rate within a company is never less than minimum wage and never exceeds $19.25 per hour. The pay rate field would be checked by the edit routine to determine if it lies between the two values. Pay rates outside this range are unusual and should be printed out for manual investigation. Where the parameter limits are not included in the program, but are input on a control card or via CRT during each run, the company should establish controls to ensure the correct limits are used for each run.

The *validity check* technique establishes that input data is valid by comparing it to a master file or to predetermined tables of valid codes (e.g., part numbers, job classification, division numbers, state abbreviations, prices, etc.). This type of edit routine can usually be found in an inventory system. At some point, when a shipment of parts arrives, an input transaction is created using a part number. The edit routine will take this number and compare it to a table of all valid part numbers. If an agreement is made, the transaction will pass for further processing; if not, the transaction will be rejected and referred to manual investigation and correction. The effectiveness of an existence check depends on the construction of the reference fields being checked. Errors of transposition and miscoding should not cause a match with another valid number. For example, if there are ninety-nine valid product codes numbered from 01 to 99, any error in the product code will be undetected by the validity check. The reference fields should be constructed in such a way that an error in the field is likely to result in a nonexistent code. Among the techniques used in the construction of reference fields are

1. *Use of check digits:* An additional digit is added to the field. One of several check digit techniques could be used such as modulus 10.
2. *Use of alphanumeric codes:* Only certain possible combinations of letters and digits are used, thus increasing the possibility that errors will not match with another valid code.

The analyst must be aware that rejections can arise out of an existence check because of either an error in the transaction, or incomplete data on the master file or table. As a result, the analyst must ensure that there are procedures to handle both possibilities.

In some situations, control over accuracy may be achieved by an overall manual review of the final output for correctness. Normally, the analyst would only apply this technique where the information printed out is sensitive and of relatively low volume (e.g., the executive payroll or a list of passwords). This technique is seldom a satisfactory accuracy check by itself, but it can be used to supplement the other techniques described.

Errors frequently arise during the initial recording of a transaction. By preprinting certain information fields on a document, such errors can be avoided.

It is possible to use a punch card, OCR (Optical Character Recognition), OMR (Optical Mark Sensitive Recognition), or MICR (Magnetic Ink Character Recognition) forms, thus making the input form machine readable and avoiding the errors inherent in keying techniques.

The payment advice form sent to customers is a familiar example of what is often called a "turnaround document," one which is output by the computer for subsequent input. However, turnaround documents also are used to amend maintenance data. When the maintenance data is first set up, the complete contents of the record are printed out. The output document is filed until an amendment is necessary. The field or fields that need to be changed are crossed out and the new values entered beside them. This process reduces the chance that incorrect values will be written down and increases the effectiveness of supervisory approval of the document. The change fields are keyed and input to the computer and the new output document is checked back one for one with the input and then filed for use next time. Turnaround documents are often used with the current data matching programmed checks technique.

It is also possible for the program to indicate on the output those fields that were changed, thus increasing the effectiveness of the one-for-one check. Some turnaround documents are punched cards, OCR, or OMR forms; only changed information need be recorded and the chance of keying errors is reduced.

Key verification is a common technique to control the conversion of information from input documents into machine-readable form. It is in common use, particularly for punched cards, and involves a repetition of the keying process by a separate individual using a machine that checks the key strokes with what was originally punched on the card. The technique is usually of limited control value, particularly in the absence of other techniques discussed above since it only covers the actual keying process.

PROCESSING CONTROLS: COMPLETENESS

In designing the processing controls for a new system, the analyst must consider both the completeness and accuracy of processing. Once it has been established that all data is input to the system, it must be determined if sufficient controls exist to ensure that all the input data accepted by the system is carried successfully through intermediate processing and recorded on the appropriate computer files. This would be similar, in a manual system, to establishing that transactions are all recorded in the books (accounting records) or subledgers.

The analyst has several techniques available to choose from when establishing processing completeness. The techniques selected should fit well within the organization, be consistent with the new system design, and provide the most coverage. They can be considered under the following headings.

1. Use of input controls.
2. Specific updating controls.
3. Reliance on controls over a previous application.
4. Simultaneous update.

Some controls over input completeness also can provide coverage of the completeness of processing updates. To accomplish this, however, the input control must be exercised at or after the update process. Input controls that involve the checking of printouts, such as one-for-one checking and agreement of control totals, will help to ensure processing completeness if the output report used for checking is produced at or after updating.

The typical computer job stream is broken down into several distinct phases. Figure 9-4 shows the processing stream for a relatively simple job. It illustrates the point that to ensure processing completeness the input must agree to the output after the appropriate files have been updated. At points A, B, and C, the reports of processed transactions are produced before the processing has updated the appropriate files. As a result, the reports may be effective in establishing input completeness, but they will not establish that a complete update has occurred. At point D, the report has been produced after updating, and through the use of controls such as sequence checks and

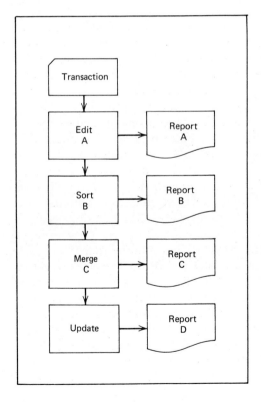

FIGURE 9-4 Shows the processing stream for a relatively simple job.

computer matching, coverage of both input completeness and processing completeness can be provided.

In cases where the input controls being utilized do not provide sufficient coverage to ensure that all transactions update the appropriate files, the analyst may have to rely upon the use of specific update controls. This technique involves use of control totals generated by the computer during, or prior to, the run in which the input control is exercised. The control totals then are matched either manually or automatically to the total of the items written to the appropriate files. It may not always be possible, however, to match the control totals directly because

1. Control totals may have been summarized because several input batches were combined for subsequent processing.
2. The total being controlled may change as the result of a particular intermediate processing step. For example, a hash total of quantities may change to a dollar total as a result of the pricing of shipments.
3. Items may be rejected at an intermediate processing step.

There are two specific techniques used to control the processing completeness, that is, a manual or computer reconciliation. The manual reconciliation technique is used often in conjunction with input batch controls, and controls processing updates for completeness by establishing a total of the items accepted by the computer. The total of the items accepted can be manually reconciled with a computer-produced total subsequent to the processing update. In using this technique, the computer establishes and prints out a total for accepted data. This total is recorded manually and reconciled to the amount updated which then is printed out on a subsequent report produced after updating. The computer reconciliation technique monitors processing through to the updating stage by having the computer internally reconcile totals computed by the programs and written on a control record in each computer file. As each individual item is processed, it is accumulated, and the total of all items must agree with the total on the control record. This activity continues until the end of the processing update is reached.

In some computer-based systems, all transactions are first summarized on independent control files. Periodically, these totals are balanced by an independent set of programs in such a way that it can be determined that the correct files have been used and that all transaction files have been combined for processing. The only manual procedures required are those that determine that the programmed reconciliations are not bypassed by unusual events such as system failure, and that the control files are brought forward properly each time the system is closed down and restarted. To be completely effective as a control over processing updates, the initial total used either in the manual agreement of computer-established totals or the run-to-run internal reconciliation should be established during or before the run when the input completeness control is exercised. There must not be a gap in the sequence of programs or events.

As systems become more integrated, and a file output from one system becomes input to another system, the analyst must consider placing some reliance on the controls being exercised within the originating system. For example, when a file of accepted shipping reports is output from an inventory application, it may be used as an input to the accounts receivable system. In these circumstances, the input completeness controls within the originating system should be designed to provide additional assurance as to the input completeness for the subsequent system. Specific update controls, as described in the previous section, are necessary to ensure that all transactions accepted by the first application are passed through the second update. If the system is designed to allow only certain transactions to be passed through, the programmed selection procedure must operate to ensure that all required transactions are in fact selected.

If the system is designed in such a way that two or more files are updated at the same point during processing, simultaneous updating, the analyst can use effectively the same input and processing update controls for each file.

PROCESSING CONTROLS: ACCURACY

Once the analyst has selected the techniques to be used in establishing processing completeness, the next step is to decide on a compatible approach for establishing processing accuracy. In general, controls on the accuracy of processing are designed to ensure that data is carried properly through intermediate processing and accurately updated to the correct generation of the appropriate computer file(s).

In selecting the control techniques that will be relied upon for processing accuracy, the analyst's earlier decisions about controls come into full play. Some of the control techniques already described can be utilized to ensure that data is processed accurately through to the master files. If the analyst has made a careful selection earlier, he or she will have obtained the broad range of coverage with the fewest number of control techniques. The techniques that can be carried forward include

1. The *one-for-one checking* technique can ensure the accuracy of processing updates if the checking is carried out on a report that reflects the information actually placed on the file.

2. The *computer matching, existence checks*, and *double matching* techniques can be used to ensure that information is updated to the correct record/account.

3. The technique of *reconciling control totals* to the update report can be used to ensure that the totaled fields accurately update the appropriate files. This

technique alone does not ensure that the correct records/accounts were updated, however.

These procedures do not cover all of the requirements for establishing processing accuracy. As a result, some reliance normally must be placed on the accuracy of the logical procedures programmed into the system to ensure the accurate processing of some or perhaps even all of the significant data fields. To justify this, the analyst must ensure that the system has been tested thoroughly.

In a system relying heavily upon master files that reside on magnetic tape, the files are processed according to generations. The current master file is processed with the input transactions to create a new master file. The current master file then becomes the old master file, or previous generation, and the new master file then becomes the current master file for subsequent processing. The danger the analyst must protect against is that, if the most recent master file is not used, the transactions processed since the last correctly updated master file will be lost.

When a *control account* technique is used at input, no additional control is necessary to ensure that processing updates have been performed on the correct generation of the master file. This technique involves recording all types of transactions that eventually should reach the master file in a manual account. The cumulative totals of this account are carried forward from period to period. Periodically, the balance of the control is matched against the accumulation of individual balances in the master file. If the wrong generation of the master file is used for an update, the control account and master file balance do not agree.

In cases where the control account approach is not used, the principal technique available to ensure that the correct generation of the master file has been updated is the check on brought-forward total technique. This technique establishes that the correct generation of the master file has been updated by reconciling manually the total of the master file currently updated to the balance of the master file at the end of the previous update run.

AUTHENTICITY

All new systems must include controls to ensure that processing is applied only to transactions that are correct and that have occurred in accordance with management's intentions. These types of controls also can provide some evidence that previous processing (computer and clerical) is both accurate and error free. Authenticity can be established by ensuring that all transactions are authorized properly by responsible supervisory officials, or by using program logic to check transactions against criteria set by management. As a rule, there are normally more documents and/or procedures that require authorization in the computer environment. To be effective, these systems must depend upon a highly structured environment. When structure and definition of procedures are increased, authorization usually appears at more points in the system. For example, individual transactions, batches, corrections to rejections, overrides of normal procedures, and maintenance data amendments may each require specific authorization.

The timing of authorization procedures are of critical importance to the analyst in designing controls for new systems. In many cases, data is authorized when it is first introduced into the system, rather than at the time it is used for processing. For example, the credit manager will conduct a credit check on a new customer and make a decision regarding the credit limit the company is prepared to extend. The credit limit is then loaded onto the customer's master file record. From that point on, the logic programmed into the system uses the credit limit field to determine if any order received from the customer exceeds the credit limit. Under these circumstances, the credit manager's authorization has become an integral part of the system and does not involve manual authorization of each order prior to processing.

In some computer systems, authorization continues in many cases to be a conventional manual procedure. It is important that the structure of control ensures that only authorized data is processed. For authorization to be effective, it is necessary to structure procedures in such a manner that manual authorization is either

1. Carried out or rechecked on documents *after control* has been established over completeness and accuracy, or
2. Carried out or rechecked *after processing*, using printouts.

An authorization carried out *prior* to establishing control over the documents may not ensure that only authorized data is submitted for processing. For example, if vendor invoices are authorized for payment before the establishment of input controls (such as batch totals) and there is no subsequent checking of authorization, unauthorized documents could be input for processing. The system may be such that once put into the system, invoices are paid automatically without further scrutiny (as would often take place in a manual system).

The analyst must ensure that the controls within the system properly handle data that is rejected. In most cases, rejected data is handled by correcting the erroneous transaction. Unless properly controlled, this could result in previously authorized data being subjected to unauthorized change. The analyst must consider the extent to which authorizations need to be repeated or checked after correction of errors.

The use of programs for checking authenticity as an integral part of computer processing is becoming quite popular. The "authorization by program" technique can be handled in two ways.

1. Invalid input is rejected or reported for followup based upon some predetermined authorization variable. For example, sales orders often are tested for

credit against a predetermined credit limit held in the customer's master file, and fixed price discounts held on the master file are given to customers based upon the accumulated sales to date or the volume of a particular order. It is important to establish that reports of items rejected, or reported for followup, are in fact received by a responsible individual and appropriate action is taken.

2. Output is generated under certain conditions. For example, it is common to find inventory systems generating reports of recommended buying quantities, or even purchase orders themselves, based upon economic order quantities when a minimum stock level is reached. This represents a standard management decision which is activated when the computer establishes that a particular set of circumstances exist.

In both these methods, a programmed procedure within the system is based upon predetermined management criteria, thus leaving the need for manual intervention out of the day-to-day decision process. More control is required, however, over the original authorization used as the criteria in the system. It is very important that adequate controls exist over the amendments to the maintenance data used by the program (e.g., credit limits, minimum stock levels) and that there is adequate user followup to exception reports. Lack of control may cause multiple and cumulative errors. It also is important for the analyst to note that most systems allow for "overrides" (e.g., manual changes to recommend purchase orders); therefore, overrides must be controlled.

MAINTENANCE CONTROLS

The analyst must consider the need for maintenance controls in designing new systems. As far as manual records are concerned, most companies rely upon the controls inherent in the continual scrutiny of day-to-day accounting procedures. In computer systems, however, records held in computer files can be changed erroneously without leaving any visible evidence that the change has been made. As a result, the analyst must recognize that controls are required for both maintenance data and transaction data, to ensure that stored data remains.

1. *Correct:* The rigid structure of controls over the completeness, accuracy, and authenticity of input and update is not bypassed and records erroneously changed.

2. *Current:* Unusual data requiring action is identified and so kept up-to-date.

Some of the control techniques that the analyst might employ to ensure that all data remains correct in the computer file are

1. The manual reconciliation of file balances with control accounts is a useful technique. This technique involves regular agreement of the accumulation of individual balances in the computer file with a manual control account. It usually is applied to transaction data, but it also can be used for standing data if a "hash" total control account is maintained. An example of applying this technique to standing data is often found in payroll applications. Suppose one standing data field on each record is the hourly rate of pay used in the program to calculate gross pay. The payroll clerk would maintain a manual record of the total of all pay rates. Periodically, this hash total of pay rates would be reconciled with the total of individual pay rates on the payroll master file. This technique helps to ensure that data in the fields totaled is maintained completely on the file, but it does not guard against compensating errors between accounts, nor does it control the accuracy of other fields.

2. The reconciliation of file totals by the computer each time the file is updated is another technique to consider. This technique uses the computer to reconcile the file to a separate control record contained within it, each time the file is used. For example, the payroll master file may contain a trailer record storing the total balance of the file. Each time the file is processed, the computer adds the individual records and matches the accumulated total to the total stored on the trailer record. If there is agreement, the processing will continue; otherwise, a report will be produced announcing the out-of-balance situation. This technique normally is insufficient by itself to ensure proper maintenance of the file; it only establishes that no records are lost within each program. In order to be effective, this technique must include
   a. Manual reconciliation of the opening balance to the closing balance of the previous use of the file, or
   b. Controls within the EDP function over access to data files.

3. The computer recognition of control files is a technique that also can be used. This technique differs from the previous one, in that an independent control file is maintained within the computer system. On a regular basis, this control file is matched, by a separate set of programs, to the results of transaction processing, even to the extent of reconciling several totals. It provides greater assurance as to the maintenance of data than the reconciliation of trailer records described above. This is partly because independent programs are used and partly because the control may reconcile the results of processing by many programs. In order to be effective, it must include
   a. A reconciliation of the opening balance against the previous closing balance each time the system is stopped and restarted (perhaps each morning).
   b. Immediate followup of any discrepancies reported by the reconciliation.
   c. Adequate controls over access to data files within EDP (these are discussed later as part of integrity controls).

4. The regular printout of the computer file for manual review is a technique that involves the one-for-one verification of accounting records. The master file would be printed out record-by-record and checked manually against source data or other information to determine if the stored records are accurate.

In completing the design of the new system controls, the analyst must ensure that, where applicable, important data is kept on a current basis. As a result, stored information must be reviewed regularly to establish that it is up to date. This is important particularly in the case of maintenance data. The principal techniques available for the analyst include

1. The regular printout of the master file for manual review is a verification technique which is carried out to determine if an amendment should be made to the data because of changing circumstances since the last update.

2. The use of exception reports is a technique that uses the computer to interrogate each record in order to establish if a situation exists that might require manual investigation and an amendment to the data. The technique is valid for both standing and transaction data. For example, this technique might be used to review the currency of maintenance data with regard to prices. Assume the company is in an inflationary period and prices are expected to change at least once a year for the majority of products. The computer would interrogate each record in the inventory file to determine if the price has changed in the last twelve months. Where there has not been a price change, the computer would report the item for manual investigation. Regarding transaction data, the computer may review the inventory file for slow-moving items (e.g., items showing no activity for six months). Another situation commonly found is the aging of accounts receivable. The old accounts are investigated manually and appropriate action taken.

The analyst has now completed the design of the new system and its related environmental and application controls. Use the matrix approach in Appendix II to identify, document, and evaluate the appropriate controls. Chapters 10 through 12 provide some further insight into the economics of the system as well as selling the idea and the system implementation.

SELECTIONS FOR FURTHER STUDY

The subject of designing and auditing system controls has been of growing concern within the last decade. While much has been written on the problem, few works are available that can assist the analyst in solving the problem. Most of these are shown below.

1. *Auditor's Study and Evaluation of Internal Control in EDP Systems.* New York: American Institute of Certified Public Accountants, 1977.

2. *Computer Audit Guidelines.* Toronto, Canada: The Canadian Institute of Chartered Accountants, 1975.

3. *Computer Control Guidelines.* Toronto, Canada: The Canadian Institute of Chartered Accountants, 1973.

4. CONNOR, JOSEPH E., and BURNELL H. DEVOS, JR. *Guide to Accounting Controls.* Boston, Mass.: Warren, Gorham & Lamont, Inc., 1979.

5. *Control Objectives.* Hanover Park, Ill.: EDP Auditors Foundation for Education and Research, 1977.

6. FITZGERALD, JERRY. *Internal Controls for Computerized Systems.* 506 Barkentine Lane, Redwood City, Calif.: Jerry FitzGerald and Associates, 1978.

7. JENKINS, BRIAN, and ANTHONY PINKNEY. *An Audit Approach to Computers: A New Practice Manual.* London, England: The Institute of Chartered Accountants, 1978.

8. MAIR, WILLIAM C., DONALD R. WOOD, and KEAGLE W. DAVIS. *Computer Control and Audit, 2nd edition.* Altamonte Springs, Fla.: The Institute of Internal Auditors, Inc., 1976.

9. PORTER, W. THOMAS, and WILLIAM E. PERRY. *EDP Controls and Auditing, 2nd edition.* Belmont, Calif.: Wadsworth Publishing Company, Inc., 1977.

10. *Systems Auditability and Control Study.* Altamonte Springs, Fla.: The Institute of Internal Auditors, Inc., 1977.

QUESTIONS

1. Name four developments that have triggered considerable interest in the subject of internal controls.
2. What are the major control components of a data processing system?
3. Once the major control components that may impact the system have been identified, what is the next step?
4. What does the final step in the control matrix approach involve?
5. What are the major application control concepts?
6. What major application control concepts can be satisfied by the one-for-one checking technique?

7. To what does one-for-one checking refer?

8. What does the batch control total technique involve?

9. How can authenticity be established?

10. Why are maintenance controls used?

SITUATION CASES

CASE 9-1

A newly hired systems analyst was assigned to design the controls for an on-line, real time, order entry system that was being implemented by KZQ Corporation. The firm had been experiencing some difficulty in processing the increasing volume of orders on a timely basis. At times the backlog had resulted in order processing delays of up to two weeks. Although the vice-president of sales was very anxious to see the new system implemented, he insisted that the controls be designed carefully.

The analyst decided to use the matrix approach to design the environmental controls for the system. As a first step, the analyst identified what she felt were the major control components to be designed. Her list included (1) input controls, (2) output controls, (3) program/computer processing controls, (4) physical security controls, and (5) system software controls. Next, for each control component, she listed the resources/assets to be protected on the left side of a blank matrix, and the concerns/exposures across the top. The analyst then tried to think of a number of control techniques that might apply to each component and made a comprehensive list numbering each one. Next the analyst reviewed each matrix and placed the number of a control in each box where she felt it could help protect a resource/asset from a concern/exposure. After completing each matrix, the analyst reviewed it and circled the number of each control she felt would be easiest to implement. She then went to the systems manager and reviewed her findings.

QUESTIONS

1. Did the analyst examine all the appropriate control components? If not, what else could she have considered?

2. What else could the analyst have considered when selecting the control techniques to recommend?

Key Punch Incorporated, a company specializing in providing data entry services to customers, decided to develop a new computerized payroll system. The current system was manual and took too long to process. In addition, some problems had been noted including lost time cards and inaccurate processing.

An analyst was assigned the task of designing the new payroll system controls. First, the analyst decided to identify the most appropriate input controls. For completeness, he decided to recommend using batch control totals. The totals would be established by each of the company's user departments. For accuracy, the analyst decided to recommend relying upon key verification. Next, he set out to identify the most appropriate processing controls. The analyst decided to recommend that users do a one-for-one check of the payroll processed report against their retained copy of the time sheets. He felt this would cover both accuracy and completeness. In order to ensure the authenticity of the time sheets, the analyst decided to require the user department supervisor to initial them. He wrote a report containing the recommended controls and submitted it to the systems manager for approval.

QUESTIONS

1. Identify any additional control concepts that could be applied to this situation.

2. How could the analyst have made a more efficient choice in selecting control techniques?

FOOTNOTE

1. The following book contains over 650 specific controls in these nine areas. See *Internal Controls for Computerized Systems* by Jerry FitzGerald, 506 Barkentine Lane, Redwood City, Calif. 94065: 1978.

CHAPTER 10

ECONOMIC COST COMPARISONS

The quality of life depends in large measure upon man's wise use of scarce resources. The systems analyst must learn to compare the costs and benefits of systems resources, while management must evaluate these costs and benefits in terms of their probable effect on the quality of the organization.

GENERAL CONCEPTS

Estimating the cost of a system is much more complex than estimating the cost of a new piece of equipment. Many variables and intangibles are involved. But estimating a system's cost is necessary to decide whether implementation is justifiable. Some of the questions that must be considered include

1. What are the major cost categories of the overall system?
2. What methods of estimating are available?
3. Can all costs be identified and accurately estimated?
4. Which benefits cannot be estimated in dollar terms? (Very few!)
5. What criteria will management use when evaluating these estimates?

TWO CONCEPTS OF COST ANALYSIS

The first concept is the *planning* type of cost analysis. It is a method based on the analysis of the *opportunity costs* of using a resource for one purpose rather than another.

Suppose that the firm has 10,000 square feet of unused (empty) floor space in its main plant. To determine the opportunity cost of alternate uses for the space, management would develop a list of all the uses to which this floor space could be put. For example, it could be used for

1. Office space.
2. Manufacturing a new product.
3. Expanding the manufacture of an existing product.
4. A new computer center.
5. Leasing to an outside firm.
6. Nothing . . . let it sit empty.

A systems analyst would now estimate the return on the investment for all six of the above uses to which the floor space could be put. Suppose the return on investment turned out as follows.

DOLLARS RETURNED PER YEAR FOR EACH $1000.00 INVESTED	USE TO WHICH THE FLOOR SPACE WILL BE PUT
$180.00 or 18% return on investment	Expanding the manufacture of an existing product
110.00 or 11% return on investment	A new computer center
60.00 or 6% return on investment	Office space
55.00 or 5.5% return on investment	Manufacturing a new product
−10.00[1] or −1% return on investment	Nothing . . . let it sit empty
−50.00[2] or −5% return on investment	Leasing to an outside firm

The opportunity cost of expanding the manufacturing of an existing product is $70.00 compared with the next best alternative which is to locate a new computer center in that empty space ($180.00 minus $110.00 equals $70.00). If the firm chooses to locate the computer center in this space, it will be foregoing $70.00 return for each $1000.00 investment required.

The second concept of cost analysis is the *budgeting* type of analysis. It is based on the *cash-flow* concept, which refers to the amount of money that will be required for a particular project and the dates when that money will be needed. The firm can then budget or set aside the required funds so they will be available when needed.

Cash flow also involves the *speed* of collection of the firm's incomes and the *amount* of collection. In other words, a firm cannot adequately plan budgets unless it knows when it will collect its incomes, the amount it will collect, and when it will have to pay out funds.

Finally, the analyst should try to determine the similarities or differences in the final estimates that are caused by differences in the systems requirements versus those caused by peculiarities in the different methods used to estimate the costs in each instance.

PAYBACK PERIOD

A criterion that frequently is used to judge the profitability of a system is the payback period. For example, if a new computer-based information system costs $700,000 and is expected to yield $100,000 per year, its payback period is 7 years (before taxes). The payback period is defined as the number of years required to accumulate earnings sufficient to cover its costs.

The payback period criterion ranks projects in terms of number of years to payback. Two factors that must be examined closely are the rate at which funds flow in, and taxes. Two investments (A and B) each have a 4-year payback period.

PROJECT	ANNUAL RETURN FIRST YEAR	ANNUAL RETURN SECOND YEAR	ANNUAL RETURN THIRD YEAR	ANNUAL RETURN FOURTH YEAR	TOTAL RETURN
A	$10,000	$10,000	$10,000	$10,000	$40,000
B	25,000	5,000	5,000	5,000	40,000

Clearly, project B is preferred because a larger amount of the investment is recovered during the first year. Project B returns the investment money more promptly.

The second factor concerns taxes. Corporate income taxes lengthen the payback period. The payback period calculator is

$$\underline{\text{before taxes}} \qquad \underline{\text{after taxes}}$$

$$P = \frac{I}{R} \qquad P = \frac{I}{(1 - T)R}$$

where P = payback period, I = investment, R = average annual return on investment, and T = corporate tax rate in percent. For example, if the investment required for a project equals $22,000 ($I$ = 22,000) and the average annual return is $3800 ($R$ = 3800) and the corporate tax rate is 48% (T = 0.48), then the payback period (after taxes) is

$$P = \frac{22,000}{(1 - 0.48)3800} = \frac{22,000}{(0.52)3800} = 11.13 \text{ years}$$

The same project has a before taxes payback period of 5.79 years.

MARGINAL EFFICIENCY OF INVESTMENT

The marginal efficiency of investment (MEI) is the rate of return that a potential new system is expected to earn after all of its costs (except interest expenses) are recovered.

For example, consider a new system that required $10,000 for installation and equipment, $30,000 in operating expenses, and was expected to return $41,000 at the end of the *one year* of its life. The MEI of this system is 10% because after covering the operating expenses (noncapital costs), the firm had enough left over to replace the original $10,000 equipment cost plus a net return of $1,000 (1,000 ÷ 10,000 = 10%).

The investment may be made if the MEI is greater than the rate of interest that a bank charges the firm to borrow.

$41,000   estimated benefits at end of year
−30,000   operating expenses during the year .

$11,000
−10,000   equipment and installation cost

$ 1,000   return on investment

The above example assumes that the system will be obsolete at the end of one year. In order to consider a longer life span, the following formula should be used.

$$\text{Fixed investment in dollars} = \frac{\text{Return over operating expenses (1st year)}}{(1+r)^1} + \frac{\text{Return over operating expenses (2nd year)}}{(1+r)^2} + \ldots + \frac{\text{Return over operating expenses in last year } (N\text{th}) \text{ of the asset's life}}{(1+r)^n}$$

where r equals the M.E.I.

For example, if a new system has a fixed investment of $10,000 and if at the end of the first year of its life it is expected to yield a return on investment of $5,500 and at the end of the second year a return on investment of $6,050, its M.E.I. equals 10%.

$$\$10,000 = \frac{\$5,500}{(1+r)^1} + \frac{\$6,050}{(1+r)^2}$$

Solving this for r, we get $r = 10\%$. This means that if r (M.E.I.) is greater than the current interest rate, the investment may be made. The financial people in the firm should be consulted to get the specific interest rate which the firm must pay to obtain

221

funds. If that is not available, the M.E.I. might be compared with the "prime rate" that banks charge their best customers.

COMPARE CURRENT SYSTEM WITH THE NEW SYSTEM

Another method is to prepare an economic cost comparison between the current system and the proposed new system. *First*, estimate the projected useful life of the proposed new system. This is the common basis by which the old and new systems can be compared. Estimate a reasonable life span for your proposed new system since no system lasts forever. It might collapse from advances in computer technology or from overuse as company sales increase.

Second, calculate the proposed new system's operating costs during the projected useful life of the system (detailed example of costs in next section).

Third, calculate the present system operating costs over the same projected useful life. In this way an accurate cost comparison can be made between the present and proposed systems. It is likely that the present system will not be able to function over the projected useful life of the new system. Estimate when the current system will fail to operate adequately or when it might collapse completely.

Fourth, compare both operating costs (present and proposed) over the projected useful life of the new system. Calculate the fixed investment costs, including any one-time implementation costs. Resources and costs are generally divided into three categories.

1. *Implementation costs* are one-time outlays to create new capability.
2. *Investment costs* are nonrecurring outlays to acquire new equipment.
3. *Operating costs* are recurring outlays required to operate the system.

For both the current and the proposed new system, the approximate costs can be estimated from the various documentation sheets, reports, or summaries that were prepared as the systems study progressed through its phases.

DETAILED EXAMPLE OF COSTS

For both the current system and the proposed new system, the following should be determined.

Salaries

The current wage rate for those people operating the system can usually be obtained from the personnel department, but be careful! Rate of pay for nonunion jobs is sometimes a well-guarded secret. Settle for estimates if the exact amount is not readily divulged.

In projecting costs do not forget to figure salary increases for personnel in future years. Always include the firm's cost of fringe benefits. Fringe benefits are generally in the area of 20%, that is, company costs (contributions) are twenty cents per dollar of wages paid. If the payroll or personnel department said that the company paid 20% in fringe benefits, the wage rate would be multiplied by 0.20 to get the amount the company contributes, per hour, toward fringe benefits.

$$\begin{array}{ll} \$7.65 & \text{wage rate per hour} \\ \times 0.20 & \text{fringe ratio} \\ \hline \$1.53 & \text{fringe cost per hour} \end{array}$$

The actual payroll cost is $9.18 per hour (7.65 plus 1.53) for this particular job classification. The actual payroll cost is then multiplied by 40 hours per week to get the weekly cost and then by 52 weeks to get the annual cost.

$$\begin{array}{ll} \$ \quad 9.18 & \text{per hour} \\ \times 40 & \text{hours per week} \\ \hline \$ \quad 367.20 & \text{per week} \\ \times 52 & \text{weeks per year} \\ \hline \$19{,}094.40 & \text{per year} \end{array}$$

Assuming this is the annual cost of one computer programmer, and six of them are needed in the proposed new system, the annual cost is $114,566 for the first year (6 × 19,094.40 = 114,566). For the second year, this figure should be increased by the expected salary increases. If the firm averages 7.5% in increases, increase the $114,566 by 7.5% when calculating the costs of the system during the second year of its existence. Repeat this entire operation for each different wage rate in the current and new system in order to get the total salary costs.

Space

The cost of floor space should be included. Many firms prorate the cost of floor space on a dollars-per-square-foot basis. Assume, for example, that the cost of floor space is prorated by the accounting department at $14.50 per square foot. Calculate the square footage of space required by the current and by the proposed new system, then multiply by $14.50 in order to assign a cost to the facilities required for each system.

If space is rented in order to house the system, then the rent is the cost of floor space. Carefully consider the location for the new system. Is it expensive or inexpensive? In a multiplant firm, different buildings have different prorated floor space costs. The new system or computer should be located in a convenient location.

Supplies and Inventories

Evaluate each operation within the system as to the needs and costs of different supplies. Supplies are used to operate the system and should not be confused with inventories. One supply might be paper for the computer's high-speed printer, whereas inventory costs have three classifications: raw materials with which to make the product, work-in-progress, and finished goods inventory.

If the system study involves inventories, a calculation should be made to determine whether inventories will increase or decrease because of the new proposed system. For example, if inventories are decreased by $10,000, there is a savings. Multiply the $10,000 reduction by the rate of interest that the firm must pay to borrow from a bank.

$$\begin{array}{ll} \$10,000.00 & \text{reduction in inventory} \\ \underline{\times 0.12} & \text{interest rate} \\ \$ \;\; 1200.00 & \text{annual savings} \end{array}$$

The proper interest figure can be obtained from the financial personnel in the firm. If the system increases inventories, it is an additional cost to be added to the system. The calculation is the same.

Overhead

Salaries, space, supplies, and inventories have already been taken into account, but there might be some other *indirect* cost items that should be taken into account. If the new system reduces overhead, it is a savings; and if the new system increases overhead, it is an additional cost. Overhead items include

1. Janitorial and maintenance services.
2. Plant protection.
3. Insurance.
4. Property taxes on a building or land.
5. Heat, light, and power.
6. Specialized services, the cost of which is borne by all departments (company library, duplicating services, etc.).

Implementation Costs

Add one-time implementation costs to the overall cost of installing the new system. Implementation costs include

1. Moving costs for equipment and people.
2. Refurbishing costs.

3. Costs for locating electrical outlets and phones.

4. Furniture costs.

5. Cost of file conversions, for example, manual to computerized or computerized to a new computerized system.

6. Cost of removing the current system.

Investment Costs

Investment costs include the cost of purchasing, renting, or leasing new equipment or facilities for the proposed system.

BASIC NONECONOMIC BENEFITS . . . INTANGIBLES

Noneconomic benefits are those benefits which are difficult to estimate in economic terms. The cost of effort to make the estimate may exceed the value of having the estimate. A list of some noneconomic benefits the analyst should look for are

1. Faster response time to inquiries from prospective customers . . . better public relations.

2. Increased employment stability . . . higher employee morale.

3. More accurate delivery promises to customers.

4. Greater stock availability . . . fewer stock-outs.

5. Improvement in quality of product or quality of service . . . fewer rejects.

6. Positive effect on other classes of investments or resources, such as more efficient use of floor space or personnel.

7. Better managerial control of the organization.

8. Effective cost reduction, for example, less spoilage or waste, elimination of obsolete materials, and less pilferage.

9. Greater flexibility in dealing with a changing business environment.

10. Future cost avoidance.

The effectiveness of a system is determined by time and cost required to operate the system versus the benefits derived from the system. Do not base the success of a system solely on economic values. Show management the other important benefits (intangibles) of your proposed new system.

CONSIDERATIONS IF A COMPUTER IS REQUIRED

If the proposed new system is recommending a larger, more powerful computer or a new computer, the following must be done prior to completing the economic section of the final system report. (See Chapter 2, Type B feasibility study, for a detailed description on this analysis.)

1. Gather together the costs of the computer and the peripheral equipment.
2. Determine the supplies cost: magnetic tape, disk packs, floppy disks, standard paper and special forms for high-speed printer, and so forth.
3. Calculate the expected costs for purchasing software packages, or outside programming assistance from a consulting firm or directly from the hardware vendor.
4. Evaluate maintenance costs.
5. Consider costs of converting forms and files for use with the computer.
6. Check the facilities costs: refurbishment of the space, air conditioning, electrical, plumbing, raised floors, special fireproof storage areas, and furniture.
7. Consider special computer training costs for personnel.
8. Calculate one-time computer installation costs.

As a further example see Figure 10-1 which is a checklist of cost-benefit analysis factors.

The main point to be remembered in calculating and assembling cost data is to be conservative. Do not exaggerate and do not compare extreme figures. Use fair and reasonable averages since you will probably be required to defend your estimates. You do not want to back yourself into a corner with figures that are pushed to extremes. You will be facing financial people (accountants and financial analysts), and these people tend to be careful and conservative in their approach to evaluating the future. Be prepared to speak their language!

SELECTIONS FOR FURTHER STUDY

Since the installation of a major new system might easily bankrupt a company if misjudged or mishandled, this phase of your system study is critical. The overall study of business economics would be beneficial. In general, you will find cost analysis information under every subject but cost analysis. These include cost accounting, cost benefit analysis, cost effectiveness, and value analysis. The periodicals indexes mentioned in Chapters 8 and 18 may be used. More sophisticated economic theory will be

COSTS

Direct costs
—Computer equipment
—Communications equipment
—Common carrier line charges
—Software
—Operations personnel costs
—File conversion costs
—Facilities costs (space, power, air conditioning
   storage space, offices, etc.)
—Spare parts costs
—Hardware maintenance costs
—Software maintenance costs
—Interaction with vendor and/or development
   group
—Development and performance of acceptance
   test procedures and parallel operation
—Development of documentation
—Costs for backup of system in case of failure
—Costs of manually performing tests during a
   system outage

Indirect costs
—Personnel training
—Transformation of operational procedures
—Development of support software
—Disruption of normal activities
—Increased system outage rate during initial
   operation period
—Increase in the number of vendors (impacts
   fault detection and correction due to "finger
   pointing")

BENEFITS

Direct and indirect cost reductions
—Elimination of clerical personnel and/or
   manual operations
—Reduction of inventories, manufacturing,
   sales, operations, and management costs
—Effective cost reduction, for example, less
   spoilage or waste, elimination of obsolete
   materials, and less pilferage
—Distribution of resources across demand for
   service

Revenue increases
—Increased sales due to better responsive-
   ness
—Improved services
—Faster processing of operations

Intangible benefits
—Smoothing of operational flows
—Reduced volume of paper produced and
   handled
—Rise in level of service quality and perfor-
   mance
—Expansion capability
—Improved decision process by providing
   faster access to information
—Ability to meet the competition
—Future cost avoidance
—Positive effect on other classes of invest-
   ments or resources such as better utilization
   of money, more efficient use of floor
   space or personnel, and so forth
—Improved employee morale

FIGURE 10-1 Cost-benefit analysis factors.

located through the *Index of Economic Articles.* This index is outstanding in its arrangement and coverage, including both periodical articles and chapters in books. Its primary disadvantage, however, is that its publishing time is very slow.

Several books are Vancil's *Financial Executives Handbook,* Black's *Accounting for Business Decisions,* and Schweyer's *Analytic Models for Managerial and Engineering Economics.* The latter is recommended for applying economic theory and analytic methods as aids in the solution of practical business problems. Martino's *Dynamic Costing,* and Bond's *Price and Cost Proposals* may be more to the point in many instances.

1. BATLIWALLA, MINOO R. *Investment Decision: Capital Budgeting with the Aid of the Discounted Cash Flow Technique.* New York: Asia Publishing House, 1978.

2. BLACK, HOMER A., et al. *Accounting in Business Decisions, 3rd edition.* Englewood Cliffs, N.J.: Prentice-Hall, Inc., 1973.

3. BUSSEY, LYNN E. *The Economic Analysis of Industrial Projects.* Englewood Cliffs, N.J.: Prentice-Hall, Inc., 1977.

4. DUDICK, THOMAS S. *Profile for Profitability: Using Cost Control and Profitability Analysis.* New York: John Wiley and Sons, Inc., 1972.

5. HARPOOL, JACK D. *Business Data Systems: A Practical Guide.* Dubuque, Iowa: William C. Brown Company, 1978.

6. HINRICHS, HARLEY. *Systematic Analysis: A Primer on Benefit-Cost Analysis and Program Evaluation.* Santa Monica, Calif.: Goodyear Publishing Company, 1972.

7. *Index of Economic Articles in Journals and Collective Volumes.* Homewood, Ill.: Richard D. Irwin, Inc., 1961–   .

8. JENSEN, RANDALL W., and CHARLES C. TONIES. *Software Engineering.* Englewood Cliffs, N.J.: Prentice-Hall, Inc., 1979.

9. MARTINO, R.L. *Dynamic Costing.* New York: McGraw-Hill Book Company, 1970.

10. PEARSON, ROBERT. *Cost Effective Decision Making.* Ann Arbor, Mich.: Masterco Press, Inc., 1973.

11. SANDERS, DONALD H. *Computers in Business.* New York: McGraw-Hill Book Company, 1975.

12. SASSONE, PETER A. *Cost-Benefit Analysis: A Handbook.* New York: Academic Press, Inc., 1978.

13. SCHWEYER, HERBERT E. *Analytic Models for Managerial and Engineering Economics.* New York: Reinhold Publishing Company, 1964.

14. VANCIL, RICHARD F. *Financial Executives Handbook.* Homewood, Ill.: Richard D. Irwin, Inc., 1970.

15. WILLOUGHBY, THEODORE C., and JAMES A. SENN. *Business Systems.* Cleveland, Ohio: Association for Systems Management, 1975.

16. WILSON, R.M. *Cost Control Handbook.* New York: Halsted Press, 1975.

QUESTIONS

1. List some of the questions that must be considered in estimating a system's cost.

228

2. Name the two concepts of cost analysis.
3. What must a firm know to adequately plan budgets?
4. What criterion is used frequently to judge the profitability of a system?
5. How does MEI relate to economic cost comparisons?
6. When comparing operating costs over the projected life of the new system, name the three categories into which resources and costs are generally divided.
7. What are "investment costs"?

SITUATION CASES

CASE 10-1

An analyst for Burger Manufacturing was trying to determine whether the company should lease a 10,000 square foot warehouse next to the factory or to maintain the present 10,000 square foot warehouse located four blocks away. He had done the following: defined the problem, made an outline for the study, gathered general information, viewed the interactions, made the effort to understand the present system, defined the requirements, and designed the new proposal. He was now faced with the problem of cost comparisons and has worked up the following schedules for the next three years.

Salaries The analyst determined that the supervisor of either warehouse system would receive an annual raise of $35.00 indefinitely. The warehousemen will receive annual raises of 10% but have a ceiling rate of $4.00 per hour. The secretary will also receive an annual raise of 10% but has a $2.50 per hour ceiling. With this information, a salary schedule for the present and proposed system was developed as follows.

	PRESENT SYSTEM		
	YEAR 1	YEAR 2	YEAR 3
Supervisor			
Monthly salary	$700.00	$735.00	$770.00
Months per year	×12	×12	×12
Total annual salary	$8,400.00	$8,820.00	$9,240.00

Warehousemen			
Hourly wage	$3.00	$3.30	$3.63
Hours per week	×40	×40	×40
Weekly salary	$120.00	$132.00	$145.20
Weeks per year	×52	×52	×52
Annual salary	$6,240.00	$6,864.00	$7,550.40
Number of workers	×4	×4	×4
Total salary cost	$24,960.00	$27,456.00	$30,201.60
Secretary			
Hourly wage	$2.00	$2.20	$2.42
Hours per week	×40	×40	×40
Weekly salary	$80.00	$88.00	$96.80
Weeks per year	×52	×52	×52
Annual salary	$4,160.00	$4,576.00	$5,033.60

PROPOSED SYSTEM

Supervisor			
Monthly salary	$700	$735	$770
Months per year	×12	×12	×12
Annual salary	$8,400.00	$8,820.00	$9,240.00
Warehousemen			
Hourly wage	$3.00	$3.30	$3.63
Hours per week	×40	×40	×40
Weekly salary	$120.00	$132.00	$145.20
Weeks per year	×52	×52	×52
Annual salary	$6,240.00	$6,864.00	$7,550.40
Number of workers	×3	×3	×3
Total salary cost	$18,720.00	$20,592.00	$22,651.20
Secretary			
Hourly wage	$2.00	$2.20	$2.42
Hours per week	×40	×40	×40
Weekly salary	$80.00	$88.00	$96.80
Weeks per year	×52	×52	×52
Annual salary	$4,160.00	$4,576.00	$5,033.60
Total Salary Costs			
Present system	$37,520.00	$40,852.00	$44,475.20
Proposed system	$31,280.00	$33,988.00	$36,924.80
Difference	$6,240.00	$6,864.00	$7,550.40

Space After making the salary comparisons, the analyst needed to know the cost of space for the present and proposed systems. He based his costs on space usable as storage only. He found the present 10,000 square foot warehouse has 300 square feet used as office space and 700 square feet is used as dock space. Under the proposed system, all 10,000 square feet can be utilized as storage space. The analyst also determined that lease costs for both warehouses would increase by 10% during the analysis period. The following schedules were developed.

PRESENT SYSTEM			
	YEAR 1	YEAR 2	YEAR 3
Monthly lease cost	$30,000.00	$33,000.00	$36,300.00
Storage space (square feet)	10,000	10,000	10,000
Monthly cost per square foot	3.00	3.30	3.63
Months per year	×12	×12	×12
Annual lease cost per square foot	$36.00	$39.60	$43.56

PROPOSED SYSTEM			
	YEAR 1	YEAR 2	YEAR 3
Monthly lease cost	$50,000.00	$55,000.00	$60,500.00
Storage space (square feet)	10,000	10,000	10,000
Monthly cost per square foot	5.00	5.50	6.05
Months per year	×12	×12	×12
Annual lease cost per square foot	$60.00	$66.00	$72.60
Total Space Costs			
Present system	$36.00	$39.60	$43.56
Proposed system	$60.00	$66.00	$72.60
Difference	($24.00)	($26.40)	($29.04)

Supplies The analyst determined that one copy of the receiving form could be eliminated under the new system. In addition, two file cabinets and one desk would no longer be needed but the cost associated with these items was so negligible, the analyst felt it was not of any importance to the study. Because the secretary was needed in both systems, his personal supplies would still be used.

Inventories Due to the added storage space available under the proposed system, the analyst determined the following added inventory costs.

	YEAR 1	YEAR 2	YEAR 3
Additional storage space (square feet)	1,000	1,000	1,000
Monthly cost per square foot	$5.00	$5.50	$6.05
Added monthly cost	5,000.00	5,500.00	6,050.00
Months per year extra space utilized	×12	×12	×12
Annual added cost	60,000.00	66,000.00	72,600.00
Cost to borrow added funds	×0.06	×0.06	×0.06
Added annual cost	$3,600.00	$3,960.00	$4,356.00

Implementation The new warehouse would be leased to meet the firm's specifications plus the changeover would not be made until the present warehouse lease expires. Because of this, the only implementation costs would be those of moving the merchandise. This cost would amount to approximately $300.00.

Noneconomic Benefits The analyst concluded that the following nonquantitative benefits would be derived from the move.

1. More centralization.
2. Fewer stock-outs.
3. More efficient use of space.
4. Better managerial control.
5. Less spoilage and waste.
6. Higher employee morale.

QUESTIONS

1. When the proposed system's salary costs were computed, were all of the old system costs included? If not, did this create a low or high estimate for the proposed system?
2. When estimating the present system's cost per square foot, was the correct square footage figure used? If not, what figure should have been used?
3. Did the analyst fail to consider a major area depicted in the chapter as an area of vital concern when making cost comparison? If so, what area?

CASE 10-2

Robert Prescott, owner and president of Universal Machine Parts, Inc., had given substantial thought as to the efficiency of the present system for processing an increasing volume of paperwork. In recent months, there had been numerous complaints from customers, and from various department heads within the company, that the required paperwork was taking too long to be processed. Upon conferring with the various department heads experiencing the problem, he concluded that a systems study must be made. Mr. Prescott hired a systems analyst to try to correct the problem.

The analyst began by making a "Type A" feasibility study. After completing the study, she discussed two alternatives with the president, that is, (1) to expand the present manual system, or (2) to computerize the company to give it better overall management control. They decided jointly that before any major decisions could be made, a full systems study should be conducted.

To begin, the analyst defined the problem and wrote an outline for the study. She then gathered some general information about the existing system and studied the interactions between the various departments, and developed a complete understanding of the existing system. After defining the new system requirements, the analyst designed two alternative new systems. One would expand the existing manual system, while the other would be a complete new computerized system. The analyst was now ready to make an economic cost comparison of the proposed new systems.

The analyst began by calculating the cost of a computerized system. She gathered together the cost of the computer, the peripheral equipment, and the cost of one-time installation. The analyst added in the cost for special training and assistance from the computer manufacturer. Finally, she calculated what it would cost to convert the existing forms and files for use with the computer and what it would cost for supplies.

Next, the analyst calculated the costs for expanding the present manual system. She determined what it would cost for added personnel, office space, supplies, and equipment.

The analyst then calculated the cost for added floor space that would be required, and the added cost of overhead for each of the proposed new systems. These overhead expenses would be for the added janitorial services, insurance, property taxes, and utilities.

The salary cost of employees was then calculated by the analyst. She determined the number of employees who would be involved in the area under study, and added in the estimated number of new employees that would be required for each new system. The analyst estimated how much each employee would be earning annually, and combined the total annual salaries of all the employees. The analyst then took this annual figure, and multiplied it by the projected useful life of each system, to get the total salary costs for each of the new proposed systems.

Having calculated the tentative costs for each system, the analyst made an estimate of the useful life of each new system relative to the growth rate of the company. She figured each proposed system would have a projected useful life of ten years, with a projected five-year (after taxes) payback period. The marginal efficiency of investment (MEI) was then calculated, and a considerable difference was found. Her calculations showed that the MEI for a computer system would be 27%, while the expanded manual system would be only 8%.

Finally, to conclude her comparison of the proposed systems, the analyst tried to establish some noneconomic benefits. Items such as faster response time to inquiries from customers, increased employee stability, more accurate delivery to customers, and improved quality of products were considered. Also, she felt there were such factors as effective cost reduction, less spoilage and waste, and better managerial control of the organization that should be considered, if the system was to meet the future needs of the company.

In comparing all the costs, the analyst concluded that a computer system would be more costly, but that it would be more beneficial to the company in the long run. With all this data compared and evaluated, the analyst was now ready to present her findings to Mr. Prescott and his financial advisors for a decision and implementation.

QUESTIONS

1. In calculating the cost for a computerized system, what did the analyst fail to consider?

2. Where did the analyst make errors in calculating the salary costs for the proposed new system?

FOOTNOTES

1. Negative dollars returned because the firm must pay property taxes.
2. Leasing to another firm might interfere with your own business, giving a negative return.

CHAPTER 11

SELLING THE SYSTEM

An important phase in the development of any useful product is the communication of that product's utility to prospective users. The systems analyst has the responsibility for presenting the systems proposal to management in a clear and objective manner, but with enthusiasm for the benefits that will be derived from its use.

BACKGROUND KNOWLEDGE

The systems analyst must realize that in the final presentation an *idea* is being sold to management, not a product. A system is an abstract idea. It rides the thin line between the tangible and the intangible. Management usually has a desire to improve the company's systems, but normally there also is fear of expensive failure. When the analyst replaces fear with reassurance, management's desire is free to express itself and the new system can be accepted.

The analyst should mentally ask many questions in relation to selling the new system to the management of the area under study. Has management had any previous experience with *your* systems studies, and was that experience favorable? Does management have a need for this specific new system? Is management aware of the need? Does management harbor any fears or prejudices of a specific or general nature? Does the management in the area under study have the authority to buy or accept the new system? Is management able to implement the changes called for in the new system? To whom is the management of the area under study accountable (who else needs to be convinced about the new system)?

After answering the above questions the analyst should think about the meanings of a few key words. Whether a written or a verbal report is used, the five key words for successful interaction with people are

1. *Empathy:* mentally entering into the feeling or spirit of other people in order to understand them better.
2. *Tact:* ability to say or do the right thing without offending other people.
3. *Rapport:* a meeting of the minds, or absence of friction.
4. *Sensitivity:* the quality of being readily affected by other people, responding to their feelings.
5. *Integrity:* basic honesty and moral uprightness.

Selling management on a new system is a continuous effort and it should be carried on throughout the entire systems study. The analyst's enthusiasm and positive mental attitude can help sell the system in the beginning. (We assume in this chapter that the analyst genuinely feels a new system is needed and that it will be a beneficial change for the firm.) As the study progresses and each individual's personality unfolds it takes empathy, tact, rapport, sensitivity, and integrity to continue the selling job. And finally, after the design phase, it takes the optimum combination of written reports and verbal presentations to negotiate a change from the old to the new.

BASIC OBJECTIONS TO OVERCOME

The analyst will be better prepared during this phase if basic objections to the proposed system are considered ahead of time. Think about them, weigh their importance, and prepare answers to them. Any experienced analyst can tell you that your hands will be full trying to answer the *unexpected* objections. Anticipating and being previously prepared for as many objections as possible has several important advantages. First, it can help you determine, prior to the final presentation, if any items were overlooked. If so, these omissions can be corrected immediately. Second, it helps you *feel* prepared because you know you are prepared. And third, your obvious preparedness probably will be recognized and appreciated by management, thus inspiring their confidence in your overall design.

Be prepared, then, for the obvious objections to the proposed system. Basic objections usually follow these lines.

1. The cost is too high, or it appears too low for what the system is claimed to be able to do.

2. The performance is not good enough, or it is more than is required at this point in time.

3. The new system does not follow the goals, objectives, and policies of the firm or the area under study.

4. The response time or processing time is either too slow or too fast with respect to the other operations within the firm.

5. The system is not flexible enough. If changes are made in other areas this system will collapse, and the investment will be wasted.

6. The quality, capacity, efficiency, accuracy, or reliability of the new system may not meet the criteria of the management personnel involved.

7. Certain management personnel may dislike the analyst's motives, personality, or presentation methods.

Why should the analyst have to sell the new system to a company that has paid for it, wants it, needs it, and will profit by it? The basic reason is that you have to *prove* to management that it needs the new system and will profit by it. Money is a scarce resource and management usually has more projects on which to spend it than there is money available. Therefore the analyst has to convince management that this new system is more important than the other projects that are being considered.

Another reason that selling is required is the fact that what the analyst offers is change. Change often is seen as a threat by some personnel. The threat of change may develop insecurity and fears of job loss. When this happens there will be opposition to the analyst and to the new system that was developed. Only when management agrees that the new system is needed and decrees that it will be implemented will some of these people cooperate. Management, in other words, must impose change. In some cases, change occurs only through directive.

GAINING ACCEPTANCE THROUGH THE WRITTEN REPORT

The final written report is the most important report of the entire systems study. All system studies require a final systems report. Up to this point the only other *required* reports were the feasibility study report and the problem definition report. The other writings were not reports; rather they were just summaries of some phase of the system. Presented below is an outline of what might be put into a final written report. It is by no means a definitive statement on the subject since the contents and outline may vary a great deal depending upon the study. It is simply a guide to the basic contents the analyst should consider for inclusion. Amplifying comments follow the outline.

Final report

I. Summary of the full systems study.
  A. Statement on events leading up to decision to conduct the study.
  B. Statement of the problem to be solved.
    1. Subject.
    2. Scope.
    3. Objectives.
  C. Statement of the major recommendations only, and justification of the new system.
  D. Review of the current system and the new system requirements.
  E. Highlights of the operations and control responsibilities within the new proposed system.
  F. Review of cost and implementation schedules.

II. Body of the report.
  A. Description of the existing system (this section may not be required in some studies).
    1. Brief description of the existing system and how it was used.
    2. Purpose of the existing system.

3. The "product" of the existing system and whether it served management effectively.

4. Major control points and responsibilities within the existing system.

B. Description of the new proposed system.

1. The "product" of the new system.

2. Major controls points and responsibilities within the new system. Are they equal to or better than the control framework within the existing system?

3. Display of all reports or forms to be used in the proposed system. Emphasize elimination of nonessential data.

4. Systems flowchart.

5. Any or all of the documentation of the new proposed system. Any not included here can be put in the appendix.

6. Full list of recommendations.

7. List of time schedules.[1]

   a. Overall elapsed time required to install the proposed system.

   b. Chargeable time, for example, number of work hours required to install the proposed system.

   c. Operating times for the new proposed system.

8. List of personnel requirements.

   a. Personnel required to install the proposed system.

   b. Personnel required to operate the proposed system.

C. Section on economic cost comparison (see Chapter 10).

1. List of the estimated useful life of the new proposed system (one method enumerated in Chapter 10).

2. Cost of the old system over the estimated useful life (from 1 above).

3. Cost of the new system over the same estimated useful life (from 1 above).

4. Other areas of cost to be covered.

   a. Salaries.

   b. Space.

   c. Supplies.

   d. Inventories.

   e. Overhead.

      f. Computer time.

      g. One-time implementation costs.

      h. Cost of capital . . . investment.

    5. Noneconomic benefits . . . intangibles.

  D. A definite, straightforward, and positive statement relative to why you believe the planned system should be installed.

III. Appendix.

  A. Prior reports (when available).

    1. Feasibility study.

    2. Problem definition.

    3. Summaries of the various phases.

    4. Letters and memoranda.

  B. Any items that do not fit into the body of the report.

    1. Charts.

    2. Graphs.

    3. Tables of data.

    4. Notes.

The preceding outline can be used as a guide when assembling the final report. In the Summary (Section I) all that is required is a couple of paragraphs for each item A through F. In the Body (Section II) the analyst should be very detailed. This section should answer all possible questions about the new proposed system and the economics involved; it may include a description of the existing system so a comparison can be made. The Appendix (Section III) should contain anything that is too bulky or appears to be out of place in the other sections of the report.

There are also a number of possible *faux pas* that the analyst should take into consideration. First, try not to send out a report full of spelling and grammar errors. If you cannot correct it yourself, have someone else do it. Second, do not mail the report to the *key* management personnel. Hand it to them personally and include a cover letter pointing out that a meeting about this report has already been arranged. Arrange a meeting, through each manager's secretary, a week or two in the future. Third, do not have numerous copies mailed to everyone involved. Instead, decide who needs a copy and give them one. Check your list carefully. If one manager receives a copy, another manager also may have to receive a copy (like a dinner party where some people cannot be invited without offending others). The offended manager of today may turn out to be the key person in your next systems study!

Data should be treated consistently. If a report begins by comparing costs with revenue in a particular way, that method of comparison should be continued throughout the report. Financial matters, especially estimates, should always be reported conservatively. This idea of conservatism assures management that it is, in fact, receiving representative cost data. Of course, a conservative estimate is much easier to defend than an estimate that is biased toward your own vested interests in the proposed system.

Finally, in writing the final systems report, the analyst should

1. Be clear and effective.
2. Be brief; verbose reports usually do not get read.
3. Use the active voice because it portrays enthusiasm.
4. Use short declarative sentences.
5. Check the spelling.
6. Demonstrate good grammar and sentence structure.

GAINING ACCEPTANCE THROUGH THE VERBAL PRESENTATION

The verbal presentation can be just as important as the final written report, although it is not always required. Psychologists tell us that almost 80% of our impressions come to us through our eyes; therefore anything that can create a positive impression for the audience is advantageous to the analyst. The use of visual aids to demonstrate important points is an especially valuable technique.

But before deciding upon the techniques of presentation, the analyst has to determine what information is going to be presented. Supposedly, the written report serves well as the basis for the verbal report; but, of course, there is nothing so boring as an analyst who *reads* the report, so the analyst will want to devise a whole new approach at this time.

A suggested outline for the verbal presentation is shown below. As with the other reports, the outline may vary with the study; but this can serve as a basis. Some approximate times are included for each of the five sections; but these, too, can be adjusted to fit the circumstances. Amplifying comments follow the outline.

Verbal presentation

I. Introduction ($\frac{1}{12}$ of time allowed).

A. State the problem.

B. Explain how the systems study progressed.

C. Explain the division of the lecture, for example, what follows the introduction.

II. The body of the presentation ($\frac{1}{2}$ of total time).

    A. Describe, very briefly, the existing system.

    B. Describe the new proposed system.

    C. Present economic cost comparisons.

       1. Be careful not to make unsupported claims.

       2. Be conservative.

    D. Make recommendations.

III. Summary of the body of the presentation ($\frac{1}{12}$ of total time).

IV. Open discussion ($\frac{1}{4}$ of total time).

    A. Answer management's questions.

    B. Keep lively discussions going; it is more profitable not to cut off a lively question session just because the allotted time has elapsed.

V. Conclusion ($\frac{1}{12}$ of total time).

    A. Summarize the points of agreement that were brought out during the open discussion.

    B. Make a positive statement as to what you are going to do next, such as get more data, install the system, start programming, or conclude the project.

The *introduction* is descriptive and is only concerned with preliminaries. For example, introduce yourself and the subject of your presentation. Explain the purpose of the meeting and make a statement of the problem. Explain how the systems study began, how it progressed, how the presentation will be made, and how much time will be devoted to each section of the presentation. In closing the introduction, the main points that will be covered during the body of the presentation can be noted.

The *body of the presentation* is the heart of the lecture and demonstration. Try to be concise since the audience will lose interest if it is too long. It is desirable to use management terminology rather than systems analyst terminology since management tends to be more interested in *what* is being done rather than *how* it is being done. Use positive concepts, such as cost reduction, faster response time, greater accuracy, more consistency, better control, and so forth. Emphasize the benefits of the new proposed system, not its special features. And although it is tempting to forget them, do not omit the major drawbacks. Management must have the most reliable information available, and only by your being honest can they make a good decision. (It may be good to keep in the back of your mind that what is bad for the firm will also be bad for you.)

In the summary, the *body* of the report is summarized. The analyst should again stress the benefits of the new proposed system. The *open discussion* is where the pressure is on the analyst. The unexpected objections will crop up here, so this is where the proposed system will have to be defended. (It is perhaps desirable to remember that if it is truly a good system, it *is* worth defending!)

Finally, in the *conclusion* the analyst summarizes any points of agreement. The presentation should then close with a positive statement on what the next step will be.

THE PRESENTATION ITSELF

Before giving the verbal presentation, the analyst has certain opportunities that will help assure success. Learn about the individuals who will attend the presentation because the more you know about them, the more the presentation can be tailored toward their interests. The technique of dropping suggestions, prior to the presentation, is very valuable. Subtle hints can be made to department managers beforehand. This is not to suggest that the analyst should be secretive about the study results. To the contrary, surprise endings are not welcome! Do ask for advice, though. It is a compliment to the person you ask.

The management involved usually decides who will attend the presentation, but the analyst should express a desire that certain personnel be present. Attempt to have present the personnel who have the most to gain from the new system and those personnel who are the most involved. The name and title of everyone who will attend the presentation should be known. Be aware, too, of what their function is within the current system and what their function will be within the new proposed system. Prepare a "benefit-list" for the *key* persons who will attend. List the items within the new proposed system that will benefit each of these people the most.

As for the actual presentation, it should be set up early. Whenever possible, make a trial presentation with another systems analyst, using a tape recorder so you can hear how you sound. Make sure, for instance, that the verbal presentation is coordinated with any charts or graphs. Know the mechanics of any equipment to be used during the presentation. Avoid memorizing the presentation, though. A memorized presentation often seems insincere, and if a question arises halfway through, it is too easy to lose composure. Make sure you understand what you are saying! Above all, be sure that YOU are sold on the values of the new system.

During the presentation, speak from the "you" or "we" standpoint, rather than "I." In other words, do not refer to the proposed system as "my" system. After all, it is management's system and management will be paying for it. Explain all charts, graphs, or illustrations, remembering to reveal each visual aid only as it is discussed, not before. When visual aids are shown beforehand, they tend to detract from a presentation since the audience begins concentrating on them rather than on what is being said. Along the same line, speak to the audience, not to the wall or the visual aids. In fact, you should be speaking *with* the audience, not *at* them. Have a conversation with the audience so each individual will feel involved. Maintain that all-important ingredient, eye contact. And since fumbling with notes or visual aids also is distracting to the audience, any fumbling should be avoided. In summary, the best advice that can be given for the presentation is this: *You are supposed to be a professional, so act like one.*

Throughout the discussion period remember the words empathy, tact, rapport, sensitivity, and integrity. Anyone asking a question in front of a group of other people generally has a good reason for asking it, no matter how poorly it is phrased. The person asking the question deserves the courtesy of the most accurate reply the analyst can give.

In every group, however, there is generally at least one person who is outspoken in objecting to whatever is being proposed. To deal with such an objector can be a very delicate matter. You may wish to view the objection from one of these two viewpoints.

1. If the objection is of minor importance, it is usually better to concede the point and make whatever revision is requested. This approach demonstrates to those in attendance that you are flexible, and it may pave the way for future cooperation.

2. If the objection would require major changes in the system, tactfully explain to the objector why the system is designed as it is in this area. Explain that you cannot make the changes without consulting others, gathering more data, or doing considerable redesign work. Tell the opposition that you will investigate this aspect and report back later on this point.

VISUAL AIDS

A visual aid is any device used to illustrate what the analyst is talking about. It may be notes on the blackboard, an object, a model, photographs, charts, graphs, tables, diagrams, lists, maps, flowcharts, or any other device the analyst selects to describe something.[2]

Flip charts are a common device used in business presentations. They are nothing more than a set of graphs, tables, or other visual aids drawn on paper and clamped together onto an easel-type device. As the speaker finishes discussing the top sheet of paper, it is "flipped over" to the next illustration.

Flip charts can be used to help the presentation proceed in a step-by-step fashion. When using charts, remember to write the verbal presentation first and then determine where to include charts and graphs. Each chart or graph should be limited to one major idea, and it should be easy to read. Any sentences on the chart or graph should be short and simple. As mentioned previously, keep the charts and graphs covered until the presentation reaches that point. Before giving any detailed explanations, read any major writing or figures on the chart to the audience.

A chalkboard is another type of visual aid. When using a chalkboard, know beforehand what you are going to draw. Keep the drawings simple, and if talking while drawing, raise your voice when facing away from the audience. Clean the board before presenting a new point and be neat. Take a trial run to look at your own handwriting from the point of view of the audience; perhaps you should print the message rather than write it.

Overhead projectors are easier to use than many other visual aids, but they do require considerably more prior preparation. Better audience control becomes possible with this type of visual aid. With a colored pen the analyst can add to the chart or graph being projected. For example, the analyst can check off or color in certain areas of the graph or chart as the presentation progresses.

Films and slides do not allow the analyst to vary the presentation to fit specific personalities in the audience. Explain the contents of the film just before it is shown. After the film, reserve some time for discussion. Films generally tend to be better for training (during the system implementation phase) than they are for selling the system to management.

ENDING THE PRESENTATION

At the end of the presentation, the analyst should attempt to get management to agree to implement the new system. Try to get management to say, yes, they want the system. There are three approaches that may be useful for getting management to agree to the implementation. The appropriate one may be used during the conclusion of the verbal presentation.

1. The ask-them-to-buy approach. Boldly ask management to accept the new proposed system.
2. The inducement approach. Conclude by pointing out that some outside event will happen if the proposed new system is not installed. Or offer a special inducement for acting now. (Be careful, though, not to let this strategy sound like a threat!)
3. The secondary-question approach. Get management to agree on something of secondary importance such as whether implementation should start on the first of next month or next week.

Under most circumstances, the first strategy will be the most appropriate. It is positive, but still respects management's prerogative in making a decision.

The second strategy is appropriate only if some impending event genuinely threatens the well-being of the organization. The analyst, while being realistic, must be careful not to exaggerate the significance of the event.

The third strategy should only be used with great caution for it borders on disregard for management's prerogative and ability to make a decision. In this strategy the analyst closes on the assumption that the case is already won—a posture which can be very irritating to line managers.

Although it is desirable to close the presentation on a positive note, for example, a point of agreement, the analyst should assess carefully both his own position and that of the audience before choosing a method of ending the presentation. Above all, the analyst must be thoroughly convinced on the proposed system before attempting to sell it to management. Otherwise any reservations should be explained clearly along with the various alternatives.

SELECTIONS FOR FURTHER STUDY

As we indicated in this chapter, many a good design has been lost because the analyst did not sell the system tactfully. Trying to tell management what is best for them must be done with respect for their prerogative and ability to make the final decision. To sell the system and to have management get on your bandwagon is truly a *coup*.

The systems analyst should keep in mind that once the system design is completed and one is convinced of its utility, one must be prepared to sell the idea to the managers who need it. For this task some of the techniques used by professional sales people may be examined.

If you have never had any selling experience, there are at least 150 books available on salesmanship and persuasion. You might begin with the very general, but highly successful, *How To Win Friends and Influence People.* Cash and Crissy's 12-volume *Psychology of Selling* covers all aspects of the subject including appeal, motivation, and resistance.

Selling the system may be broken down bibliographically into writing the formal report, writing your speech, organizing your visual aids, conference techniques, and the oral presentation.

There are many books available on business report writing. Among these are Brown's *Effective Business Report Writing*, Gallagher's *Report Writing for Management*, and Robinson's *Writing Reports for Management Decisions.*

The art of speech writing is covered adequately in books on public speaking such as Sandford and Yeager's *Effective Business Speech.*

The organization of visual aids follows the writing of your speech. Carroll's *How To Chart Data* illustrates different charting techniques for use in solving multivariable factor problems, and as aids to use when presenting management with complex problems and answers. Schmid's *Handbook of Graphic Presentation* is a highly regarded volume which the system analyst can use to improve a presentation of charts, graphs, and diagrams.

Since the systems analyst is often in the position of running the meeting, some conference techniques should be learned. First and foremost, Robert's *Rules of Order* should be studied for the basics of parliamentary procedure.

For some people, the most difficult stage is the verbal presentation itself. It might be mentioned here that if your public speaking background is minimal, either taking a speech course or joining Toastmasters International would probably help you gain confidence. There are many books to help you with the actual presentation, such as Wilcox's *Oral Reporting in Business and Industry* or Jay's *New Oratory*, which includes planning, delivery, and visual aids, but remember, there is nothing like DOING IT!

Throughout this whole procedure, remember that keeping a positive mental attitude is your best asset!

1. BAIRD, JOHN E., JR., and SANFORD B. WEINBERG. *Communication: The Essence of Group Synergy.* Dubuque, Iowa: William C. Brown, Company, 1977.

2. BARBARA, DOMINICK A. *How to Make People Listen to You.* Springfield, Ill.: Charles C. Thomas Publishing Company, 1974.

3. BETTINGHAUS, ERWIN P. *Persuasive Communication, 2nd edition.* New York: Holt, Rinehart and Winston, Inc., 1973.

4. BROWN, LELAND. *Effective Business Report Writing, 3rd edition.* Englewood Cliffs, N.J.: Prentice-Hall, Inc., 1973.

5. CARNEGIE, DALE. *How to Win Friends and Influence People.* New York: Simon and Schuster, Inc., 1977.

6. CARROLL, PHIL. *How to Chart Data.* New York: McGraw-Hill Book Company, 1960.

7. CASH, HAROLD C., and W.J. CRISSY. *Psychology of Selling.* Flushing, N.Y.: Personnel Development Associates, 1958–

8. GALLAGHER, WILLIAM J. *Report Writing for Management.* Reading, Mass.: Addison-Wesley Publishing Company, 1969.

9. HARPOOL, JACK D. *Business Data Systems: A Practical Guide.* Dubuque, Iowa: William C. Brown Company, 1978.

10. JAY, ANTONY. *The New Oratory.* New York: American Management Association, 1971.

11. JENSEN, RANDALL W., and CHARLES C. TONIES. *Software Engineering.* Englewood Cliffs, N.J.: Prentice-Hall, Inc., 1979.

12. MAMBERT, W.A. *Effective Presentation: A Short Course for Professionals.* New York: John Wiley and Sons, Inc., 1976.

13. MAMBERT, W.A. *Presenting Technical Ideas: A Guide To Audience Communication.* New York: John Wiley and Sons, Inc., 1968.

14. ROBERT, HENRY MARTYN. *Rules of Order, Revised.* New York: William Morrow and Company, 1971.

15. ROBINSON, DAVID. *Writing Reports for Management.* Columbus, Ohio: Charles E. Merrill Publishing Company, 1969.

16. SANDERS, DONALD H. *Computers in Business.* New York: McGraw-Hill Book Company, 1975.

17. SANDFORD, WILLIAM P., and WILLIAM H. YEAGER. *Effective Business Speech, 4th edition.* New York: McGraw-Hill Book Company, 1960.

18. SCHMID, CALVIN F., and STANTON E. SCHMID. *Handbook of Graphic Presentation.* New York: Ronald Press, 1978.

19. VAN FLEET, JAMES. *How to Put Yourself Across to People.* Englewood Cliffs, N.J.: Prentice-Hall, Inc., 1971.

20. WIKSELL, WESLEY A. *Do They Understand You: A Guide to Effective Oral Communication.* New York: Macmillan Company, 1961.

21. WILCOX, ROGER P. *Oral Reporting in Business and Industry.* Englewood Cliffs, N.J.: Prentice-Hall, Inc., 1957.

22. WILLOUGHBY, THEODORE C., and JAMES A. SENN. *Business Systems.* Cleveland, Ohio: Association for Systems Management, 1975.

QUESTIONS

1. Name four problem areas the analyst should analyze when selling the new system to the management of the area under study.

2. Is selling management on a new system a one-time effort? Why or why not?

3. Name three important advantages to anticipating and preparing for as many objections as possible.

4. Name three *faux pas* that the analyst should try to avoid.

5. Name six things the analyst should consider in writing the final systems report.

6. How should the analyst respond to an objection to a proposal during the verbal presentation, if that objection would require major changes?

7. In ending the verbal presentation, what is the "inducement approach" for getting management to agree to the implementation?

SITUATION CASES

CASE 11-1

The Reed Manufacturing Company had received numerous complaints from its customers regarding delays in the delivery of its products. The vice-president of Sales requested the Systems Department to conduct a system study and to develop a system that would reduce the delivery time of its finished products.

The study was assigned to a junior analyst. After conducting preliminary work and gathering all the data, he designed a new system that would include the use of a computer. The analyst knew that the Accounting and Production Departments did not have authority to buy the new system. As a result, his presentation would have to be geared toward top management. The analyst thought it would be a good idea to prepare answers to some of the basic objections he knew management would raise against the proposed system.

Next, he read some of the current literature on how to prepare a final report to refresh his memory in this area. The analyst knew that the final report was an important phase in selling the system. All areas of the written report were composed perfectly as suggested by the literature he had read. The analyst had remembered that the summary section required only a few paragraphs and explained details in the body of the report.

In the report, he decided to increase the estimated cost of installing the system. The analyst figured everyone thought a computerized system was expensive, so he decided it would be better to overestimate the cost in order to make it appear that the company spent less than planned. His secretary typed the report and after checking it for correct grammar and spelling, copies were made and sent through the message center to everyone involved except one of the vice-presidents. The vice-president was unintentionally left off the list of people who should receive a copy of the report.

With the help of his fellow analysts, preparations were made for the verbal presentation. He decided to use an overhead projector for key points that needed to be emphasized. Because the analyst was relatively new with the company, his supervisor advised him on the personalities of management who would attend the presentation. The supervisor told him it would be difficult to convince the vice-president of Production. This vice-president was against change and believed in the philosophy, "What was good for the company ten years ago is good for the company today." Unknown to the junior analyst, this was the vice-president to whom he had forgotten to send a written report. The analyst then prepared a "benefit list" for the key personnel who would be present, pointing out the savings in time and money. He personally checked out the conference room for voice projection and all equipment that would be used. Everything was in good working condition. He went over the presentation using a tape recorder and discovered he should be more forceful in his delivery. At this point, the analyst felt he was ready to make the presentation. He asked the secretary to schedule the meeting in two days.

Meanwhile, the vice-president of Production read a copy of the written report that another vice-president had lent to him. He was furious to learn that everyone else in top management had received a report. The vice-president made inquiries and found out the date of the meeting. Then he obtained a copy of the written report and did extensive research into all aspects of the report. Any discrepancies would be used to sabotage the analyst's system.

On the day of the presentation the analyst felt great. The analyst thought to himself, "Finally, I have a chance to display my talents." The analyst introduced himself and the subject of the presentation. The outline and allotted time to be spent on each segment was explained to the group. During the presentation he primarily used management terminology, but at times reverted to using systems analyst phrases and terminology. The new system was referred to by the analyst as "management's system" emphasizing cost reduction, faster response, increased performance, high accuracy, more flexibility, and more consistency. He made good use of the overhead projector, underlining key points and benefits each department would receive from the new system.

Finally, the meeting was opened to questions. The offended vice-president of Production immediately asked, "Why should we spend this large sum of money on a computerized system when we could hire additional help for less money?" He also stated that the extra help would solve the existing problems. The analyst explained that the new system was based on Reed's long-range planning and would be able to handle an increased load in the system without any additional cost for the next five years.

Next, the vice-president of Production wanted to know why the estimates were higher than the total he had received from several computer companies. The vice-president was sure he had sabotaged the presentation. The analyst called upon all the skills he had learned and replied that he had anticipated phase two to end in a few months and that prices of equipment and labor would increase approximately 5%. The analyst had included this increase because, if adopted, the system would take at least a year to install. There were no further questions so he summarized all the points of agreement that were brought out during the discussion period. He closed the meeting and assumed management would accept the new system because it would reduce costs, processing, and delivery time.

Management told the analyst that they would inform him of their decision on the new system in approximately two weeks.

QUESTIONS

1. Describe how the analyst could have improved the way he handled the financial and report distribution aspects of this case.

2. What were the major weaknesses in the analyst's verbal presentation?

3. How could the analyst have improved his approach to closing the meeting and selling the project to management?

CASE 11-2

The management of Sliz Corporation, manufacturers of electronic and stereo equipment, noted a consistently gradual loss of profits and market share over the previous two years. A systems analyst was asked to determine the problem and suggest a workable solution. After discussing the problem Sliz Corporation was facing with key personnel and studying the goals and objectives of the company, the analyst determined that the marketing strategy of the company could not keep pace with changes in the marketplace and the recent influx of imported merchandise. An on-line time-sharing computer processing system was suggested that could process the variety of data required to develop an updated marketing strategy and forecast for each new model year.

When the analyst felt that her solution would effectively meet all the requirements of Sliz Corporation, she prepared the final written report. This included the statement of the problem (subject, scope, and objectives), the major recommendations, and a justification of the new system. She reviewed the current system, the new system requirements, highlighted the operations of the new system and reviewed the cost and implementation of the new system. In preparing the written report, the analyst attempted to be brief and concise, describing the new system in the active voice and using correct spelling and grammar.

A copy of the written report was delivered personally to all key personnel with a cover letter pointing out that a future meeting already was arranged to discuss the report. The analyst then set the meeting date for ten days in the future.

At the meeting, the analyst began the verbal presentation by explaining the problem, and describing the existing system and "her" new proposed system. All key personnel attended including those in upper management and those who would operate the system. A "benefit list" was prepared for each member of management explaining the special features of the new system by using management terminology. The analyst went on to discuss the new system explaining the economic costs, comparing them conservatively with those of the existing system. She then closed by explaining to management that continued market share deterioration and loss of profits will continue rapidly if her new system was not accepted.

A disagreement surfaced during the open discussion because some attendees felt that the costs were far too conservative and not representative of the real costs involved in implementing this system. The analyst quickly reviewed her notes, stated that her costs were realistic according to all available sources, and then abruptly cut short the discussion by asking for other questions. Management replied that it would notify the analyst of its decision within the week.

QUESTIONS

1. In what sense did the analyst misuse the "benefit list" in her efforts to sell the new system to management?

2. What error did the analyst continually make when referring to the proposed system?

3. What should the analyst have done to improve the closing of her presentation rather than stressing the impending danger if her system was rejected?

4. What could the analyst have done to more effectively handle the objections that were raised?

CASE 11-3

A manufacturer of toys had previously enjoyed numerous successful products. Consequently, its brand name was quite well known. In fact, its latest toy was currently number one in the market. Even though sales were up, however, the company had experienced a drop in profit percentages because of the high production costs of their latest toy. This prompted the Board of Directors to request a feasibility study on improving the production system. The Board decided that an improved system was possible and assigned the company's senior analyst to conduct a full systems study of production. The analyst was told that he had until the scheduled start of production of the company's "new" toy to report his findings (some two months away).

The analyst conducted a very thorough systems study. First he defined the problem, then outlined the study, obtained general background information, and learned the interactions between the production system and the other systems of the company. Next he familiarized himself with the existing system, determined the requirements for the new system, and then designed the proposed system. After performing a comparison of the economic costs of the new system as opposed to the existing one, all he had left to do before implementation was to sell the system to management. Time was running short, however, so the analyst decided to concentrate the rest of his time on his final written report and forego the preparation of a verbal report. He knew that the final report was the most important report of the entire systems study. Since management thought a new system was feasible anyway, he figured that it would not take much effort to verbally persuade them to implement the new system.

The analyst presented the written report, starting with a summary of the full systems study, followed by a detailed body of the report which included, first, a description of the new proposed system, second, an economic cost comparison of the new and existing systems, third, a description of the existing system, and fourth, a straightforward positive statement on why he believed the planned system should be installed. The written report concluded with an appendix which consisted of prior reports and items that did not fit into the rest of the report. The analyst made sure that the report was clear, effective, and brief. He wrote in an active voice, using short declarative sentences; and both he and a secretary checked for spelling, good grammar, and sentence structure. In other words, the written report, like the rest of the systems study, was a very commendable job. The analyst was proud of the new system and was quite confident that he could sell it to management even though a verbal presentation had not been prepared.

At the presentation the next day the analyst recognized most of the people but did see a few new faces. What startled him most was the fact that two friends, whom he had counted on for support, were not there. Using the written report as an outline for the presentation, the analyst first introduced himself and the subject of "his" proposed system. He followed with the body of the presentation in which he emphasized the "special features" of "his" new system. Preprinted charts, graphs, and diagrams would have been ideal for the presentation; but he had not prepared any that were big enough for this purpose. Without an open discussion, he concluded the presentation by summarizing the points that would benefit "his" new proposal. The analyst ended the presentation without using any kind of strategy for closing; he assumed that management would install the new system. When finished with the presentation, the analyst was quite surprised at the objections that were being raised. Because he had done such a thorough job of the systems study, except for the verbal presentation, all of the objections were minor. Nevertheless, he was not about to let someone change "his" proposed system when he had spent so much time preparing it.

QUESTION

1. How could the analyst improve on selling the proposed system?

FOOTNOTES

1. Notice that some of the time now being estimated pertains to the time required to install the new proposed system.

2. There are numerous volumes on technical writing, most of which include a section on graphic presentations. For specific examples, the reader is referred to *Basic Technical Writing* by Herman M. Weisman (Columbus, Ohio: Charles E. Merrill Publishing Company, 1974), now in its third edition.

CHAPTER 12

IMPLEMENTATION, FOLLOW-UP, AND RE-EVALUATION

Finally, after analysis, design, and approval, the phase that everyone has been working for arrives: the conversion of the conceptual idea to the reality of the empirical system. Implementation emphasizes peopleware. This chapter tells the analyst how to implement the new proposed system.

THE IMPLEMENTATION PROCESS

The implementation process begins *after* management has accepted the new system. Implementation consists of the installation of the new system and the removal of the current system. It involves hardware (machines), software (computer programs, procedures, forms), and peopleware (personnel). The implementation phase is often the longest, costliest, and most difficult part of the systems job. Naturally, the time chosen for the changeover to the new system should be a time when the system is *not* involved in heavy operations. A slack period should be chosen if possible.

During implementation of the system, problems that had not been anticipated during the study and design often appear. Solutions to these problems usually require modification to the original design. The analyst should be willing to accept changes where necessary, but should prevent extreme distortions of the original design.

IMPLEMENTATION OF A MANUAL SYSTEM

The steps involved in the implementation of a new system can be very complex and demanding. To make implementation proceed as smoothly as possible, the analyst should work out an implementation plan. The plan should specify who will do what and when they will do it. The best method is to prepare detailed instructions for the supervisors to follow during the system implementation in their departments. The detailed instructions should outline the responsibilities, time schedules, and operating instructions that are required in order to properly implement the new system. Various items that will help the analyst in organizing the plan are To Do lists or an outline (see Chapter 3), Gantt charts and flowcharts (see Chapter 13), and job procedures (see Chapter 17).

Actually, the analyst should think of implementation from the very beginning of the project in order to keep the system oriented to all user levels. The users of the new system will have to be taught new methods. Benefits of the new system should be pointed out early to enable the users to understand why their cooperation is needed. They also should be made aware of both the limitations and capabilities of the new system so they can accept it in realistic terms. This avoids later disappointments when users learn the system cannot handle something they had assumed it could handle. The workings of the new system and its procedures should be explained thoroughly to them.

The personnel also have to be trained to operate the new system. The analyst should assist in the training function and, whenever possible, train each employee individually. The analyst may have to train in small groups, however, dividing similar tasks into separate training groups.

The analyst operates in a staff capacity and, as such, operates without line authority. The analyst, therefore, cannot *order* anyone to do anything. If line authority is required, the manager of the department where the implementation is taking place should be the one to give orders. Since the analyst can only request, methods which will secure the cooperation of the operations employees are vital to the system analyst's success. The analyst should learn the art of dealing with the operations employees in such a way as to secure their goodwill and cooperation right from the start of the project. Learning to deal with employees who have extreme aversion to change or who are obstinate in their attitude toward cooperating is especially important. Needless to say, getting the job done without line authority of any kind can be tricky; but the competent analyst does it.

Once implementation has begun, the design of the new system should be frozen except for essential modifications. *One central control* point should be established to handle these modifications. This central control, whether it is an individual or a group of people, needs to have authority to authorize any system changes which must be made. The central control is responsible for making a permanent record of the change, informing all affected personnel of the change, and the prompt issuance of revised written procedures as soon as possible after the change is made. Usually all forms used in the new system must be on hand before implementation can begin. The central control should also approve any form modifications since changes may affect more than one group of users.

A time schedule must be estimated for the system implementation. First take into account the overall elapsed time from the start of implementation to its finish. Calculate chargeable time to determine how much it will cost to implement the new system. Then decide *how* the new system will be implemented. There are three basic approaches to starting up the new *manual* system.

1. All at once. All operations are started up at the same time (one-for-one changeover).
2. Chronologically, and in sequence through the system. Start with the first operation and move through to the last operation.
3. In predetermined phases. Similar areas within the system are started up at one time and then other areas are started up.

The analyst should help the operating personnel in modifying old data so it will fit the new system. New data sources may have to be located or cross-reference files started. In a system change where both old and new files must be maintained, the analyst should try to help the personnel make the transition with as little frustration as possible.

For example, if the old system files were indexed by customer name and the new system files are by customer number, the two files are not compatible. The analyst can assign customer numbers and refile the entire old system or have two files: an old file by customer name, and a new file by customer number for all of the current business. If all subsequent reports were to contain *only* customer number, a cross-reference file would be required to use the old-system files. The operating personnel would have to look up the customer number in the cross-reference file to get the customer name so they could use the old-system files. In time, the old-system files would become outdated and would no longer be used because all the current business would be filed in the new customer number file.

The old system must definitely be phased out. Operating personnel must learn to depend on the new system and forget the old one. By some preplanned date, only the new system should be operating. Do not allow a mixture of the old and the new systems unless that mixture was part of the original systems design or implementation plan. In a manual system the changeover is usually made without the benefits of parallel conversion (see Glossary). When the operating personnel are busy with one manual system (the old one), they cannot be expected to simultaneously operate both the old and the new manual systems. Even if temporary employees were considered, a physical space problem often precludes parallel conversion of a manual system.

In summary, the systems analyst, when installing a *manual* system, does the following.

1. Develops a plan for implementation.
2. Assists line management in understanding their commitment to support the new system, that is, their responsibilities toward the new system.
3. Prepares any required documentation, such as written procedures or instruction manuals.
4. Orients the operating personnel and assists in their training.
5. Coordinates the installation of the new system and the phaseout of the old system.

IMPLEMENTATION OF A COMPUTER-BASED SYSTEM

If the new system is to be computerized, the systems analyst has a whole new set of problems with which to contend. All the steps to be followed in the implementation of a manual system are still followed, only they are much more complicated. For a small computerized system, one analyst may coordinate the whole project. For very large systems, however, a team of analysts may work together, either from the design phase through implementation, for example, or from the very beginning of the study. If a team approach is utilized, the individuals making up the team should exhibit certain personal qualities. Among these are experience, thoroughness, good logic, patience, tact, the ability to communicate, and above all, they must be practical.

The major new ingredients in the implementation of a computer-based system are more sophisticated hardware and more complicated software. In a small system the analyst might now change hats and become a computer programmer, too, although a more successful system is usually developed when the systems analyst is one person and the programmer is another person. Whenever professional programmers are working on the job, the analyst should advise them on the new system and not on the details of their programming techniques. They, too, are professionals.

It is advisable not to start any extensive programming until management fully approves the system and orders that it be implemented. Prior to management's acceptance, only small test programs should be undertaken. Never spend the firm's money on programming until the system has been formally accepted. This usually means that programming commences during the implementation phase.

Before discussing the programming portion of the implementation phase, it should be mentioned that if this is a new computer installation, an even larger set of problems exists. The selection of a computer was discussed in the Type B feasibility study in Chapter 2.

Whether the analyst or a programmer does the work, the programming should now begin. The programming proceeds in the following sequence (amplifying comments follow the list).

1. Study the design of the new system to become knowledgeable on how the system operates.
2. Develop the format of the outputs and then the inputs.
3. Develop a program flowchart (see Chapter 13) of the program's step-by-step operations.
4. Determine file layouts.
5. Code the program.
6. Desk check.
7. Program test.
8. Systems test.
9. Document the program.

The programmer must study the design details of the new system in order to understand how the system should operate. The analyst who performed the study should be available for advice. Together they can decide which programming language will be used, for example, COBOL, APL, and so forth.

The analyst and programmer analyze the new system requirements and break it up logically into cycles and jobs. The frequency of processing must be determined and the appropriate cycles identified, such as daily cycle, weekly cycle, monthly cycle, quarterly cycle, semiannual cycle, and annual cycle. The next step is to identify the job requirements to meet the needs of each cycle. Each job is then broken down into its basic components, that is, programs required.

For each program, the next step is to lay out the format of the outputs and inputs. First the format of the output reports should be considered. Lay them out based on the information they will contain and on how they should look when they come out of the computer. The outputs and inputs were already designed during the system study. Next determine how the inputs will be put into the computer. Where does the input data come from? Who submits it? How will it be converted to machine-readable format? At this point the programmer should know what the output reports will look like and what the input data will look like.

Now develop a program flowchart. A program flowchart delineates the logic of the program. It portrays the step-by-step sequence of the program's instructions, and each decision-point is spelled out. The program flowchart is a pictorial model of the program. (Flowcharts are discussed in Chapter 13.)

The file layouts should be determined next. The computer-based files can be stored in any of four layouts. *Sequential* files are in sequential order by their keys,[1] for example, from the lowest to the highest part number. *Index-sequential* files are in sequential order by their keys also, but an index is created so specific data can be accessed directly without searching sequentially through the file. *Random-access* files are in an order prescribed by a mathematical formula, and can be accessed directly. *Partitioned* files are those in which various areas are partitioned into unique file areas for your specific data only.

Coding the program follows. If the program flowchart is complete, the coding should be easy. Coding is nothing more than writing the computer program in whatever programming language was chosen. After coding is completed, the program is keypunched into machine readable format.

Desk checking is next and it is the most overlooked step in the sequence of programming. Desk checking consists of checking for keypunch errors and checking the keypunched program code against the program flowchart. The program is followed through *mentally* with some simple input data. The object is to see if the program logic is in proper sequence and if the output is what is expected.

The program test is made by actually trying out the program on the computer to see whether it will compile and run. Rarely does a program run perfectly the first time. Removing the errors to get the program to run correctly is called debugging the program. Test data should be run in the program to see if the program logic develops as it should. Besides normal data, input some extreme and invalid data to make sure the program will not fail under extreme conditions or when someone puts invalid data into the system. All logical paths in the program must be tested during this phase.

After each separate program has been tested individually, they should all be put together to test the entire system. This phase, called the systems test, is necessary. Individual programs often perform perfectly, but fail when joined together in a system. The important point is that the entire system should be tested thoroughly before actual installation in the company. This is achieved by running test data through the system prior to the real thing.

And finally, document the program. Documentation is necessary for communication of program characteristics to persons other than the programmer, and for the programmer's future reference as well. People come and go, but the programs stay. Documentation provides the written and charted explanations necessary to familiarize new personnel with the program. The documentation package for a computer program consists of the following.

1. In-line comment or note statements in the program coding itself. For every two or three program coding instructions, there should be a comment that explains what these instructions do and how these instructions interact with other related areas in the program.

2. One program flowchart for each program; and if there are multiple programs, an overall flowchart showing how the various programs interact.

3. For sections of complicated logic or calculations, detailed flowcharts or narrative descriptions of the activity should be prepared.

4. Pictorial layouts of the files, the outputs, and the inputs.

5. Program run book that contains operating instructions, for example, brief program narrative, computer setup information, tapes, disks, carriage control tapes, special printer forms, or any special restart procedures in case of failure prior to normal program end.

At this point the analyst and the programmer together should be ready to implement the new computer-based system. Now the decision whether to implement using a parallel conversion or a one-for-one conversion must be finalized. In a parallel conversion, the old and the new systems run simultaneously for at least one cycle (see Glossary) using current data. It is costly, but it is the safest method because the old system keeps operating until the new system has proven its accuracy and reliability. In a one-for-one conversion, the new system is installed with simultaneous removal of the old system. For protection during a one-for-one conversion, emulators may be used to process the old programs by the old procedure using current data on the computer.

Actual implementation of the new system can begin at this point using either a parallel or a one-for-one plan, or some blend of the two. It is advisable that the systems analyst and the programmer observe the following basic principles during any implementation.

1. Avoid disrupting the day-to-day business activities during the implementation process.
2. Do not require excessive overtime work during implementation.
3. Inform management of all changes in the implementation method or time schedule.
4. Do not give demanding orders; you are functioning as advisory staff, not as a line manager.

PURPOSE OF THE FOLLOW-UP

During the follow-up phase the analyst determines how good the system design really is by returning to observe the actual operation of the installed system. The analyst finds out if the operating personnel are using the new system with its formal procedures or if the operating personnel have started their own informal procedures. Many observations are required during follow-up. Management wants to know if the objectives of the new system are being accomplished. Management also wants to know whether any anticipated cost savings are being realized, whether the new system is providing the information or "product" that is required, and whether the day-to-day working schedules are being maintained.

The analyst, during the follow-up, should make sure that all parts of the new system are actually operating and that minor activities or operations have not been overlooked. The operating personnel should be observed to be sure they have not reverted back to the old system! Operating costs should be examined to see if they are in line with the estimates and whether the facilities are really adequate, for example, equipment, space, number of operating personnel, and so forth.

The programmer also should be active at this point to determine if each program is operating correctly and if the program's output is really adequate. Both the programmer and the analyst should collect a list of employee complaints for evaluation and possible corrective action.

WHY RE-EVALUATION IS REQUIRED

The primary reason for re-evaluation of the system after it has been installed is to make whatever changes are needed for the refinement and improvement of the new system. It may be necessary to redesign some portions of the system, and revise some of the original recommendations.

Review the list of employee complaints, and evaluate the efficiency of the work flow of your new system. Seek out more opinions from the operating personnel. Determine when peak loads occur, the quantity of paperwork flowing through the system, the accuracy, the utility, and the timeliness of the outputs. Using the *evaluation criteria* from the new system requirements phase of the systems study, develop some performance standards. Use the evaluation criteria to monitor the day-to-day performance of the system as well as for possible re-evaluation and change. Do not be afraid to change the system you designed. It is better that *you* do it than someone else who might use the opportunity to your detriment.

FINAL SYSTEM DOCUMENTATION

The final documentation package requires the pulling together of all the documents that were prepared along the way. It is a *mandatory* step. If certain documents were bypassed originally, they must be prepared now. The final system documentation consists of the developmental documentation that was used in defining and analyzing the problem, the documents used to control the whole project, the documents that formally describe the new system, and the documentation that actually will be used during the operation of the system. In other words, it consists of all the records showing how the system was designed and how it is to be operated.

A suggested format for the final documentation package is shown in the following outline.

Final system documentation

  I. Introduction and table of contents.
    A. Name of the system.
    B. Why the system was developed.
    C. Purpose and objectives of the system.
    D. Who uses the system.
    E. Where the system fits into the company.
    F. Any other pertinent remarks.

 II. Explain, in narrative, how the system operates and include the following.
    A. A system flowchart of the overall system.
    B. Any required documentation sheets that are necessary to explain the system, such as

      1. Area cost sheet.

      2. Documented flowchart and documentation section.

      3. Systems requirements.

      4. Input/output sheet.

      5. Decision table.

      6. Equipment sheet.

      7. Personnel sheet.

      8. File sheet.

III. Show how the computer programs fit into jobs, how jobs fit into cycles, and how cycles fit into the overall system. Describe, in narrative, what each program, job, and cycle does. Include the following.

  A. A computer listing showing the statements of each program.

  B. In-line comment or note statement. For every two or three program coding instructions, put in a comment that explains what these instructions do and how these instructions interact with the rest of the program.

  C. One program flowchart for each program and an overall flowchart showing how the various programs interact with the rest of the program.

  D. Flowcharts and detailed narrative descriptions of complicated logic or calculations within each program.

  E. Pictorial layouts of the files, the outputs, and the inputs.

  F. Program run book that contains operating instructions, for example, brief program narrative, computer setup information, tapes, disks, carriage control tapes, special printer forms, or any special restart procedures in case of failure prior to normal program end.

IV. Summarize the implementation plans and the results of the implementation.

V. The appendix contains all other documentation. Nothing is thrown away.

  A. Feasibility study report.

  B. Problem definition report.

  C. Outline of the study.

  D. Summaries of various phases.

      1. General information on the area under study.

      2. Interactions between the areas being studied.

      3. Understanding the existing system.

      4. Definition of the new system's requirements.

    5. Designing the new system.

    6. Economic cost comparisons.

E. Final written report.

F. Verbal presentation.

G. Implementation plan.

H. Any significant notes made by the analyst or programmer.

The complexity and level of detail will vary with the particular system or firm involved. However, the report should be sufficiently complete to enable the reader to understand the essential characteristics outlined above. Such information can become extremely valuable in the future when the inevitable changes are required in the system.

SELECTIONS FOR FURTHER STUDY

When large-scale electronic data processing installations first came into existence, there were a few things written on administering the EDP function. A search of the literature reveals a very positive outlook: come get on the computer bandwagon where wonderful things are in store for you.

At that point in time the few voices in the wilderness who advocated systems and procedures for EDP were quite weak and went relatively unnoticed. Gradually in the late 1965 and 1966 period, a few articles began appearing that indicated some monsters had been created. Most of the problems (e.g., no software documentation, empire building, and exorbitant costs) were blamed on everything imaginable including the hardware manufacturers. Few companies were willing to admit that costs had been underestimated, the program had been underplanned, or that no one had seen to it that proper documentation had been kept. The cost estimating problem showed up fairly quickly; but the other problems were more subtle and took time.

Finally, in 1967 and 1968, the books by Greenwood (*Managing the Systems Analysis Function*), Canning (*Management of Data Processing*), Cleland (*Systems Analysis and Project Management*), and Rothery (*Installing and Managing a Computer*) appeared and people began to notice that the implementation and follow-up phase was important.

To avoid the pitfalls of these early systems analysts, it is recommended that you peruse the latest systems texts. Hopefully, you will be better prepared than were the early analysts.

Computing Technology, Inc. (Reference 16 below), in a survey of programming documentation practices at seven Federal agencies, found that (1) programmers lack the basic ability to write for the needs of others, (2) documentation is necessary to prevent anarchy but its biggest defect is its rigidity, and (3) documentation problems are fewer in EDP organizations where the design, implementation, and maintenance functions are performed by a single task force. London's *Documentation Standards* is a handbook designed for the person who is establishing a new or revising already existing programs. Tomlin's *Managing the Introduction of Computer Systems* is one of the few texts on implementation.

1. *Auditing for Systems Improvement.* Cleveland, Ohio: Association for Systems Management, 1973.

2. BASIL, DOUGLAS, and CURTIS W. COOK. *The Management of Change.* New York: McGraw-Hill Book Company, 1974.

3. CLELAND, DAVID I., and WILLIAM R. KING. *Systems Analysis and Project Management, 2nd edition.* New York: McGraw-Hill Book Company, 1975.

4. DONALDSON, HAMISH. *A Guide to the Successful Management of Computer Projects.* New York: Halsted Press, 1978.

5. GREENWOOD, FRANK, and LEE GAGNON. *Assessing Computer Center Effectiveness.* New York: American Management Association, 1977.

6. GUEST, ROBERT H., et al. *Organizational Change Through Effective Leadership.* Englewood Cliffs, N.J.: Prentice-Hall, Inc., 1977.

7. HARPOOL, JACK D. *Business Data Systems: A Practical Guide.* Dubuque, Iowa: William C. Brown, 1978.

8. *How To Develop and Conduct Successful In-Company Training Programs.* Chicago: Dartnell, 1979.

9. *How To Prepare an Effective Company Operations Manual.* Chicago: Dartnell, 1979.

10. JENSEN, RANDALL W., and CHARLES C. TONIES. *Software Engineering.* Englewood Cliffs, N.J.: Prentice-Hall, Inc., 1979.

11. LONDON, KEITH R. *Documentation Standards, 2nd edition.* New York: Van Nostrand Reinhold Company, 1974.

12. MARLOW, HUGH. *Managing Change.* New York: International Publications Service, 1976.

13. MORGAN, JOHN S. *Managing Change: The Strategies of Making Change Work for You.* Houston, Tex.: Gulf Publishing Company, 1972.

14. ORGANIZATION FOR ECONOMIC COOPERATION AND DEVELOPMENT. *The Evaluation of the Performance of Computer Systems.* Washington, D.C.: The Organization, 1974.

15. SANDERS, DONALD H. *Computers in Business.* New York: McGraw-Hill Book Company, 1975.

16. SCOTT, BERNARD E., et al. *Survey of Computer-Program Documentation Practices at Seven Federal Government Agencies, Final Report.* Paramus, N.J.: Computing Technology, Inc., March 1967 (PB-175 701).

17. SCOTT, M.F. *Project Appraisal in Practice.* New York: James H. Heinemann, Inc., 1976.

18. TAYLOR, BERNARD, and GORDON LIPPITT. *Management Development and Training Handbook.* New York: McGraw-Hill Company, 1975.

19. TOMLIN, ROGER. *Managing the Introduction of Computer Systems.* New York: McGraw-Hill Book Company, 1971.

20. WILLOUGHBY, THEODORE C., and JAMES A. SENN. *Business Systems.* Cleveland, Ohio: Association for Systems Management, 1975.

QUESTIONS

1. What does implementation involve?
2. What should the implementation plan specify?
3. What type of capacity does the analyst operate in?
4. When one central control point is established to handle essential modifications, for what should it be responsible?
5. If a team approach is used for implementation of a computer-based system, what personal qualities should the team members exhibit?
6. In what sequence does the programming normally proceed?
7. Computer-based files may be stored in how many layouts? Name them.
8. What is the most overlooked step in the sequence of programming?
9. Of what does the documentation package for a computer program normally consist?
10. What does the analyst examine during follow-up?
11. What is the primary reason for re-evaluation of the system after it has been installed?

SITUATION CASES

CASE 12-1

The Goshen National Bank has just completed a Type B feasibility study. Their analyst, Ms. Karen Fuller, is working on the installation of their new computer. The new system will allow the tellers to inquire directly about customer balances to the head office computer and display the results on the local terminal. Since there is very little seasonal change in the amount of work performed in the bank, management and the analyst both agreed that the actual implementation could begin immediately.

The analyst prepared a plan for the implementation which outlined all responsibilities, made the time schedules, and listed the instructions necessary for proper implementation. In order to do this correctly, she made a Gantt chart, daily and overall To Do lists, flowcharts, memos, and job procedures. After identifying various jobs to be done, she ordered the staff in the area to perform them. Ms. Fuller did not feel that a parallel changeover was required and, therefore, called for implementation to be made on a one-for-one basis.

All personnel were trained in the actual operation of the system. The analyst first met with all of the supervisors to discuss the operations in detail. She then personally met with the tellers at all the branches to discuss the details of concern to them. To be sure that all the people working with the new system would be familiar with its limitations and capabilities, she explained these during the visits. The response time for an inquiry and answer had been set at a maximum of five seconds.

Management agreed that the analyst was quite efficient and that any suggestions or possible alterations in the new system should go through her. Ms. Fuller, in turn, resolved to issue any changes in procedure to all concerned.

After studying the design of the new system, COBOL was chosen as the programming language to be used. Final approval was received on the inputs and outputs which had been devised during the design segment of the study. The analyst then proceeded to develop a complete step-by-step flowchart of each program. It was decided that the files should remain in sequential order by account number. The analyst requested that a programmer be assigned to write the program. After the cards were punched, she personally desk checked the cards for any logic or syntax errors. The result of all program and systems tests were favorable.

Having completed the implementation of the new system, Ms. Fuller awaited a further date when she would do a follow-up and re-evaluation of her work. The new system appeared to be running smoothly and both management and the tellers seemed quite pleased.

QUESTIONS

1. In what way does Ms. Fuller exceed her rights in the department under study and jeopardize her rapport with management?

2. What did Ms. Fuller fail to do which may affect the future understanding and success of her system?

Jim Barber, systems analyst for the machinery division of the Delano Corporation, has been requested to resolve a problem with the "Job Order Cost Sheet" form. Management requested that all charges be separated by department, but wanted all the information listed on one form.

The analyst defined the problem and all the other steps necessary to solve it. He successfully sold the solution to management and was now ready to implement the new system. He had designed a form that incorporated all the old information with the inclusion of department separation. The purpose for the change was to reduce reporting errors, facilitate recording, and provide management with accurate and concise reports on job costs.

Mr. Barber began the implementation phase by planning who would do what and when. He determined that the approved form should first be printed, distributed to the user department (namely Accounting), and held until July 1 which was two months away. A slack period normally occurs in job orders in July, allowing a period for transition to the new form. On June 15, he planned to start an orientation program for the managers in charge of the clerks who would be using the form. He also scheduled times for the clerks to be given a dry run on the new system. Training the clerks, in groups of two, was to be the responsibility of the analyst; however, the training was to be carried out by the supervisors.

The analyst planned to have any problems such as inaccurate or incomplete department reporting of job costs (i.e., materials requisitions, time tickets, scrap summaries, and spoilage reports) reported to him during the dry runs. By the final implementation date, the analyst hoped to have all users entering the information correctly. The date for project completion was set as August 1. This allowed four posting cycles before evaluating the acceptance and use of the form. He already had rewritten the job description for the clerks using the form, and planned to update the manuals on July 1 when the new form was to be introduced.

On June 1, the forms arrived and were held until June 15, when Mr. Barber began instructing the supervisors on their responsibilities and the time schedule. Next the analyst distributed a supply of the new forms to appropriate departments. The supervisors held instructional classes for the clerks and provided test data and noted the results. These test forms were delivered to Mr. Barber on June 25 for his evaluation. On June 26, he made a tour of the Accounting Department, updated the manuals, and checked requisitions and other input forms for consistency in reporting.

On July 28, the analyst again visited the Accounting Department to check the results, review complaints, and see if any of the old forms were being used. He found that most clerks enjoyed using the new form and felt that it saved time in posting. Mr. Barber found that one clerk was using the old form sporadically. He informed the supervisor who had the old forms removed and the files changed. No other changes seemed necessary at the time so the analyst started compiling his material for documentation. The implementation was complete.

QUESTIONS

1. During the implementation phase, what factors were overlooked or not specified by the analyst?

CASE 12-3

A newly hired member of the Systems Branch at Community Hospital had been assigned to do the final documentation on the "patient accountability system" which was designed by her predecessor. The former analyst had completed the implementation, follow-up, and re-evaluation, but was transferred before a final documentation package of notes he had prepared during the study could be compiled.

After several days of hard work, the new analyst submitted to the systems manager what she considered to be a complete documentation of the new system. In looking over the new analyst's work, the supervisor found the following.

The introduction stated that the name of the new system was "Patient Accountability System—Community Hospital," and that the system was developed primarily to eliminate duplication of effort on the part of several hospital departments and to effect greater patient control. It related the purpose and objectives of the new system, who utilized it, and just where it fit into the hospital. To close the introduction, the analyst included a few pertinent remarks meant to further introduce the reader to the system.

After the introduction, the analyst began the body of the documentation package with a brief narrative description of how the system operated. In this section, she also included many of the documents that the former analyst had prepared. These included an area cost sheet, documented flowcharts with documentation sections, and a list of the system requirements.

She also included a list of the system's inputs and outputs, all decision tables, listings of equipment and personnel, and a file sheet. All remaining reports and documents of any importance and notes prepared by the first analyst were placed in the appendix. Having included what she considered to be all pertinent documents, the new analyst continued the documentation with a description of the data processing aspects of the new system.

The data processing section began with a lengthy narrative on what function each computer program in the new system actually performed. In it she explained how one program gave an alphabetical listing of all patients in the hospital and that another would list all patients who have been in the hospital 90 days or more. The analyst continued by explaining the six remaining programs in the system in a similar manner.

She then closed this section with summaries of the implementation plans for the new computer-based system and how they were carried out.

The remainder of the report was comprised of an appendix that contained documents pertinent to the system but not already included in the main body of the documentation. In the appendix, the analyst also included summaries of the various phases of the study that had been written by the previous analyst. These included notes on general background information, interactions between areas under study, a description of the old system, notes on design of the new system, economic cost comparisons, and notes on the implementation, follow-up, and re-evaluation.

Having reviewed the analysts' work, the manager asked her to come into his office to discuss the documentation package. The following conversation took place:

"What you have done is fine," began the supervisor, "but where are the rest of the documents and reports your predecessor prepared before leaving?"

"All of the material that I consider pertinent is either in the body or appendix of the package," answered the analyst.

"But where are the feasibility study, the problem definition report, the outline of the study, and the final written report submitted by the previous analyst?," asked the supervisor.

"Since the system has been implemented and is running smoothly," replied the analyst, "I see no reason why these should be included. They served their purpose at one time but are not relevant now."

"Fine, but what about data processing?," asked the supervisor. "All you have done here is given an explanation of the programs. Did you consider whether someone might need some of the material you left out?"

"Yes," answered the analyst, "I did think about those things and that is why I chose to omit them. All we do here at the hospital is punch the data cards. We have nothing to do with the actual data processing. All that is done somewhere else, by specialists. Besides, if anyone wants that information, it will surely be available at the computer center."

"Yes," agreed the supervisor, "that information *might be* available if needed, but just like the rest of the material you left out, I want to see it included when you redo this job."

With that the analyst returned to her desk to correct the omissions in the documentation package.

QUESTIONS

1. Why should the systems manager have insisted upon the inclusion of such documents as the feasibility study and problem definition reports in the final documentation package?

2. What other documents were omitted from the final documentation package?

3. The analyst undoubtedly believed in her reasons for omitting a large portion of the material on the data processing work. How would you convince her to include more documentation in this area?

4. What additional material would you, as the supervisor, expect to find in the data processing section once the analyst has redone the documentation package?

FOOTNOTE

1. The key is whatever the data are filed by or accessed by, for example, part number, name, or whatever the programmer chooses.

THE TOOLS OF SYSTEMS ANALYSIS

Part Three of this book contains the basic tools that a systems analyst will use. These chapters can be used in any sequence with any of the previous chapters, as required.

CHAPTER 13

CHARTING

If a picture is worth a thousand words, a chart can be worth even more —for a chart can portray the logic of a situation, as well as the positioning of the situations' parts. For this reason, charts are one of the primary tools of the systems analyst.

WHAT IS CHARTING?

Charting is a graphical or pictorial means of presenting data. Charting takes the flow of work and makes a picture of it. Charts can be used to illustrate statistical data, locations of desks or equipment, relationships between people and jobs, sequences of events, work flow, organizational structure, and planning or implementation schedules.

The primary use of charting is for communication and documentation of the system. Charting also is used during feasibility studies, problem definition, understanding the existing system, defining new system requirements, design, cost comparisons, final report, and implementation.

GENERAL CATEGORIES OF CHARTS

Charts and graphs can be separated into four categories. The first is *activity charting*. In activity charting, the analyst is pictorially summarizing the flow of work through the various operations of a system. Flowcharts are the best example of activity charting. The second category is *layout charting*. Layout charting pictures the physical area under study. Layout charts show the locations of work areas and equipment before and after the new system design. A layout flowchart actually combines activity and layout charting because the work flow is shown as well as the physical locations of work areas and facilities.

The third category is *personal relationship charting*. Charts in this category depict lines of authority, job responsibilities, or job duties. Good examples are work distribution charts and organization charts. The fourth, and last, category is *statistical data charting*. These charts convert statistical data into meaningful statistical information by graphic portrayal. The analyst can be original in illustrating statistical data. Whatever format conveys the picture in a simple, easy-to-read, and factual manner is acceptable. Columnar tabulations or graphs are examples of statistical data charting. The acronym ALPS might help you remember Activity, Layout, Personal, and Statistical charts.

GENERAL CHARTING GUIDELINES

Charts and graphs can be used as a communication aid and visual device for the presentation of a new system to management. Charts also provide a means of analysis and evaluation by providing an overall picture of the current system or the proposed system. The area cost sheet from Chapter 6 is a good example of a cost picture of the current system. Charts show what goes on where, and allow comparisons of efficiency, timeliness, and costs between the old and new systems. Grouping data together with charts assists the analyst in spotting duplications, bottlenecks, redundant operations, and many other system peculiarities. Finally, charts can serve as aids in isolating problems caused by imbalanced work assignments.

Before developing a chart, the analyst should know what standard charts are appropriate to the application at hand. This will require a familiarity with flowcharts, Gantt charts, work distribution charts, organization charts, and others.

The analyst must first learn the basic kinds of charts which have developed over the years, and their usual applications. Then it becomes possible to devise creative variations in charting methods and applications. Following this general discussion we will present the specifics of the more useful charting methods.

The analyst should, depending on the type of chart, determine the flow of work, job duties involved, responsibilities, and the organization involved. These items can be ascertained during the phase understanding the exiting system and they may be modified during the phase designing the new system.

Try to picture the system as it really is, when charting the existing system. Remember that the formal organization, with its formal written job procedures, provides only one picture of the system. The informal organization, with its informal and unwritten job procedures, provides another view of the system. These formal and informal procedures may or may not be the same, and neither may be totally correct. The point to be made is that the system should be pictured as it actually is when the existing system is being portrayed.

Regardless of what the written procedures say, or what some opinions are about how things *should* be, find out and chart what the system is *actually* doing. The actual system idiosyncrasies usually can reveal the important reasons for having things as they are. The analyst can then study these reasons to facilitate a more realistic approach to redesigning the system.

SIX FLOWCHARTS

A flowchart is a graphic picture of the logical steps and sequence involved in a procedure or a program. Flowcharts help the analyst or programmer break down the problem into smaller workable segments and aid in the analysis of sequencing alternative paths in the operation. Flowcharts usually bring to light new areas of the problem that need further study and evaluation. Many laborsaving or timesaving ideas can come from a flowchart.

When developing flowcharts the analyst or programmer should observe the following guidelines.

1. Flowcharts are drawn from the top of a page to the bottom and from left to right.
2. The activity being flowcharted should be carefully defined and this definition made clear to the reader.
3. Where the activity starts and where it ends should be determined.
4. Each step of the activity should be described using "one-verb" descriptions, for example, "prepare statement" or "file customer statement."
5. Each step of the activity should be kept in its proper sequence.
6. The scope or range of the activity being flowcharted should be carefully observed. Any branches that leave the activity being charted should not be drawn on that flowchart. A connector symbol should be used and that branch put on a separate page, or omitted entirely if it does not pertain to the system.
7. Use the standard flowcharting symbols.

Systems Flowchart

A systems flowchart shows the overall work flow of the system. It is a pictorial description of the sequence of the combined procedures that make up the system. A systems flowchart shows *what is being done* in the system. All of the flowchart symbols are listed at the end of this section. A very simple systems flowchart is portrayed in Figure 13-1.

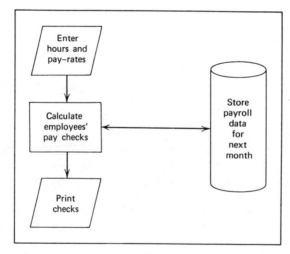

FIGURE 13-1 Systems flowchart.

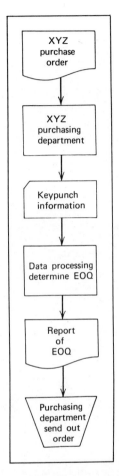

FIGURE 13-2 Systems flowchart.

Another example of a systems flowchart might be as follows: The XYZ Company has found that it can purchase a raw material at a cost of $40 per order. The company has a 10% carrying charge on average inventory. They expect to use $20,000 of the raw material within the next year. We want to determine the Economic Order Quantity (EOQ). The systems flowchart portrays only "Data processing determine EOQ" (Figure 13-2). The actual calculations appear in Figure 13-4 (program flowchart).

Program Flowchart

This flowchart is derived from a systems flowchart. A program flowchart is a detailed explanation of *how* each step of the program or procedure is actually performed. It shows *every* step of the program or procedure in the exact sequence in which they occur. Programmers use program flowcharts to show the step-by-step sequence of

279

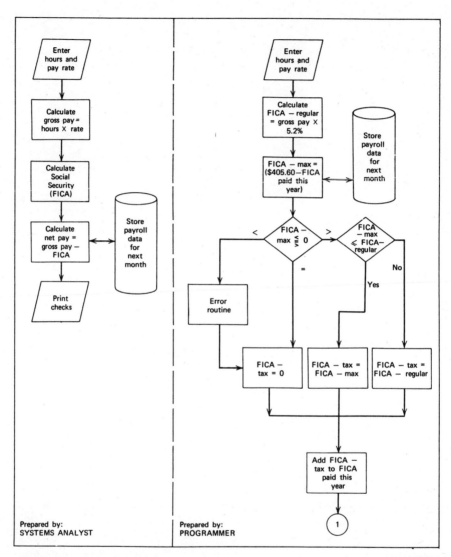

FIGURE 13-3 Program flowcharts.

instructions of the computer program. Systems analysts use program flowcharts to show the step-by-step sequence of job tasks within a procedure or operation. Two very simple program flowcharts are shown in Figure 13-3. The one on the left is a program flowchart prepared by a systems analyst to show details of procedural operations. The one on the right is a program flowchart prepared by a programmer to show details of computer program operations.

The reader should note that the systems flowchart in Figure 13-1 was further expanded by the systems analyst into a program flowchart as in Figure 13-3, left side. The programmer then expands the detail into the program flowchart as shown in Figure 13-3, right side.

Another example of a program flowchart is that in Figure 13-4, which shows the detailed, step-by-step procedure of the processing function of the systems flowchart from Figure 13-2. This is the data processing procedure the XYZ Company uses to determine the EOQ.

Layout Flowchart

A layout flowchart shows the floor plan of an area. It also may show the flow of paperwork or the flow of goods as well as the location of file cabinets or storage areas. Layout flowcharts are used effectively to show the before and after layouts of the area under study. The analyst draws the area as it is in the existing system and traces in the flow of work. For comparison, the analyst draws the area as it will be in the new proposed system and traces in the more efficient flow of work (Figure 13-5).

It is advisable to use scale model layouts. Draw the outside boundaries of the area under study in $\frac{1}{4}$-inch to 1 foot scale, and cut paper models (top view) of the equipment needed in the area, to the same scale. Draw in all permanent interior walls and other permanent features. Then play a game with the equipment cutouts by moving them about in order to get the most efficient layout.

Observe simple office common sense. The manager will not want the secretary 200 feet away. The manager's assistant and other supervisory personnel should be located in such a way that it is obvious that they are the supervisors. Those who need partitioning around their desk or a door to their office should get it, unless the economic situation prohibits it. It is usually a mistake to give one supervisor partitioning and a peer none.

Do not make the error of forgetting to provide electric wall sockets for those personnel who have electric calculators, and so forth. Remember to arrange for telephones to be repositioned. Each employee should keep the same extension number, if at all possible.

The layout must meet local fire safety and other emergency procedure requirements. City hall can usually provide the needed information.

Schematic Flowchart

A schematic flowchart is similar to a systems flowchart in that it represents an overview of a system or an overview of a procedure. Instead of using only the standard flowchart symbols, the analyst includes small pictures of the computer, peripherals, forms, or other equipment used in the system.

The principal use of schematic flowcharts is communication of the system to persons who are unfamiliar with conventional flowchart symbols. Using pictures

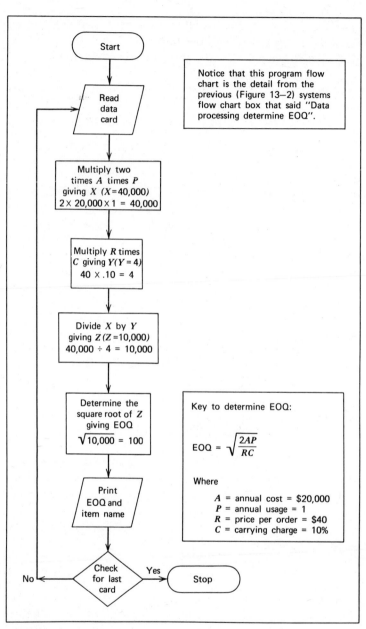

FIGURE 13-4 Program flowchart.

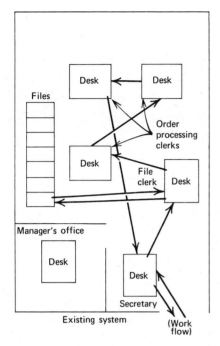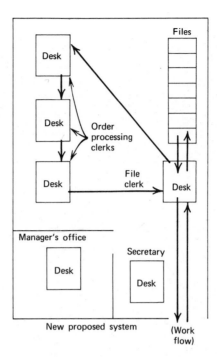

FIGURE 13-5   Layout flowchart.

instead of symbols saves the time it would require the person to spend learning the abstract symbols before being able to follow the chart. Pictures reduce the possibility of the viewer receiving an erroneous view of the system because of confusion over the symbols. The pictures make things easier for the viewer. This presents a more favorable opinion of your work, which is especially important when presenting the system for management approval.

Paperwork Flowchart

A paperwork flowchart traces the flow of written data through the system. Its primary use is to trace the flow of forms and reports through the system. It presents a clear picture of the form or report itself as well as where each copy ends its life, such as in a file cabinet or wastebasket, and so forth. In Figure 13-6, a form is traced through its life cycle.

Process Flowchart

A process flowchart is an old industrial engineering charting technique that breaks down and analyzes successive steps in a procedure or system. Process flowcharts use their own special five symbols (see Figure 13-7).

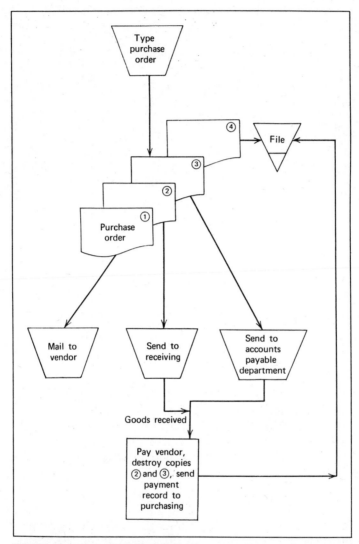

FIGURE 13-6 Paperwork flowchart.

A process flowchart lends itself to industrial engineering where the industrial engineer is studying and improving manufacturing processes. In systems analysis it can be used effectively to trace the flow of a report or a form. See Figure 13-8 for a filled-in process flowchart, then compare Figures 13-6 and 13-8.

The standard flowcharting symbols are shown in Figure 13-9. The shape of each recommended symbol appears on the left. On the right is its meaning with an example.

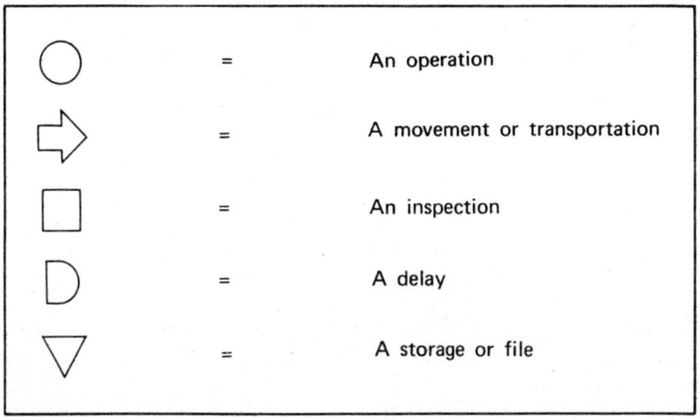

FIGURE 13-7 Process flowchart symbols.

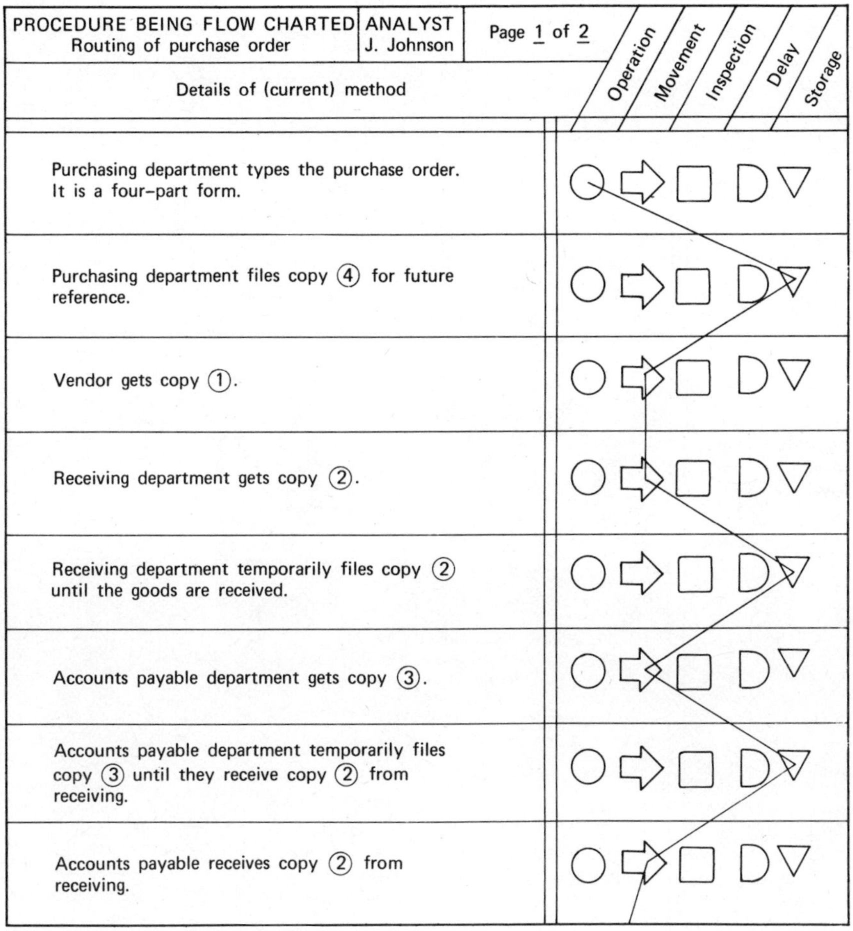

PROCEDURE BEING FLOW CHARTED Routing of purchase order	ANALYST J. Johnson	Page 1 of 2	Operation	Movement	Inspection	Delay	Storage
Details of (current) method							
Purchasing department types the purchase order. It is a four–part form.			○	⇨	□	D	▽
Purchasing department files copy ④ for future reference.			○	⇨	□	D	▽
Vendor gets copy ①.			○	⇨	□	D	▽
Receiving department gets copy ②.			○	⇨	□	D	▽
Receiving department temporarily files copy ② until the goods are received.			○	⇨	□	D	▽
Accounts payable department gets copy ③.			○	⇨	□	D	▽
Accounts payable department temporarily files copy ③ until they receive copy ② from receiving.			○	⇨	□	D	▽
Accounts payable receives copy ② from receiving.			○	⇨	□	D	▽

FIGURE 13-8 Process flowchart.

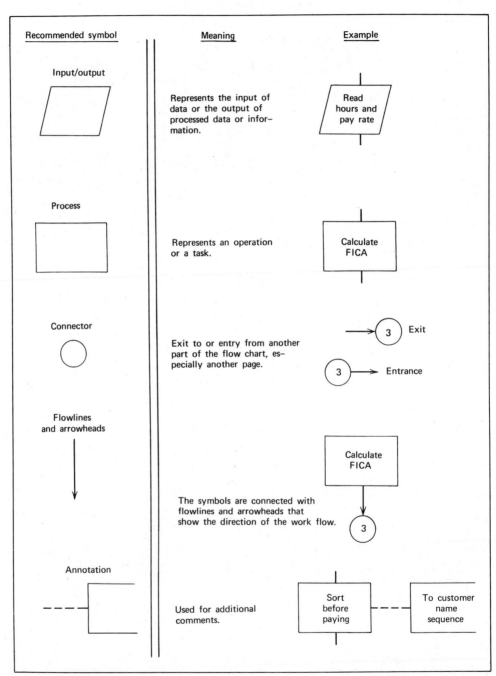

FIGURE 13-9   Standard flowchart symbols. These symbols conform to the International Organization for Standardization (ISO) draft recommendations on flowchart symbols for information processing.

Recommended symbol	Meaning	Example

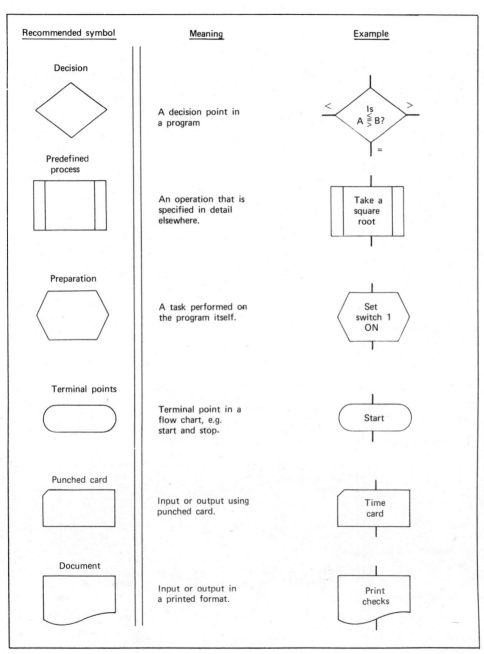

Decision — A decision point in a program

Predefined process — An operation that is specified in detail elsewhere.

Preparation — A task performed on the program itself.

Terminal points — Terminal point in a flow chart, e.g. start and stop.

Punched card — Input or output using punched card.

Document — Input or output in a printed format.

FIGURE 13-9 Continued.

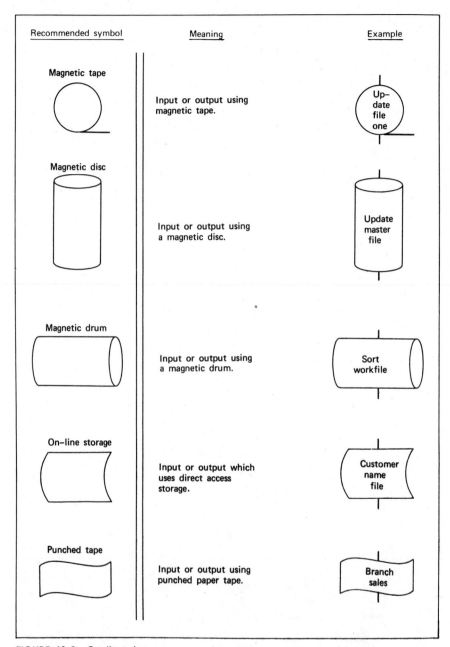

Recommended symbol	Meaning	Example
Magnetic tape	Input or output using magnetic tape.	Up–date file one
Magnetic disc	Input or output using a magnetic disc.	Update master file
Magnetic drum	Input or output using a magnetic drum.	Sort workfile
On–line storage	Input or output which uses direct access storage.	Customer name file
Punched tape	Input or output using punched paper tape.	Branch sales

FIGURE 13-9 Continued.

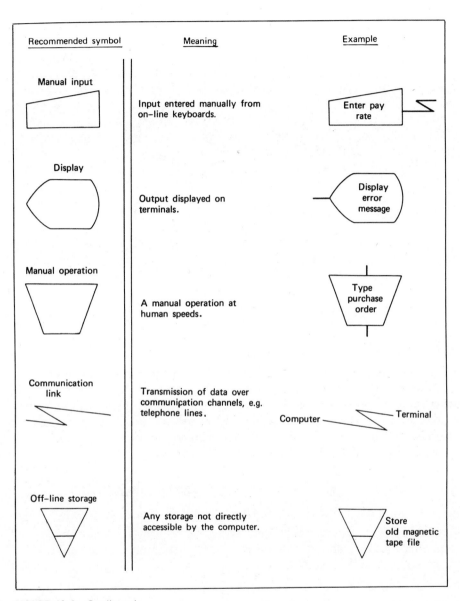

Recommended symbol	Meaning	Example
Manual input	Input entered manually from on-line keyboards.	Enter pay rate
Display	Output displayed on terminals.	Display error message
Manual operation	A manual operation at human speeds.	Type purchase order
Communication link	Transmission of data over communication channels, e.g. telephone lines.	Computer — Terminal
Off-line storage	Any storage not directly accessible by the computer.	Store old magnetic tape file

FIGURE 13-9 Continued.

GANTT CHART

Gantt charts are used as a scheduling tool. They portray output performance against a time requirement. As applied to systems work, the Gantt chart finds its greatest usefulness in planning the systems project and for scheduling its various phases. Figure 13-10 shows an example of a Gantt chart. The horizontal axis represents units of time—hours, days, weeks, months—whatever time units are appropriate. The vertical axis can have a listing of different projects, or one project with all of its phases listed. The S and C in Figure 13-10 stand for "scheduled" and "completed," respectively. For example, if a project or project phase was scheduled for May, the analyst would put X's in the S row under May 7, 14, 21, and 28. As the project is completed the analyst puts X's in the C row to show the state of completion. Figure 13-10 has the project "reprogram FICA computation" scheduled for May. It is one-half completed. At a glance the analyst can see the overall scheduled time of a project in weeks, when the project will start and end, and how much of the project is completed. It can be seen at a glance whether the project is behind or ahead of schedule by comparing today's date with the completed (C) portion of the Gantt chart.

HIPO DIAGRAMS

Hierarchy plus Input-Process-Output (HIPO) is a technique for documenting programming systems. HIPO was developed by IBM personnel who believed that if programming systems documentation was created that emphasized function, it could speed up program maintenance by facilitating the location in the code of a procedure to be modified. In addition, HIPO can satisfy the requirements of a variety of people who use documentation for many different purposes, such as

1. A manager can use HIPO documentation to obtain an overview of the system.
2. An application programmer can use HIPO documentation to determine program functions for coding purposes.
3. A maintenance programmer can use HIPO documentation to locate quickly the functions to which changes will be made.

HIPO documentation usually is prepared using the HIPO worksheet (see Figure 13-11) and the HIPO template (see Figure 13-12). Although many of the symbols are similar to standard flowchart symbols, there are some new conventions.

As a design and documentation technique, HIPO has three major objectives. The *first* objective is to provide a structure by which the functions of a system can be

PROJECT NAME	S or C	JAN				FEB				MAR				APRIL				MAY				JUNE				JULY				AUG				SEPT				OCT				NOV				DEC			
		7	14	21	28	7	14	21	28	7	14	21	28	7	14	21	28	7	14	21	28	7	14	21	28	7	14	21	28	7	14	21	28	7	14	21	28	7	14	21	28	7	14	21	28	7	14	21	28
Reprogram FICA computation	S																	X	X	X	X																												
	C																	X	X																														
	S																																																
	C																																																
	S																																																
	C																																																
	S																																																
	C																																																
	S																																																
	C																																																
	S																																																
	C																																																
	S																																																
	C																																																
	S																																																
	C																																																
	S																																																
	C																																																
	S																																																
	C																																																
	S																																																
	C																																																
	S																																																
	C																																																
	S																																																
	C																																																

FIGURE 13-10 Gantt chart.

GX20-1970-0 U/M 025 *
Printed in U.S.A.

Author: _____ System/Program: _____ Date: _____ Page: _____ of _____

Diagram ID: _____ Name: _____ Description: _____

Input

Process

Output

Extended Description

Notes		Ref.

Extended Description

Notes		Ref.

FIGURE 13-11 The HIPO worksheet.

* The number of sheets per pad may vary slightly

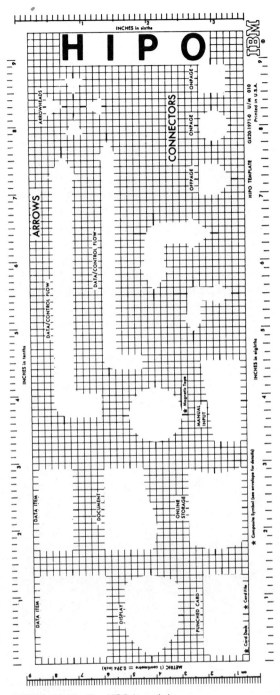

FIGURE 13-12 The HIPO template.

understood. The diagrams are organized in a hierarchy structure (see Figure 13-13), where each diagram at any level is a subset of the level above it. The *second* objective of HIPO is to state the functions to be accomplished by the program, rather than to specify the program statements to be used to perform the function. The *third* objective of HIPO is to provide a visual description of input to be used and output produced by each function for each level of diagram (see Figure 13-14).

A typical HIPO package contains three kinds of diagrams: a visual table of contents, an overview, and detail HIPO diagrams (see Figure 13-15). The objective of the visual table of contents is to show the major functions to be performed by the system and the relationships between each function (see Figure 13-13). In the visual table of contents, the top box identifies the overall function of the system. The next level of boxes break that function down into logical subfunctions. In the case of Figure 13-13, the subfunctions include calculation of gross pay and calculation of net pay. Although four to five levels in the top down structure are usually all that is necessary, as a general rule it is wise to continue to lower levels until both the designer and user completely understand the functions being described. Each box in the visual table of contents refers to a HIPO diagram, the description and identification number of which are shown in the box. For example, box #2.0 labeled "Calculate gross pay" is described further by HIPO diagram 2.0, as shown in Figure 13-14. The visual table of contents should include a legend to describe how the diagrams are to be read and what the various symbols mean.

The objective of the overview diagram (a high-level HIPO diagram) is to provide general information about a particular system. This is accomplished by describing the major functions within the system and referencing the detail diagrams necessary to expand each function to a detail level (see Figure 13-14). The overview diagram describes, in general terms, the inputs, processes, and outputs. The process section contains a series of numbered steps that describe the function being performed. In Figure 13-14, these steps describe the procedure for calculating gross pay. The input section identifies the data items used by the process section, and includes all major input items used in any lower-level diagrams. The input data items are connected to the process steps by arrows. The output section identifies the data items created or modified by the process steps, and includes all major output items shown in lower-level diagrams.

The detail HIPO diagrams contain the basic elements of the system, describe the specific functions, show specific input and output items, and may reference other HIPO diagrams as well as flowcharts or decision tables of complex logic. The detail diagrams contain an extended description section that is used to amplify the process steps and can reference the code associated with the process steps. The number of levels of detail HIPO diagrams that are required is determined by the number of functional subassemblies, the complexity of the processing, and the amount of infor-

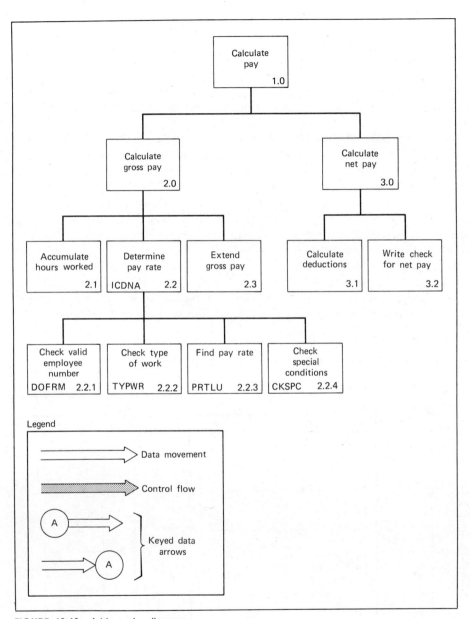

FIGURE 13-13   A hierarchy diagram.

mation to be documented. Figure 13-16 is a sample detail HIPO diagram with an extended description section. This figure is an extension of box 2.2 in the visual table of contents shown in Figure 13-13, and process step 2 in the overview HIPO diagram shown in Figure 13-14. The specific labels in the input section refer to record names (PAYMSTR) and data names (EMPNBR) to be used in the actual coding. The labels referred to in the extended description relate to program modules and segments of code.

There are three major kinds of HIPO documentation packages: the initial design, the detail design, and the maintenance package (optional). Each package contains all three kinds of diagrams, but each package has a distinct purpose, characteristics, and audience (see Figure 13-17). The initial design package is prepared by the design group at the beginning of a project. It describes the overall functional design of the project and is used for design reviews by management and user groups. The detail design package is prepared by the development group. The initial design package is used as a base to specify more detail and add more levels of detail HIPO diagrams. The resulting package then can be used for implementation and comparison with the initial design package to ensure that all requirements have been covered. During the coding process, the extended description area is expanded to include program labels and other pertinent information regarding implementation. This package is good reference material for

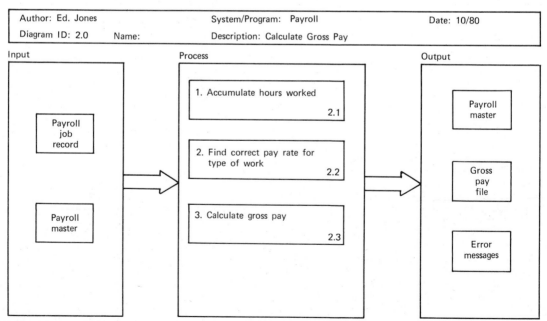

FIGURE 13-14 A HIPO diagram.

1. A Visual Table of Contents

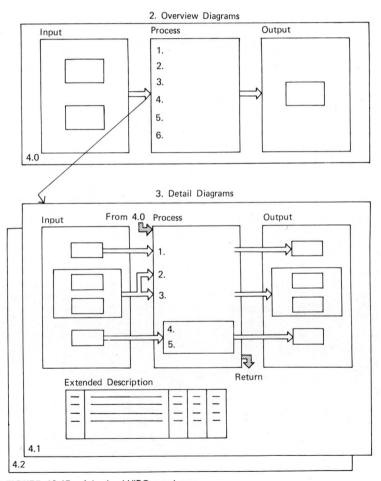

2. Overview Diagrams

3. Detail Diagrams

FIGURE 13-15   A typical HIPO package.

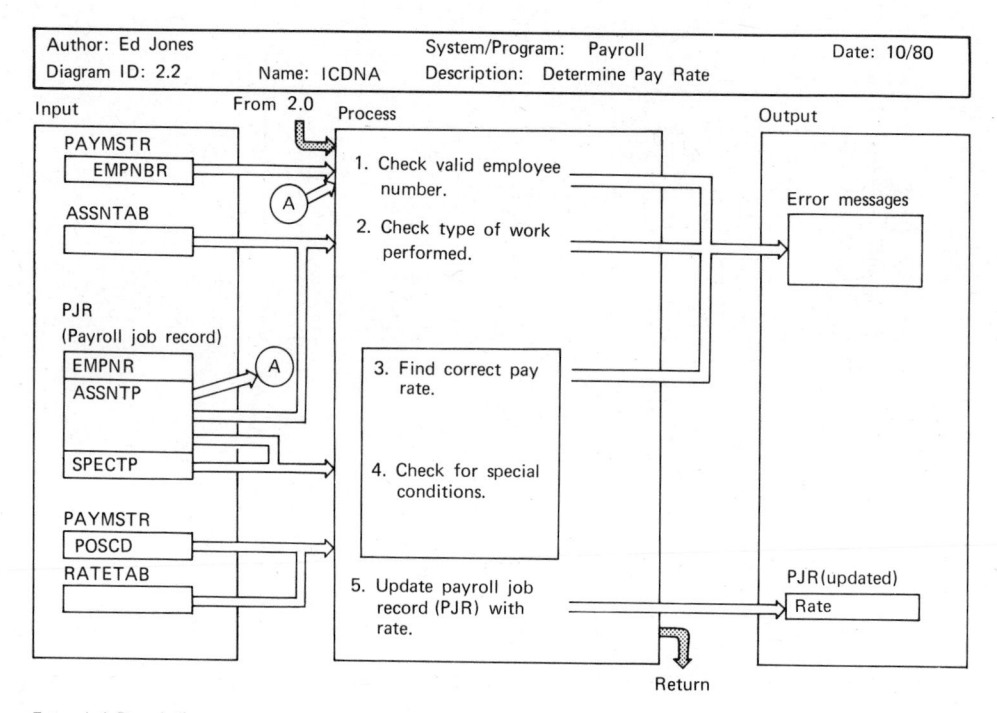

FIGURE 13-16   A sample detail diagram.

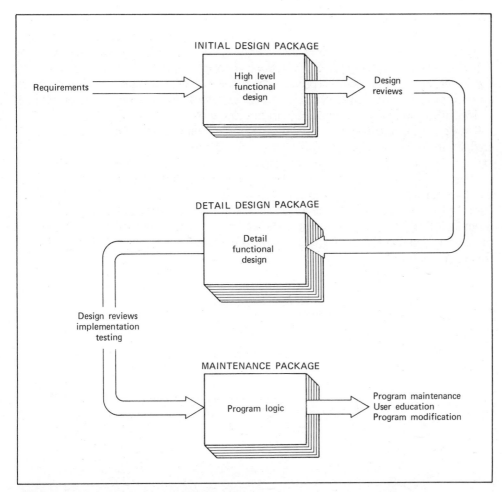

FIGURE 13-17   The three kinds of HIPO packages.

maintenance programmers, and also may be used to develop user instructions. If a maintenance package is assembled, it is used for corrections, changes, or additions to the system. It is basically the detail design package with some of the lower-level diagrams deleted.

Combining HIPO with the structured design process and structured programming can provide the analyst and system designer with powerful tools which result in a design and related programs of extreme modularity, both in function and logical structure. HIPO assumes that a system (a collection of related programs) will be organized into a hierarchical structure of functions. The structure of HIPO is well suited to a functional design made by starting at the top and subdividing into increasingly lower levels of detail such as the Structured Design Process discussed in Chapter 8. In this type of development, the functions are implemented in the same sequence as the structure. The top module contains the highest level of control logic and decisions for each program within the system, and either passes control to lower-level modules or identifies lower-level modules for in-line inclusion.

If structured programming is used during implementation, the functions are considered as single entities. The code is written in segments, each with a single entry and exit, and is created from HIPO programs. Figure 13-18 is the HIPO visual table of contents for a system that prints a Payroll Journal. Figures 13-19, 13-20, and 13-21 illustrate how detailed HIPO diagrams can be taken down to the COBOL coding level. Figure 13-19 illustrates the code for module 100. Figure 13-20 illustrates the code for module 200. Figure 13-21 illustrates the code for submodule 210. The technique of developing detailed HIPO diagrams down to the code level is applicable regardless of the language utilized.

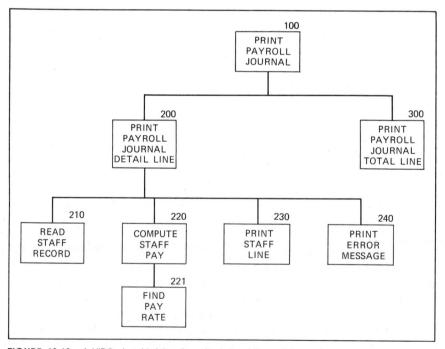

FIGURE 13-18   A HIPO visual table of contents for a Payroll Journal.

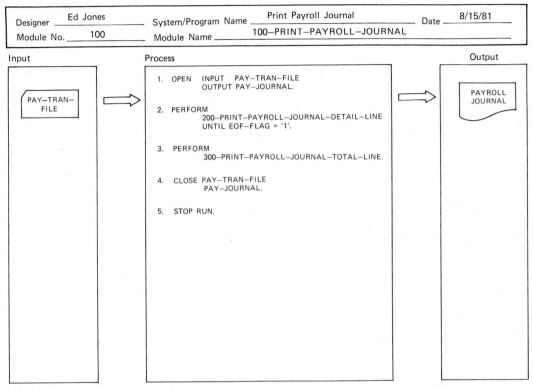

| Designer | Ed Jones | System/Program Name | Print Payroll Journal | Date | 8/15/81 |
| Module No. | 100 | Module Name | 100–PRINT–PAYROLL–JOURNAL | | |

Input

PAY–TRAN–FILE

Process

1. OPEN   INPUT   PAY–TRAN–FILE
            OUTPUT PAY–JOURNAL.

2. PERFORM
            200–PRINT–PAYROLL–JOURNAL–DETAIL–LINE
            UNTIL EOF–FLAG = '1'.

3. PERFORM
            300–PRINT–PAYROLL–JOURNAL–TOTAL–LINE.

4. CLOSE PAY–TRAN–FILE
            PAY–JOURNAL.

5. STOP RUN.

Output

PAYROLL JOURNAL

FIGURE 13-19   Code for module 100 from Figure 13-18.

WORK DISTRIBUTION CHART

A Work Distribution Chart provides an analysis of what jobs are being performed, who performs them, how the work is divided, and the approximate time required to perform each job. Work distribution charts are used often for analysis when morale problems exist because morale problems are caused frequently by uneven work distribution or some other form of imbalanced work load among the employees.

The work distribution chart for the area under study consists of a list of job duties down the left-hand column and identification of the personnel across the top of the chart. Figure 13-22 shows an example of a typical work distribution chart.

Within the appropriate box the analyst enters a short statement of what the person does in relation to the job duty. In the example in Figure 13-22, under JOE SMITH and to the right of CORRESPONDENCE the analyst entered "gives dictation" and "reviews priorities." The analyst also entered the number of hours *per week* spent on

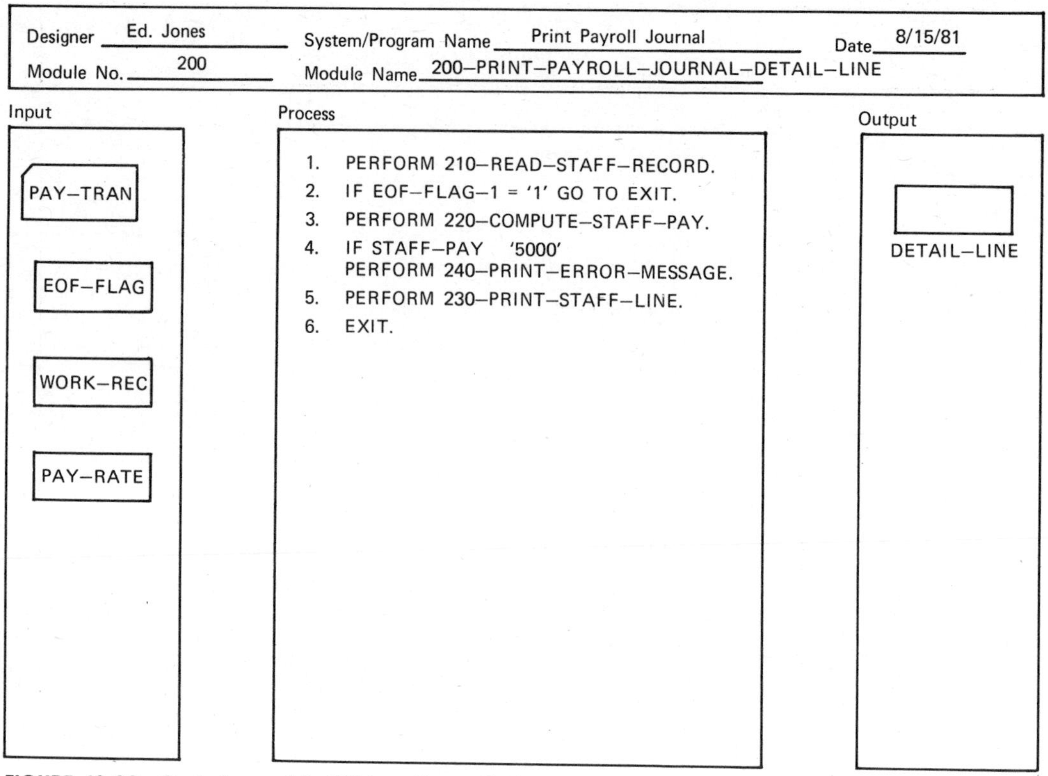

Designer	Ed. Jones	System/Program Name	Print Payroll Journal	Date	8/15/81
Module No.	200	Module Name	200–PRINT–PAYROLL–JOURNAL–DETAIL–LINE		

Input

PAY–TRAN

EOF–FLAG

WORK–REC

PAY–RATE

Process

1. PERFORM 210–READ–STAFF–RECORD.
2. IF EOF–FLAG–1 = '1' GO TO EXIT.
3. PERFORM 220–COMPUTE–STAFF–PAY.
4. IF STAFF–PAY   '5000'
    PERFORM 240–PRINT–ERROR–MESSAGE.
5. PERFORM 230–PRINT–STAFF–LINE.
6. EXIT.

Output

DETAIL–LINE

FIGURE 13-20   Code for module 200 from Figure 13-18.

those tasks. Other units of time can be used; but hours per day is too short a span to include all job tasks, and hours per month is usually too long a span for people to estimate accurately.

The analyst should sum the number of hours in each column to verify that the sum equals the number of hours in the work week for each employee. Summing the hours horizontally shows the total hours being spent per week on each job duty. The analyst should now proceed to study the chart, the work itself, and the employees to determine what changes should be made.

The study might reveal that far too much time is being spent on certain activities, or that certain employees are carrying more than their share of the load. These employees will welcome the opportunity to have their performance contrasted with the performance of others who may be doing less than their share. Such imbalanced workloads are often the source of severe employee discontent and frustration. The work distribution chart can often be used to objectively bring such problems to light.

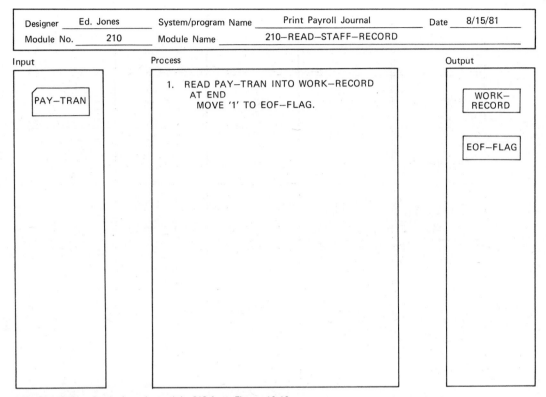

| Designer | Ed. Jones | System/program Name | Print Payroll Journal | Date | 8/15/81 |
| Module No. | 210 | Module Name | 210—READ—STAFF—RECORD | | |

Input

PAY—TRAN

Process

1. READ PAY—TRAN INTO WORK—RECORD
    AT END
        MOVE '1' TO EOF—FLAG.

Output

WORK—RECORD

EOF—FLAG

FIGURE 13-21   Code for submodule 210 from Figure 18-18.

Probably the chart's biggest limitation is the difficulty of using it where the work is highly creative or continuously changing. Even under such circumstances, however, the chart is useful for sorting out the routine tasks.

ORGANIZATION CHART

Organization charts (Figure 13-23) show the structure of the organization in terms of functional units or in terms of superior-subordinate relationships. Organization charts can be designed to portray

1. Levels of authority, from the top to bottom of the organization.
2. The important distinctions between line and staff personnel.
3. Divisional, departmental, or job functional interrelationships in the organization.

303

JOB DUTIES	JOE SMITH SUPERVISOR	CAROL JONES SECRETARY	CARL WARREN CLERK, SR.	AL JOHNSON CLERK
CORRESPONDENCE	Gives dictation, reviews priorities (10)	Typing, dictation (30)	Report writing (5)	
ORDER PROCESSING			Processes rush orders (10)	Processes new orders for tires (40)
SUPERVISION	Supervises the department (20)		Supervises ten clerks (2)	
INVOICE AUDITING			Checks the dollar totals (20)	
MISCELLANEOUS	Attends staff meetings (5)	Maintains file of orders (10)		

FIGURE 13-22 Work distribution chart.

4. The lines of formal communication channels.
5. The names and relative prestige of employees.

What the organization chart does not show is often of more interest to the systems analyst than what it does show because the chart shows only the formal organization. It does not show the *informal* organization that may have developed. Therefore it may not show the true relationships between people and between departments, and it may not show the true lines of communications. For example, in the chart shown in Figure 13-23 is Tom Taylor a relative of Phil Taylor? Think about the importance of knowing such relationships. Other things may be omitted by the organization chart. It may not show the actual degree of authority held by any one person (the person's ability to get the job done through others), and it may not show status and importance.

Nevertheless, organization charts are extremely important to the systems analyst. During the first interviews with management in the area under study the analyst should make sure that an up-to-date chart of the authority and functional relationships

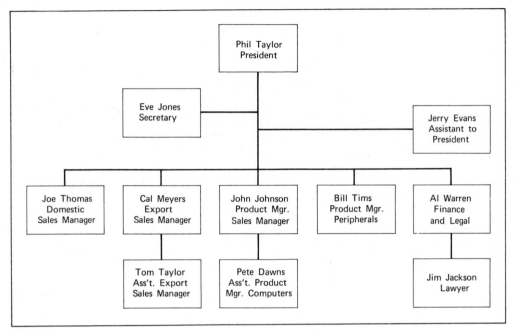

FIGURE 13-23   Organization chart.

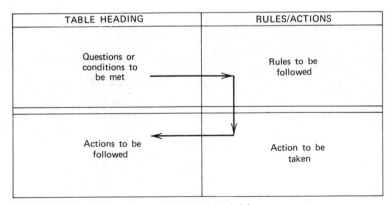

FIGURE 13-24   General format of a decision table.

of the area is obtained. Quite often the analyst will have to draw it as the manager explains the organization, but the manager should verify the accuracy of the final draft.

DECISION TABLES

A decision table is another method (besides program flowcharts) of describing the logic of a computer program. The table is divided into two main parts. The upper part contains the questions and rules, and the lower part contains the action to be taken.

In the example shown in Figure 13-24 the upper part of the decision table contains the questions and rules in the sequence in which they are to be considered in reaching a decision. The lower part (below double horizontal line) contains the prescribed action to be taken when a given condition is either met or not met. The *table heading* is the name of whatever the decision table is defining. The *questions* are the specific conditions that the analyst must choose between; it is for these questions that the decision table is constructed. The *rules* are the different situations such as YES, NO or $>$, $<$, $=$, $\geq$, $\leq$ or TRUE, FALSE or whatever rule fits the particular situation. The actions to be followed are the specific actions that the analyst will carry out depending upon the rules. The *action* to be taken shows which of the actions will be followed explicitly.

In reading a decision table the analyst first looks in the upper left corner "questions or conditions to be met." The analyst reads the first question and then reads the upper right section "rules to be followed." If the answer to the first question is yes (see Figure 13-25), then the analyst reads straight down that column to the "action to be taken" section. From there the analyst reads left to the "actions to be followed" section. In the example, the order would be approved. If the answer to the first question is no, then the analyst looks at the second question. If the answer to the second question is yes, then the analyst again reads down to the appropriate X in the "action to be taken" section,

ORDER ACCEPTANCE PROCEDURE	RULES/ACTIONS			
1. Is credit limit okay?	Yes	No		
2. Is pay experience favorable?		Yes	No	
3. Is D. & B. rating AAA?			Yes	No
Approve order	X	X	X	
Reject order				X

FIGURE 13-25   Decision table.

reads left and discovers that the order should be approved. Again, if the answer to the second question is no, then the analyst reads the third question. If the answer to the third question is yes, the order is approved; if no, the order is rejected. In summary, the analyst using a decision table like the one in Figure 13-25 would not reject an order unless a no answer was received to all three questions.

A decision table can be viewed as a method of describing the flow of data through a system. Decision tables are most often used for describing the logic of a computer program; but where numerous and complex conditions affect the logic flow in a system, decision tables can be used. Whenever there are situations involving numerous combinations, conditions, and actions, it is easy to overlook the occurrence of any one given combination. Decision tables provide a precise method of ensuring that every possible combination is included. Never allow a decision table to replace a systems flowchart, however, since a decision table should only be used to supplement systems flowcharts. One final warning, the analyst must remember that decision tables are complex and, therefore, a lower-level clerical employee may reject the use of decision tables. It is important to make sure that employees who must use decision tables can in fact understand them thoroughly. The reader who is unfamiliar with decision tables should work with some until their logic is well understood. Decision tables can be extremely useful to the systems analyst because they subtly force management into defining routine policies.

PERT/CPM

The acronym PERT stands for Program Evaluation Review Technique and CPM stands for Critical Path Method.

Throughout this section on PERT it should be noticed that it is always referred to as PERT rather than PERT/CPM. In fact, CPM is automatically included in PERT. The basic difference between PERT and CPM is that in PERT three time estimates are used while in CPM only one time estimate is used (this difference will become clear as you read further).

The first question to clarify is, what is PERT? PERT is a planning and control tool for defining and controlling the efforts necessary to accomplish project objectives on schedule. PERT is a unique method of graphically illustrating the interrelationships of events and activities required to bring a project to its conclusion. PERT was developed jointly by the United States Navy and the Lockheed Aircraft Corporation; CPM was developed by the DuPont Company.

PERT is a statistical technique. It is both diagnostic and prognostic, and it is used for quantifying knowledge about the uncertainties that are faced in completing all of the individual project activities that lead to the successful conclusion of a major systems project. It is a method of focusing management attention on

1. Danger signals that require remedial decision in order to prevent the materialization of problems.
2. Areas of effort where trade-offs in time, resources, or technical performance might increase the possibility of meeting the major schedule dates of a product.

PERT is a tool that aids the decision maker but does not make the decisions. It is a technique which the systems analyst can use to

1. Establish coordinated and definitive job activities at the lowest organizational responsibility level.
2. Determine relative importance of each activity.
3. Simulate real or proposed changes in the project and show the effect of these changes on the project.

The department manager or the project leader will want a method of organizing activities. PERT offers them a method of visually seeing what must be done. PERT will provide these management people with

1. An excellent medium for coordinating the various project tasks, particularly if the project has its various tasks separated geographically.
2. A definitive plan in which each analyst really understands each portion of the whole task, and the relative importance of each of the activities is easily determined.
3. A basis for determining the time to complete each task within the overall project as well as the total time required to complete the entire project.
4. Identification of the tasks that will delay the project if the individual tasks do not meet the planned schedule.
5. The means for rescheduling in order to reduce the total time required to meet the project's objectives.
6. The criteria for measuring the project's progress.

PERT is a management tool for defining and integrating what must be done to accomplish the project's objectives on time. It provides a precise method of planning for the development of a new product, the installation of a new computer, the planning of a major systems study, and many other projects.

PERT can be carried out either by using hand calculation methods or by using a computerized PERT program. When the analyst is using a computerized program, the input format and the specialized output format of the PERT program for whichever computer the company has, must be learned first. In this book computerized PERT programs will not be discussed because they are developed specifically for a certain computer by the computer manufacturer's software programmers.

In this book we will discuss PERT hand calculations. PERT hand calculations are often advantageous for small networks of approximately 150 events or less. Even when a computer facility is available, usable results may often be obtained by hand in a shorter overall elapsed time and at less cost than is possible by the use of the computer. The point is, a systems analyst who wants to use PERT for a relatively small network may be able to design the network and perform the hand calculations required in order to determine the critical path in less time than it would take to use the computerized PERT program. The hand calculation is particularly valuable in small companies or in operations that are performed in the field.

The first definition that the reader must know is *network*. A PERT network is the foundation of the PERT procedure. Figure 13-26 shows a PERT network.

The network consists of events and activities. All of the required events are connected by arrows that indicate the preceding and succeeding events. An *event* (A, B, C, D, or E) is the beginning or ending of an activity. An event is a decision or the accomplishment of a task (activity). An event is looked at as a milestone. Events have no time dimension and are usually represented by a circle or a box. An *activity* links two successive events together and represents the work required between these two events. An activity must be accomplished before the following event can occur. Activities are presented by an arrow.

This example of a PERT network contains five events (lettered A through E) and six activities (represented by the arrows). In summary, each event represents the ac-

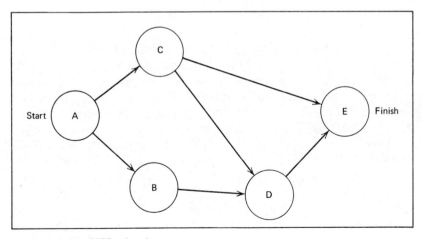

FIGURE 13-26   PERT network.

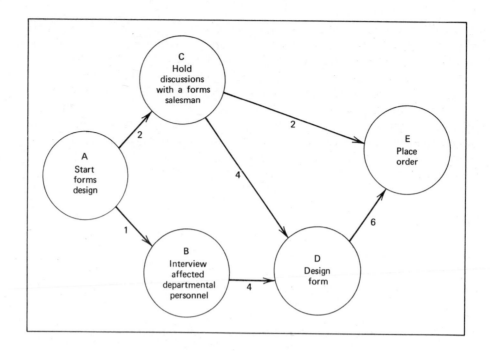

complishment of a task and each activity represents the time it took to accomplish that task. The above example expands upon the previous PERT network.

This PERT network represents an overly simplified forms design PERT chart. This PERT chart begins at *start forms design*. The analyst then branches off in two directions. One activity requires 2 weeks while the analyst holds discussions with a forms salesperson (A to C). The other activity requires 1 week while the analyst interviews affected department personnel (A to B). Branching off in both these directions implies that both of the events can take place simultaneously. For example, the analyst can hold discussions with a forms salesperson, and interview the affected department personnel. The next event to look at is D, design form. This event can take place only after the analyst has held discussions with a forms salesperson and interviewed affected department personnel. The point is, if you are drawing a PERT network and one event follows another, it implies that the preceding event must be totally completed before the succeeding event can begin. In the above example, one path takes 1 week to complete the interviews with the affected department personnel, 4 weeks to carry out the forms design, and 6 weeks to get the necessary approvals to place the order (A to B to D to E). Another path shows that it takes 2 weeks to complete the interview with the forms salesperson, 4 weeks to carry out the forms design, and again 6 weeks to get the necessary approvals to place the order (A to C to D to E).

The only activity not mentioned so far on the above PERT chart is the activity that leads from hold discussion with a forms salesperson to place order (C to E). Let us assume that after holding the discussions with the forms salesperson it is discovered that the forms company already has a preprinted form that might be suitable for use within the company. This activity shows that if you used the preprinted form that was already available, it would only take 2 weeks to get the necessary approval; and then the order could be placed. Therefore if the analyst designed the form and got the necessary approvals, it would be a 6-week activity (D to E); but if the analyst just got the approval for the preprinted form, it would be a 2-week activity (C to E). Therefore, another path is A to C to E.

$$\text{expected activity time } (t_e) = \frac{\text{OT} + 4(\text{MLT}) + \text{PT}}{6}$$

The *expected activity time formula* is used to get the expected times that were recorded below the activity arrows in the previous example. These expected times are collected by the analyst from the lowest level of supervision responsible for the completion of each activity of the project. This lowest level of supervision could be another systems analyst. The planning is done by those persons most familiar with the details of that operation of the project. This is a most important portion of the PERT method of planning and collecting times because it establishes a true graphic picture of each event of the project. The most common unit of time used is weeks, but any other unit of time may be used that fits the project.

- *Expected activity time* (t_e). The time *in weeks* calculated for an activity from the three time estimates given.
- *Optimistic time* (OT). Time estimate for an activity assuming everything goes better than expected, that is, no unforeseen problems arise.
- *Most likely time* (MLT). The time which the responsible manager or analyst thinks will be required for the job. (This is the *only* estimate in CPM.)
- *Pessimistic time* (PT). The time required if many adverse conditions are encountered, not including acts of God, strikes, power failures, and so forth.

These three estimates are obtained from the people closest to and responsible for the activity in question. The analyst inserts the optimistic time, the most likely time, and the pessimistic time into the expected activity time (t_e) formula to arrive at the expected activity time (t_e). For example, if OT = 2 and MLT = 3 and PT = 6, then t_e = 3.33 weeks.

$$t_e = \frac{2 + 4(3) + 6}{6} = \frac{20}{6} = 3.33 \text{ weeks}$$

In hand-calculated networks, time usually is written on the activity arrows as shown below.

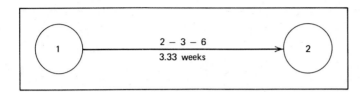

In summary, *events* are shown on the network as meaningful specific accomplishments that do not consume time or resources, but can be identified with a particular point in time. They may be such things as *program go ahead, start forms design, place order, start fabrication,* and so forth.

Activities are the time-consuming elements of the network. The activities are shown by an arrow that connects two events. The arrow indicates the beginning and the ending of the activity. The length of the arrow is unimportant.

The *expected activity time* (t_e) is the calculated time of completion for an activity that links two events. As was previously stated, PERT hand calculations are very advantageous for networks of 100 to 150 events or less. Computers should be used for larger networks. The whole objective of the PERT network is to determine the critical path for the entire project. *The critical path is the longest path through the network.* It is obtained by summing all of the activities for all possible paths. It is the shortest time to complete the entire project even though it is the longest path through the network. For example, the critical path in the previous example which started out with forms design and ended with placing the order was 12 weeks unless the management of the operation decided to use a preprinted form.

The following is a list of the various terms that will be used as we develop a larger PERT network for hand calculation.

- *Activity.* The effort required to accomplish an event measured in time. Represented by an arrow.
- *Critical path.* The longest path through the network. It is found by summing all activities for all possible paths. It is the limiting factor because the longest path, in weeks, is the earliest time that the entire project can be completed.
- *Event.* A point in time where something has been accomplished. Represented by a circle or a box.
- *Expected activity time* (t_e). The time in weeks calculated for an activity from the three time estimates given.

$$t_e = \frac{\text{optimistic} + 4(\text{most likely}) + \text{pessimistic}}{6}$$

- *Expected event time* (T_E). The sum of all expected activity times (t_e) along the longest path leading to an event.

$$T_E = \Sigma\, t_e \text{ (See Footnote 1)}$$

- *Latest allowable time* (T_L). The latest time that an event can be accomplished without affecting the date scheduled for completion of the entire project (last event).

$$T_L = T_S - T_E$$

- *Scheduled time* (T_S). The time in weeks from the starting event to the planned-for or contractually-obligated completion date (last event).

- *Slack* (S). The difference between the latest allowable time (T_L) and the earliest expected event time (T_E) for an activity to be completed. Slack may be positive (spare time) or negative (predicted slippage) or zero (project completion on schedule).

$$S = T_L - T_E$$

In order to start the example, a worksheet with seven columns is needed (see Figure 13-27). The column headings follow the worksheet.

(1)	(2)	(3)	(4)	(5)	(6)	(7)
Succ.	Pred.	t_e	T_E	T_L	T_S	S
8	$\sqrt{6}$	3.0	~~16.8~~			
	$\sqrt{7}$	1.8	17.0	18.0	18.0	1.0*
6	$\sqrt{5}$	2.2	~~9.7~~			
	4	4.3	13.8	15.0		1.2
7	$\sqrt{4}$	5.7	15.2	16.2		1.0*
	$\sqrt{3}$	6.5	~~7.7~~			
5	2	3.0	7.5	12.8		5.3
4	$\sqrt{2}$	5.0	9.5	10.5		1.0*
3	1	1.2	1.2	9.7		8.5
2	1	4.5	4.5	5.5		1.0*

Column	Abbreviation	Definition
1	Succ.	Successor event number
2	Pred.	Predecessor event number
3	t_e	Expected activity time
4	T_E	Expected event time
5	T_L	Latest allowable time for an event
6	T_S	Scheduled time for an event
7	S	Slack time

FIGURE 13-27 Worksheet.

The step-by-step method of determining the critical path and the slack by hand calculation is as follows.

1. Calculate the expected activity times (t_e) for *all* activities from the formula

$$t_e = \frac{OT + 4(MLT) + PT}{6}$$

For example, the activity between events 1 and 2 in Figure 13-28 is

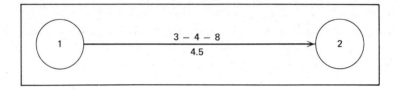

$$t_e = \frac{3 + 4(4) + 8}{6} = \frac{27}{6} = 4.5$$

2. Start with the last event (event 8 in Figure 13-28) and record the event number on the top line of the Successor column (column 1) of the worksheet (Figure 13-27).
3. Record all immediate predecessor events (events 6 and 7) in the Predecessor column (column 2).
4. Record the expected activity time (t_e) for each activity defined by each set of predecessor-to-successor event numbers. For example, record 6 to 8 and 7 to 8 (t_e) in column 3.

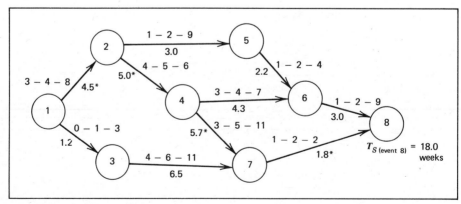

FIGURE 13-28 PERT network.

5. Record all "last events" if there are more than one. Only in very sophisticated PERT networks will there be more than one last event.

6. When all last events have been recorded, go back to the first event number, listed in the Predecessor event column, and determine the number of succeeding activities from that event. (In this case, events 6 and 7 have only one succeeding activity.)

7. If there is *only one succeeding activity*, record that predecessor event number next in the Successor event column and proceed to record its predecessor events as in step 3 above.

8. If an event has *two or more succeeding activities*, do not record it in the Successor column until it appears in the Predecessor event column as many times as there are succeeding activities.

9. In order to avoid errors, place a check (√) next to each event number in the Predecessor column as you record that event number in the Successor column.

10. Record each t_e value as in step 4. The Successor event column (column 1) should have each event recorded once except for the starting event (event 1) which should not be recorded in column 1. The first three columns of Figure 13-27 are now completely filled-in.

Calculation of the T_E (Column 4)

The computation of T_E is the sum of the time from event 1 to the predecessor event "in question" plus the activity time from the predecessor event "in question" to its successor event. For example,

$$T_E \text{ (successor)} = T_E \text{ (predecessor)} + t_e \text{ (activity)}$$

1. Start from the bottom of column 1 in Figure 13-27. The T_E for event 2 is the T_E for event 1 (0.0) plus the t_e of activity 1—2 (4.5).

$T_E \text{ (event 2)} = T_E \text{ (event 1)} + t_e \text{ (activity 1—2)}$
$T_E \text{ (event 2)} = 0.0 + 4.5$
$T_E = 4.5$

2. $T_E \text{ (event 3)} = T_E \text{ (event 1)} + t_e \text{ (activity 1—3)}$
$T_E = 0.0 + 1.2$
$T_E = 1.2$

3. $T_E \text{ (event 4)} = T_E \text{ (event 2)} + t_e \text{ (activity 2—4)}$
$T_E = 4.5 + 5.0$
$T_E = 9.5$

4. T_E (event 5) = T_E (event 2) + t_e (activity 2—5)

$$T_E = 4.5 + 3.0$$
$$T_E = 7.5$$

5. T_E (event 7) = T_E (event 3) + t_e (activity 3—7)

$$T_E = 1.2 + 6.5$$
$$T_E = 7.7$$

6. T_E (event 7) = T_E (event 4) + t_e (activity 4—7)

$$T_E = 9.5 + 5.7$$
$$T_E = 15.2$$

Note: Event times are determined by the *longest* elapsed times; therefore the T_E (event 7) = 15.2 is the limiting time. Cross out and disregard the T_E (event 7) = 7.7. Use only the 15.2 for T_E of event 7 in any future calculations.

7. Calculate the rest of the T_E values.

At this point, a preliminary evaluation can be made of the project that is being scheduled using a PERT network. The expected time to reach each of the objectives (events) is available in the T_E figure. This time can be compared with any scheduled requirements in order to determine the compatibility of this PERT network plan with the need to meet a scheduled date.

The limiting or critical path is the longest path between the last event and the starting event (event 8 and event 1). Working backward through the network from event 8 back to event 1, one can determine the critical path by summing the figures in the T_E column (column 4). For example, start with event 8 in the Successor column (column 1). Go to the Predecessor column (column 2) and note that 8—6 with a T_E of 16.8 has been crossed out; therefore, 8—7 with a T_E of 17.0 is the proper value to use. Now that you have found that 8—7 is the first leg, go to the Successor event column again and find event 7. The proper leg backward from event 7 is 7—4 because its T_E is 15.2 (you already crossed out 7—3 with a T_E of 7.7). Now go to the Successor column again and find event 4. Event 4 has a predecessor event of event 2. The T_E for activity 4—2 is 9.5. Now go to the Successor column again and find event 2. The T_E for activity 2—1 is 4.5. The point is, the critical path for the PERT network in Figure 13-28 is 17.0 weeks long. The critical path itself is event 8 to event 7 to event 4 to event 2 to event 1.

Calculation of the T_L (Column 5)

Start at the top of column 1 in Figure 13-27. A given scheduled commitment date for the completion of the last event (project completion) determines the latest allowable time (T_L) for the last event. Therefore, $T_L = T_S$ for the last event (event 8). If there is no scheduled completion time given (T_S), then let $T_L = T_E$ for the last event. In our example $T_L = T_S = 18.0$ for event 8; but if there were no T_S given, then $T_L = T_E = 17.0$ for event 8. We have assumed that $T_S = 18.0$ was given as a scheduled completion date.

1. The T_L for event 8 is $T_S = 18.0$.

2. T_L (predecessor) $= T_L$ (successor) $- t_e$ (activity)

3. T_L (event 6) $= T_L$ (event 8) $- t_e$ (activity 6—8)
    $T_L = 18.0 - 3.0$
    $T_L = 15.0$ (record this time under T_L opposite event 6 of column 1)

4. T_L (event 7) $= T_L$ (event 8) $- t_e$ (activity 7—8)
    $T_L = 18.0 - 1.8$
    $T_L = 16.2$ (record this time under T_L opposite event 7 of column 1)

5. T_L (event 5) $= T_L$ (event 6) $- t_e$ (activity 5—6)
    $T_L = 15.0 - 2.2$
    $T_L = 12.8$ (record this time under T_L opposite event 5 of column 1)

6. When an event has more than one succeeding activity there will be several T_L values. The *lowest* value should be recorded for that event. For example

   T_L (event 4) $= T_L$ (event 6) $- t_e$ (activity 4—6)
       $T_L = 15.0 - 4.3$
       $T_L = 10.7$

   T_L (event 4) $= T_L$ (event 7) $- t_e$ (activity 4—7)
       $T_L = 16.2 - 5.7$
       $T_L = 10.5$ (record the *lowest* time under T_L opposite event 4 of column 1)

7. Calculate the rest of the T_L values. It is not necessary to calculate values for the starting event (event 1).

Calculation of S (Column 7)

Slack is the latest allowable time (T_L) minus the earliest expected event time (T_E).

$$S = T_L - T_E$$

1. S (event 8) $= T_L$ (event 8) $- T_E$ (event 8)
    $S = 18.0 - 17.0$
    $S = 1.0$

2. S (event 6) = T_L (event 6) − T_E (event 6)

$$S = 15.0 - 13.8$$
$$S = 1.2$$

3. Calculate the rest of the slacks (S).

Earlier, the T_E column was used to find the critical path. The events with the least slack lie on the critical path. Also note that the slacks are all equal for the critical path. The critical path in this example (Figure 13-28) is marked with asterisks, that is, 8—7—4—2—1.

In summary, the previous example of a PERT network has shown us many things. First, it has forced the analyst to define all of the individual events that must take place from the start of a project to its completion (last event). Second, it has forced the analyst to obtain time estimates for the completion of each activity. Third, it has shown the analyst the critical path, that is, the longest path through the network. This longest path through the network is the shortest time to completion for the entire project because the project will not be complete until all of the events and activities are complete. Fourth, the analyst can now see which paths through the network have slack. The paths that have positive slack show the analyst the activities that will be done in plenty of time. In other words, those activities will not hold up the entire project. Those paths with negative slack show the analyst which activities will need close scrutiny and maybe a little extra effort in order to bring them back to zero slack so the project will not fall behind schedule. Those paths with zero slack show the analyst the activities that are right on schedule and that must be watched so they do not fall behind at all. If they do fall behind, the project will be delayed.

PERT/CPM is a very valuable tool to the systems analyst and to the manager of the systems analysts because it can be used to plan individual projects or to plan the workload of an entire department. This simple hand-calculated method can be used quickly and easily by an analyst in order to plan a complicated systems study.

In closing this chapter, it must be mentioned that there are numerous other types of charts and graphs available for use by the systems analyst. The charts presented in this chapter are the ones used most often. The analyst also should become familiar with the following.

1. *Linear responsibility charts* relate the degree of responsibility of key individuals to their various job duties.

2. *Right- and left-hand charts* and *Simo charts* are used for motion studies and time studies by industrial engineers.

3. *Operations analysis charts* are used to analyze manufacturing operations.

4. *Various graphical techniques* show the relationships between fixed costs, variable costs, and break-even points.

SELECTIONS FOR FURTHER STUDY

In 1921, when Frank and Lillian Gilbreth wrote their now-classic paper *Process Charts*, they were considered something of an oddity. Today, charting is not only accepted, but you will find both books and chapters in books wholly devoted to the subject. Two works dealing with various types of charts are Maynard's *Industrial Engineering Handbook* and Lazzaro's *Systems and Procedures*. Another volume containing a large number of examples is Willoughby and Senn's *Business Systems*. The Maynard book has an entire chapter devoted to Gantt charts.

Bohl's *Flowcharting Techniques* is designed for people who are just learning charting but who have some knowledge of FORTRAN, BASIC, COBOL, or PL/1. Schriber's *Fundamentals of Flowcharting* discusses formulation of problem-solving procedures. He feels many students learning to program have difficulty distinguishing between the procedure a computer program expresses and the grammatical detail of the language used to write the program.

There are numerous books dealing with PERT and/or CPM. Among these are the works of Horowitz, Moder, and Faulkner. The basic concepts of decision tables are set forth in IBM manual F20-8102-0. A minimum set of conventions for their use in systems analysis, procedure design, and documentation are described. You may wish to refer also to McDaniel's *An Introduction to Decision Logic Tables*, Montalbano's *Decision Tables, Program Structure, Data Structure*, or Gildersleeve's *Decision Tables and Their Practical Applications*. The concept of HIPO is discussed by Katzan.

1. BAKER, KENNETH R. *Introduction to Sequencing and Scheduling.* New York: John Wiley and Sons, Inc., 1974.

2. BOHL, MARILYN. *Flowcharting Techniques.* Chicago: Science Research Associates, 1971.

3. BYCER, BERNARD B. *Flowcharting: Programming, Software Design and Computer Problem Solving.* New York: John Wiley and Sons, Inc., 1975.

4. CHAPIN, NED. *Flowcharts.* New York: Van Nostrand Reinhold Company, 1971.

5. COOMBS, A.J., and R.F. MADGIC. *Variable Modular Scheduling.* Encino, Calif.: Glencoe Publishing Company, 1972.

6. FAULKNER, EDWARD E. *Project Management with CPM, 3rd edition.* Duxbury, Mass.: Robert Snow Means Company, Inc., 1973.

7. GILBRETH, FRANK B., and LILLIAN M. *Process Charts.* New York: American Society of Mechanical Engineers, 1921.

8. GILDERSLEEVE, THOMAS R. *Decision Tables and Their Practical Application in Data Processing.* Englewood Cliffs, N.J.: Prentice-Hall, Inc., 1970.

9. HARPOOL, JACK D. *Business Data Systems: A Practical Guide.* Dubuque, Iowa: William C. Brown Company, 1978.

10. HECHT, MATTHEW S. *Flow Analysis of Computer Programs*. New York: Elsevier-North Holland Publishing Company, 1977.

11. HOROWITZ, JOSEPH. *Critical Path Scheduling: Management Control Through CPM and PERT*. New York: Ronald Press Company, 1967.

12. *IBM Data Processing Techniques: Decision Tables; a Systems Analysis and Documentation Technique*. White Plains, New York: International Business Machines Corporation, 1962 (F20-8102-0).

13. JENSEN, RANDALL W., and CHARLES C. TONIES. *Software Engineering*. Englewood Cliffs, N.J.: Prentice-Hall, Inc., 1979.

14. KATZAN, HARRY. *Systems Design and Documentation: An Introduction to HIPO*. New York: Van Nostrand Reinhold Company, 1971.

15. LAZZARO, VICTOR. *Systems and Procedures: A Handbook for Business and Industry, 2nd edition*. Englewood Cliffs, N.J.: Prentice-Hall, Inc., 1968.

16. MAYNARD, HAROLD B. *Industrial Engineering Handbook, 3rd edition*. New York: McGraw-Hill Book Company, 1971.

17. MCDANIEL, HERMAN. *An Introduction to Decision Logic Tables, revised edition*. Princeton, N.J.: PBI Petrocelli Books, Inc., 1978.

18. MODER, JOSEPH J., and CECIL R. PHILLIPS. *Project Management with CPM and PERT, 2nd edition*. New York: Van Nostrand Reinhold Company, 1970.

19. MONTALBANO, MIKE. *Decision Tables, Program Structure, Data Structure*. Chicago: Science Research Associates, 1974.

20. PASSEN, BARRY J. *Programming Flowcharting for the Business Data Processing*. New York: John Wiley and Sons, Inc., 1978.

21. POLLACK, SOLOMON L. *Decision Tables: Theory and Practice*. New York: John Wiley and Sons, Inc., 1971.

22. SANDERS, DONALD H. *Computers in Business*. New York: McGraw-Hill Book Company, 1975.

23. SCHRIBER, THOMAS J. *Fundamentals of Flowcharting*. New York: John Wiley and Sons, Inc., 1969.

24. STERN, NANCY. *Flowcharting: A Tool for Understanding Computer Logic*. New York: John Wiley and Sons, Inc., 1975.

25. WIEST, JEROME D., and FERDINAND K. LEVY. *A Management Guide to PERT-CPM: With GERT-PDM, DCPM and Other Networks, 2nd edition*. Englewood Cliffs, N.J.: Prentice-Hall, Inc., 1977.

26. WILLOUGHBY, THEODORE C., and JAMES A. SENN. *Business Systems*. Cleveland, Ohio: Association for Systems Management, 1975.

QUESTIONS

1. Name four things charts can be used to illustrate.
2. Define activity charting.
3. Define layout charting.
4. Define personal relationship charting.
5. What is a flowchart?
6. Name five types of flowcharts.
7. As applied to systems work, where does the Gantt chart find its greatest usefulness?
8. What is the biggest limitation of the work distribution chart?
9. What does the organization chart often not show?
10. What does each of the two parts of a decision table contain?
11. For what do PERT and CPM stand?
12. What does the inducement approach include?
13. For what does HIPO stand?
14. How can HIPO satisfy the requirements of the maintenance programmer?
15. What are HIPO's three major objectives?
16. What type of diagrams are contained in a typical HIPO package?
17. Name the three kinds of HIPO documentation packages.

SITUATION CASES

CASE 13-1

Robert Swift, an analyst for a local firm, has progressed satisfactorily in his current systems study. He has defined the problem, adequately outlined the study, gathered sufficient general information, and examined the interactions of the inputs, outputs, and resources of the area under study. He now is considering the various tools that could be used to investigate the existing system.

One of the first tools Mr. Swift thought of was charting. He had very limited experience in working with charts, but had seen some and was impressed by the way other analysts had used them in their systems studies. Since he was only vaguely

JOB DUTIES	ALAN YOUNG SUPERVISOR	MABEL BROWN SECRETARY	BOB BROMAN CLERK, SENIOR	RON WHITE CLERK
Correspon-dence	Gives dictation (10)	Typing and dictation (30)		
Order Processing			Processes rush orders (10)	Processes new orders for mats (35)
Supervision	Supervises the department (20)		Supervises ten clerks (2)	
Invoice Auditing			Checks the dollar totals (25)	
Miscellaneous	Attends staff meetings (6)	Maintains file of orders (5)		
Nonjob Related Activities	Lavatory breaks, coffee breaks, etc (3)	Same (5)	Same (3)	Same (5)

FIGURE 13-29   Work distribution chart for Department C.

familiar with charting, he assumed that only two charts could help him better understand the existing system—a Gantt chart and a work distribution chart.

Mr. Swift reasoned that a work distribution chart and a Gantt chart could be used for the purpose of determining how the personnel of the area under study spent their working hours. He had drawn this conclusion because he seemed to recall that these two charts had been used for this purpose in another system study.

After doing a limited amount of research (there was an example of a partially completed work distribution chart in his notes), the analyst prepared a work distribution chart of the area under study (Department C) based on observation and conversations with the individuals involved (Figure 13-29). In preparing the chart, he determined what jobs were being performed, who performed them, how the work was divided, and the approximate time required to perform each job. Since all the employees worked on a 40-hour per week basis, he recorded the number of hours per week spent on the various tasks.

Mr. Swift, having previously involved the affected personnel in his plans, presented the work distribution chart to the area under study for their comments. He went over the chart with Bob Broman, the senior clerk, who pointed out that 5 of the 25 hours defined as being spent on "checking the dollar totals" were actually spent on "report writing." Mr. Broman was apologetic about his oversight during their earlier meeting and expressed his opinion that this correction was pertinent.

Mr. Swift, upon hearing this information, pretended to make a note of it for the clerk's benefit since he was quite satisfied that the chart was complete and was reluctant to change it. Conversations with the other employees had not resulted in any other substantially pertinent information that would require a change in the chart so he was not about to change it at this time.

Returning to his office at the end of the day, Mr. Swift prepared a "To Do" list for the proceeding day. The list included the following notations among others.

1. Prepare a Gantt chart.

2. File the work distribution chart—no corrections necessary.

The analyst planned to prepare another chart to support his work distribution chart. He decided to prepare a Gantt chart for this purpose in order to depict the job duties of each employee.

Mr. Swift's conception of how he could use the Gantt chart is represented in Figure 13-30.

EMPLOYEE NAME	CORRES-PONDENCE					ORDER PROCESSING					SUPER-VISION					INVOICE AUDITING				
	M	T	W	T	F	M	T	W	T	F	M	T	W	T	F	M	T	W	T	F
ALAN YOUNG	X	X				X	X	X	X	X	X	X	X	X	X					
MABEL BROWN	X	X	X	X	X	X	X	X	X	X										
BOB BROMAN						X	X	X	X	X	X	X	X	X	X	X	X	X	X	X
RON WHITE						X	X	X	X	X										

FIGURE 13-30   Gantt chart of daily job duties for Department C.

QUESTIONS

1. For each chart developed by the analyst, indicate whether the use was valid and explain why.
2. What additional steps should Mr. Swift have taken in preparing the work distribution chart?
3. Redraw the correct work distribution chart.

CASE 13-2

A programmer-analyst had decided to use charting techniques as a means of providing management with a concise comparison of the current system and the new system, allowing both systems to be put in their proper perspectives. She also felt the use of these techniques would help her (1) gain a thorough understanding of the existing system, (2) clearly illustrate new system requirements, and (3) identify key points in the design of the new system.

The programmer-analyst completed preparation of the management report. In the final report, the time and cost comparisons were presented in narrative form in great detail. Several managers commented that there was such a mass of numerical data that it was difficult to be certain how reliable the estimates of projected savings and time were. Some even confessed to being totally confused. The analyst decided to rewrite the report. In an effort to better illustrate the time and cost comparisons, and subsequent savings to be expected, the analyst used the payroll preparation procedure as her first example.

The analyst used a bar chart to illustrate that under the existing system, it takes two days to prepare the payroll, while under the new system it will only take six hours. (See Figure 13-31.) Next the analyst used a bar chart to illustrate that under the current system, it takes twenty-one days to process a customer complaint, while under the new system, it was expected to take only ten days. In addition, the analyst showed that the new system would raise the number of complaints that could be processed per month from 25 to 50. (See Figure 13-32.)

As a part of the report, the analyst included an organization chart to better illustrate the impact of the new system. The analyst found it necessary to prepare a new chart since the only one she could find was forty years old. To obtain the information needed for the new chart, the analyst (1) interviewed all of the top executives in the home office and each of the twelve regional managers, and (2) read through the corporation's policy and procedure manuals. She prepared a draft of the organization chart in detail and submitted it to top management for approval. In the final section of the management report the analyst provided some information on the timing and phases of the project. To present this information, she decided to use the basic format of a work

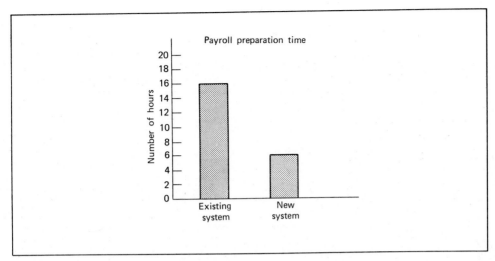

FIGURE 13-31 Payroll preparation time.

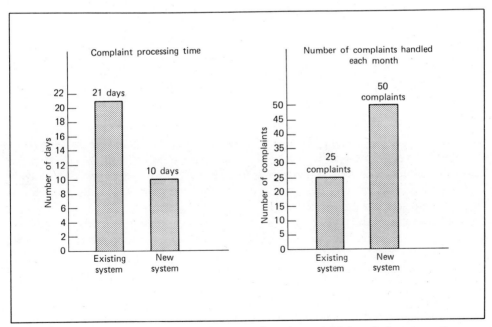

FIGURE 13-32 Complaint processing time and number of complaints handled each month.

PHASE	STARTING DATE	PROJECTED COMPLETION (TIME)	TOTAL TIME (WEEKS)	TIME SPENT (WEEKS)	TIME LEFT (WEEKS)
Payroll	Jan 14	Feb 7	4	COMPLETED	
Accounts Receivable	Feb 7	Apr 7	9	1 2 3 4 5 6789	9 8 7 6 5 4321
Cash Receipts	Feb 21	Mar 14	4	1 2 3 4	4 3 2 1
Accounts Payable	Mar 14	Apr 14	5	1 2 3 4 5	5 4 3 2 1
Commissions Receivable and Payable	Mar 28	Apr 28	5	1 2 3 4 5	5 4 3 2 1
General Ledger	Apr 21	Jul 14	12	1 2 3 4 5 6 7 8 9 10 11 12	12 11 10 9 8 7 6 5 4 3 2 1
Cash Disbursements	Jun 21	Jul 28	6	1 2 3 4 5 6	6 5 4 3 2 1

FIGURE 13-33   Selected portion of completed work distribution chart.

distribution chart. (See Figure 13-33.) She knew that this chart provided an analysis of what jobs are performed, how the work is divided, and the approximate time to complete each job. The analyst had used this type of chart to determine the workloads of the employees in each department, and felt that by improvising a little she could adequately present the information needed by management.

The analyst issued the revised report to management and they seemed quite pleased with the improvements that had been made; however, some questions arose concerning the way the project timing was shown.

QUESTIONS

1. What additional organization chart should the analyst attempt to develop besides the formal organization chart and why?

2. Management was not entirely satisfied with the analyst's portrayal of the reorganized office and work flow. How could she better present this information required by management by using one form of charting?

3. Is the work distribution chart being used correctly? If not, what type of chart would better illustrate the data?

FOOTNOTE

1. Σ means summation, for example, Σa_n means $a_1 + a_2 + a_3 \cdot \cdot \cdot a_n$.

CHAPTER 14

FORMS DESIGN

It is hard to imagine the operation of a business without the use of forms. They are the vehicles for most communications and the blueprints for most operations. This chapter describes the role of the systems analyst in forms design and control.

OBJECTIVES OF FORMS CONTROL

Somewhere in the firm there should be one person or one functional area that is responsible for forms. The objectives of centralized forms control are

1. To prevent the inception of new forms when suitable forms already exist somewhere else in the firm.
2. To offer expert advice on forms design techniques.
3. To reduce the total number of forms by consolidating similar forms already in existence.
4. To ensure a unique forms numbering system with no duplication of form numbers or revision dates.
5. To provide for proper replenishment, storage, and distribution of forms throughout the firm.
6. To consider the design costs, the cost of artwork, and the quantities to be purchased.
7. To coordinate forms design between all users of the form in question.
8. To review the usage of existing forms in order to detect changes in requirements.
9. To assure that forms will be well designed and usable for a long period of time.

The forms control area does not necessarily approve or disapprove the introduction of a new form or the modification of an existing form. When the forms control area has the authority to disapprove forms, the departments that need design assistance tend to avoid using the form expert's advice. They often fear that getting involved with the forms authorities will result in their form being altered or disapproved. The forms control function then is best organized as one which coordinates and assists rather than rules the forms requirements of the organization.

A forms control program can follow the general sequence of first getting the backing of management, then announcing commencement of the program, publicizing the objectives of the program, and collecting one sample of every form (including any written procedures on how to fill out the form). Samples of *all* forms should be collected, whether they were purchased originally from an outside vendor or printed within the firm.

The collected samples can be filed now into a *form-number file* where all forms are filed numerically by their identification numbers. (Attach any existing instructional procedures to the form before filing it.) Each form should be put into a separate file folder with all appropriate data recorded on the front of the file folder. The analyst should have some file folders preprinted similar to Figure 14-1. The form name(s) should include *all* of its names, the official name of the form plus any informal names

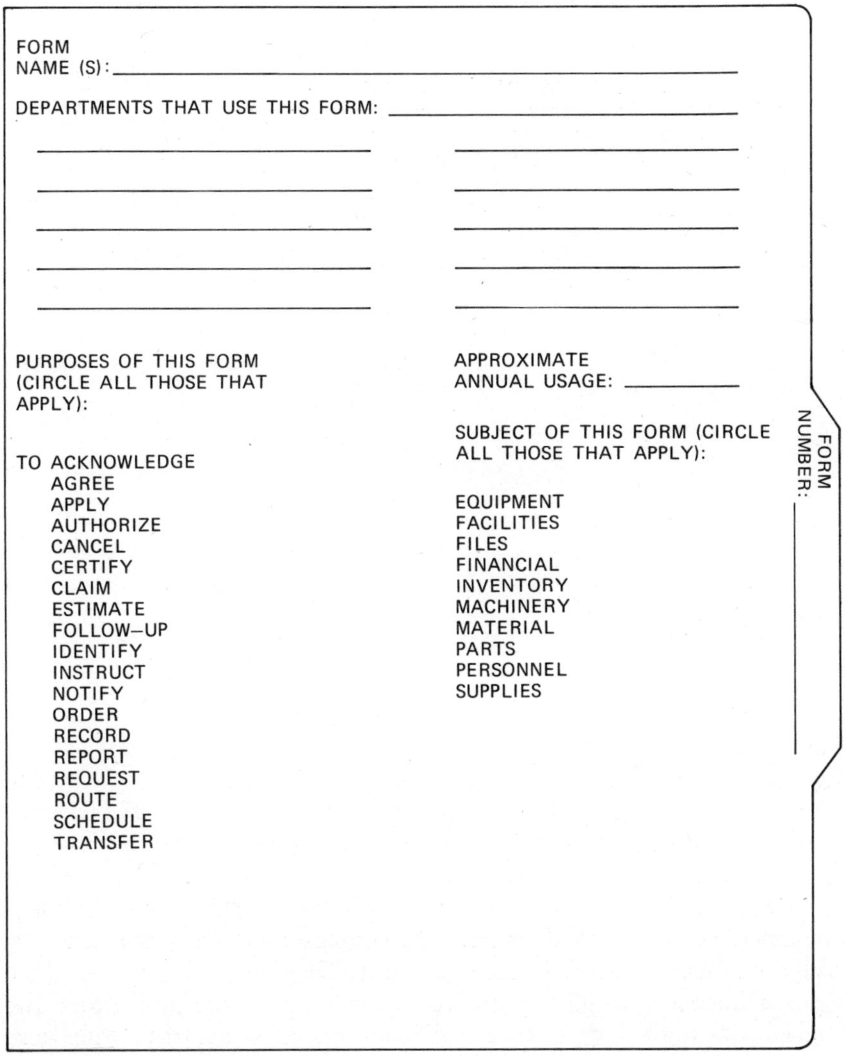

FORM
NAME (S): _____

DEPARTMENTS THAT USE THIS FORM: _____

_____   _____

_____   _____

_____   _____

_____   _____

_____   _____

PURPOSES OF THIS FORM   APPROXIMATE
(CIRCLE ALL THOSE THAT   ANNUAL USAGE: _____
APPLY):

                                     SUBJECT OF THIS FORM (CIRCLE
TO ACKNOWLEDGE   ALL THOSE THAT APPLY):
  AGREE
  APPLY   EQUIPMENT
  AUTHORIZE   FACILITIES
  CANCEL   FILES
  CERTIFY   FINANCIAL
  CLAIM   INVENTORY
  ESTIMATE   MACHINERY
  FOLLOW—UP   MATERIAL
  IDENTIFY   PARTS
  INSTRUCT   PERSONNEL
  NOTIFY   SUPPLIES
  ORDER
  RECORD
  REPORT
  REQUEST
  ROUTE
  SCHEDULE
  TRANSFER

FORM NUMBER:

FIGURE 14-1 File folder for each form.

that operating personnel use to refer to the form. List all of the *major* departments that use the form or any of its copies. Circle *all* of the purposes of the form, for example, a form may request a typewriter, authorize the movement of that typewriter, and instruct where to move the typewriter. Circle all of the subjects to which usage of the form is applied.

```
┌─────────────────────────────────────────────────────────────────┐
│                                                                   │
│   FORM                                                            │
│   NAME:              PURCHASE  REQUISITION                        │
│             _____             │
│                                                                   │
│                  FORM  NUMBER:   1261                             │
│                                  Rev. 3/79                        │
│                                                                   │
│                                                                   │
│   Also known as:                                                  │
│                                                                   │
│     1. pick sheet                                                 │
│     2. req. "pronounced wreck"                                    │
│     3.                                                            │
│     4.                                                            │
│                                                                   │
│                                                                   │
│                                                                   │
│                                                                   │
└─────────────────────────────────────────────────────────────────┘
```

FIGURE 14-2   Cross-reference file card.

The next project in the forms control program is to develop a *cross-reference file*. Enter each form name (formal and informal) with its form number on a 3 × 5 card and file them alphabetically (see Figure 14-2). This cross-reference file can be used to find a file folder for any form if you know any one of the form names. (Remember that the file folders were filed by form number.) Another method of developing a cross-reference file is to use a Key Word In Context (KWIC) index system on a computer. KWIC may be described as a method whereby the computer lists each title of a form according to all the important words in the form's title. These important words all fall into alphabetical sequence with all the important words from other form titles. The titles are matched back to the form's number so that the analyst can trace either from the key words in the form's title, or from the form's number.

The last project in setting up a forms control program is to develop a forms classification system, the *functional file*. Each form already is classified by name, user departments, purpose, and subject, as recorded on the file folder in Figure 14-1. All that remains is to set up a data retrieval system. Two simple systems are edge-notched cards and superimposable card systems. A good breakdown on these retrieval methods will be found in Dykes' *Practical Approach to Information and Data Retrieval.*

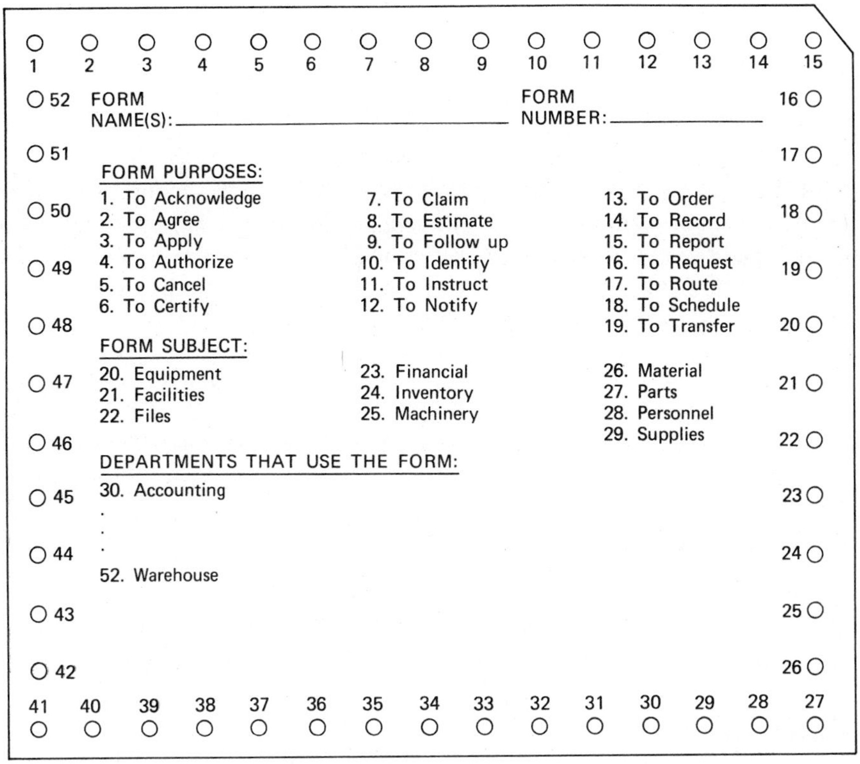

1 2 3 4 5 6 7 8 9 10 11 12 13 14 15

52 FORM NAME(S): ____ FORM NUMBER: ____ 16

51 17

FORM PURPOSES:

50 18

1. To Acknowledge	7. To Claim	13. To Order
2. To Agree	8. To Estimate	14. To Record
3. To Apply	9. To Follow up	15. To Report
4. To Authorize	10. To Identify	16. To Request
5. To Cancel	11. To Instruct	17. To Route
6. To Certify	12. To Notify	18. To Schedule
		19. To Transfer

49 19

48 20

FORM SUBJECT:

47 21

20. Equipment	23. Financial	26. Material
21. Facilities	24. Inventory	27. Parts
22. Files	25. Machinery	28. Personnel
		29. Supplies

46 22

DEPARTMENTS THAT USE THE FORM:

45 30. Accounting 23

.

.

44 . 24

52. Warehouse

43 25

42 26

41 40 39 38 37 36 35 34 33 32 31 30 29 28 27

FIGURE 14-3 Edge-notched card.

At this point you have a *form-number file* (in form number sequence) containing a copy of each form, a *cross-reference file* (by form name) on 3 × 5 cards, and now a *functional file* using, for example, edge-notched cards. This edge-notched card file is used to search out similar forms already in existence.

The analyst codes each of the characteristics, from the front of the file folder, onto an edge-notched card (see Figure 14-3). Then the analyst prepares one edge-notched card for each form in the form-number file. Whenever the analyst wants to see if the firm has an existing form that will serve a new purpose, the edge-notched cards are sorted. After finding the desired characteristics, such as user departments or purpose or subject, the analyst then goes to the *form-number* file to check visually the form in question. For a complete description on edge-notched cards see Appendix III. A summary on information storage and retrieval appears in Chapter 18.

OBJECTIVES OF FORMS DESIGN

An entirely new form often needs to be designed for a new use. When designing such a new form, begin by writing down the purpose of the form as precisely as possible. Of course, a particular form may have many purposes. As more departments become involved in the use of the form, its function becomes more complex. The forms designer should be clear on the form's purposes and should resist any additions or other complexities which might compromise its efficiency. The major cost associated with forms is not the cost of the form itself. Processing the form through the system is by far the most expensive aspect, and the most important concern of the analyst. For example, if a poorly designed form causes each of nine people who handle it to lose 30 seconds of time, then $4\frac{1}{2}$ minutes of time is lost per form. If the firm uses 7000 of these forms per year, then 31,500 minutes per year are lost $(4.5 \times 7000 = 31,500)$. This amounts to 525 hours of lost time per year $(31,500 \div 60 = 525)$ which could have been avoided by a better forms design. Bear in mind that the major objectives of forms design are to have

1. Forms that perform their function as simply and efficiently as possible.
2. Forms that can be filled out easily.
3. Forms that are legible, uncomplicated, and economically feasible.

KNOW THE MACHINES TO BE USED

It is very hazardous to attempt to design a form without knowing the type of equipment that will be used to fill it out. Will the equipment be a human hand or a computer? The standard typewriter is the basic machine used in filling out forms. The analyst should know whether the typewriters to be used have $\frac{1}{6}$-inch vertical spacing, elite type (twelve letters to the inch horizontally) or pica type (ten letters to the inch horizontally). This is important because the right amount of space must be allowed on the form for each entry. For typewritten forms the analyst also should specify the correct vertical spacing of items. Typewriters have six lines per inch vertically. Improper spacing can render a form useless because whenever the typewriter carriage returns, the form must be in a position to accept the next line of typing. If not, the typist will have to realign the form for each line; and the form will be very inefficient and time consuming.

An entire book could easily be written on the various kinds and uses of printing presses, but the analyst should be familiar with the basic processes used to print forms. In the letter press process, the form is printed from a raised surface of type that is smeared with ink. In the offset (lithography) process the finished drawing of the form is photographed onto a printing surface from which the form is then printed. Offset is a

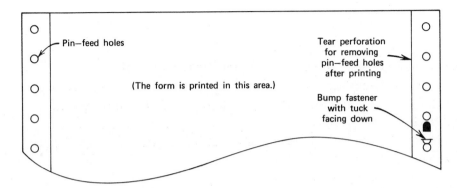

FIGURE 14-4 Line printer form.

photographic reproduction process. The familiar term "camera-ready copy" refers to the designer's finished, touched-up draft of the form which is perfect enough to be photographically copied and reproduced as the actual form. Offset is probably the most flexible reproduction method available to the forms designer.

Computer line printers are very important to a forms designer. A form for a computer line printer requires pin-feed holes so the printer's mechanism can move the forms through the printer at high speeds (see Figure 14-4). Multiple copy continuous strip forms can be fastened together with staples, crimping, or can be bump fastened. (When using bump fastening, be sure that the tuck faces down or there may be paper jams in the printer.) A form thicker than one original and five copies may give poor, hard-to-read copies. (Multilith paper, which can be used directly for offset printing, is available whenever a large number of copies are required.) Line printers usually have 120 characters horizontally across the paper (up to 160) and have six or eight lines per inch vertically. Check these figures, however, for any specific line printer. Line printer forms come in a continuous strip that is continually fed through the printer.

TYPES OF PAPER

The type of paper to be used in a form is important because the form must be able to withstand a certain amount of mishandling in addition to fulfilling a definite function. If it fails in either of these two areas, it is not a successfully designed form. See Figure 14-5.

Paper is measured in pounds. For example, 16-pound bond paper means that a ream of the particular paper weighs 16 pounds. A ream is 500 sheets of paper measuring 17 inches by 22 inches. Each sheet is usually cut into four standard 8½ by 11 inch sheets.

- *Bond paper.* Bond paper is high quality paper. Rag bonds are the highest quality because they have 25% or more rag content. Sulfite bonds are inexpensive and usually are durable enough for use in forms. Duplicating bonds are for reproduction (Ditto machines) where ink absorption is required.
- *Ledger paper.* Ledger paper is a heavier weight bond paper (24 to 44 pounds).
- *Index bristol paper.* Heavier than either ledger or bond, it is better known as card stock (60 to 220 pounds).
- *Manifold paper.* Manifold paper is a lightweight bond sometimes called onionskin (8 to 10 pounds).
- *Safety paper.* Safety paper is paper that cannot be erased without leaving an obvious mark. It is the type used for checks.

USE	PAPER	WEIGHT
Multipart forms	Sulfite bond	9–12 lbs.
Letterhead and checks	Rag bond	16–20 lbs.
Machine posted ledgers	Ledger paper	28–32 lbs.

FIGURE 14-5   Use of paper.

TECHNIQUES OF FORMS DESIGN AND LAYOUT

The designer of a form should follow certain basic guidelines in order to assure that the form is both usable and used.

When designing a form, first list all the items of information or data to be recorded on the form and the maximum number of characters to be allotted for each. Make a tabular contents list as illustrated in Figure 14-6. If anything is deleted from an existing form, explain the reason for the deletion. The contents list should be checked by the personnel who use the form in order to make sure nothing is omitted. Always have the form users check the contents list before giving them a rough draft copy of the form because the draft copy of the form may distract their attention from checking for omitted items. Instead, they may become preoccupied with layout and so on.

Give the form a descriptive title, and make the name as long as necessary to describe clearly how the form is to be used. The name should be placed at the top of the form where it can be seen readily.

Assign a new form number if the form is a brand new one, or a revision date if it is a revised form. Some firms use a form number and revision number instead of a revision date. The best placement for form number and revision date is in the extreme lower

INFORMATION OR DATA ITEM	NUMBER OF CHARACTERS
Ship to: Name	24 alphabetic
Address	24 alphanumeric
City, State	24 alphanumeric
Quantity	4 numeric
Drawing number	9 alphanumeric
Description	44 alphanumeric
.	.
.	.
.	.
etc.	etc.
.	.
.	.

FIGURE 14-6 Contents list.

left corner of the form. Be consistent by placing the form number and revision date in the same position on all forms.

Determine what the correct sequence of data should be. Lay out the data on the form either so it is most convenient for whomever fills out the form or for whomever reads the form and copies from it. This depends on the use of the form. Maybe a combination of both is desirable. If in doubt, lay out the data so it is most convenient for whomever fills out the form, especially if the form is to be filled out by typewriter.

Use boldface type or double parallel lines to make special information stand out. Screening, a light shading in one area of the form, also can be used to separate the form's sections.

Common variations of ruled lines are

Broken — — — — — — — — — — —

One-point _____

Half-point _____

Hairline _____

Double parallel _____

Provide for a continuous writing flow by the person who will fill out the form. Writing should be from left to right and from top to bottom. Boxed design is favored over caption and a line (see Figures 14-7 and 14-8). In Figure 14-7 it is perfectly clear where the data go, but in Figure 14-8 it is not as clear as you move down the form. For example, does marital status go below or above the caption "marital status"? Compare the two layouts on this point.

Whenever possible use a simple ballot box so all the user has to do is check the box applicable to the transaction (see Figure 14-9). Other important features to consider during the layout phase include

TERMINATION NOTICE	
EMPLOYEE NAME	EMPLOYEE NUMBER
ADDRESS	CITY, STATE
DATE OF BIRTH DEPARTMENT REASONS	
MARITAL STATUS	

FIGURE 14-7 Boxed design.

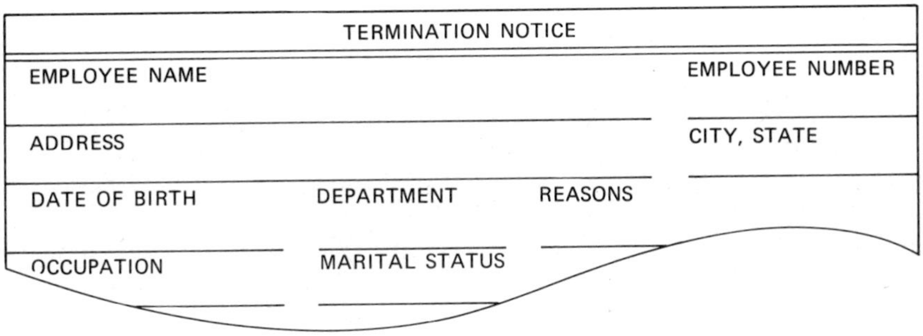

FIGURE 14-8 Caption and a line.

1. In boxes or areas for numeric data, allow sufficient space for the largest probable string of figures.
2. Place filing information near the top of the form so it can be seen easily when looking for it in a file. This varies with the type of file to be used.
3. For multiple copies use a different color paper for each copy. Also print the destination of each copy at the extreme bottom of the copy.
4. Avoid abbreviations. They might have different meanings to different people.
5. Preprint as much as possible. A consecutive document number can be preprinted. If the form is a tab card form set, a number can be prepunched into the card part of the form.
6. Consider printing the form in some ink color other than black in order to make the form stand out.
7. If possible, print the instructions on how to fill out the form on the back of the last copy of the form. Print a message on the glue margin (top of form) stating that the instructions are on the back of the form. Another method of providing instructions is to code each area with a number and have a corresponding instruction sheet. This method was used in Figure 7-8 in Chapter 7 of this book.
8. Group similar data in the same area on different forms, for example, always put form number and revision date in the extreme lower left corner.
9. Put a border around the form to give it a look of balance and professionalism.

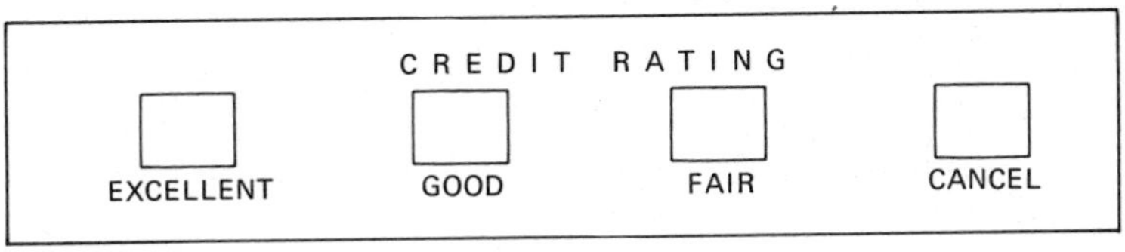

FIGURE 14-9 Ballot boxes.

FORM WIDTH

The overall width of the form is an important dimension in determining horizontal printing space. Remember that for typewritten forms elite types twelve characters per inch while pica types only ten. For handwritten forms the horizontal space per character should be ⅛ inch. The forms designer should consider the requirements for each entry on the form in terms of these constraints.

Forms costs can be reduced by confining form widths to the standard sizes of paper stock supplied by the forms printer. In addition, standard widths allow for the purchase of binding and filing supplies in standard sizes. This, of course, increases the efficiency of forms handling and filing. If the form must be mailed, consider the width of the firm's envelopes. Always try to use standard sizes. In addition, it is important to consider whether the firm uses plain envelopes or window envelopes. If the latter, the form can be designed so the addressee's name shows through the glassine window. With imagination the forms designer can even indicate by the use of arrows where the form should be folded to efficiently fit into the envelope.

For filing convenience limit form maximum size to 8½ by 11 inches. Smaller than 8½ by 11 may be acceptable, but avoid anything larger since most filing cabinets are built to accept 8½ by 11 paper size. Any larger size will have to be folded before filing or special files will have to be purchased. The analyst will have to determine if the form will be filed in cabinets, on shelves, in loose-leaf binders, in tub files, in visible files, and so forth, before determining its size. The filing method can have much to do with the required size of the form.

Vertical lines that separate boxed design are called horizontal spacing lines (see Figure 14-10). They are drawn so each one splits a printing position, that is, splits a character. If they are drawn between adjacent positions, the paper shrinkage and variations in form alignment may prevent satisfactory registration during both the printing of the form and during the filling out of the form, especially if the form is filled out with a typewriter.

Avoid horizontal spacing lines as much as possible because a typist must set a tab stop on a typewriter for each different horizontal spacing line across the forms width.

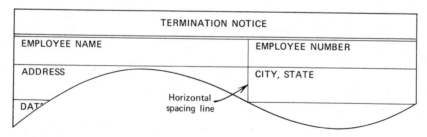

FIGURE 14-10 Horizontal spacing lines.

Try to locate horizontal spacing lines on typewriter forms at the same tab position, even though some horizontal spaces will be allotted excessive width. Avoid cutting one shorter than its minimum requirement, though.

FORM LENGTH

The overall length of the form is important in determining the number of lines of printing space. Typewriters have six lines per inch vertically, for example, $\frac{1}{6}$ inch per line. Computer line printers have either six or eight lines per inch vertically depending upon the model. Handwritten forms should have a minimum of $\frac{1}{4}$-inch line width vertically for writing. Notice that if there is $\frac{1}{3}$ inch between lines, the form is perfectly suitable for either handwritten entries or typewritten entries because a typewriter set on double spacing indexes $\frac{1}{3}$ inch. Handwritten entries can readily adapt to the larger space.

Forms costs can be reduced by confining form lengths to the standard sizes of paper stock supplied by the forms printer. The same benefits listed for standard form widths will be realized: fewer sizes of binding and filing supplies required, more convenient forms handling, and easy filing. Again, remember that most file cabinets are made for $8\frac{1}{2}$ by 11 inch paper.

CARBON COPIES AND CARBON PAPER

It may seem that cost savings should be pursued by reducing the number of carbon copies in a form to a minimum; but if there are not enough copies, the form will be ineffective and probably expensive to process. Employees will need to hand copy data from the form or expensive copy machines may be needed. Or the form may just be dumped and not used—with a loss of all the design time.

If one of the copies has to be a heavy paper stock, like ledger paper, it should be the bottom (last) carbon copy. Each carbon copy has a cushioning effect and the typing or handwriting becomes lighter and harder to read as it goes down through the copies. Use a lighter weight paper for the copies below the original. For example, for the original (top copy) use 12-pound and for the carbon copies use 9-pound paper weight.

When a multipart form with many copies (over eight) is typed, a very hard platen should be put on the typewriter. This gives more legibility to the copies. When a form is entirely handwritten, softer carbon paper should be used. A softer carbon paper transfers more easily and thus gives more legible copies.

Various methods of transferring the impression between copies are in use. The basic types of carbon paper are

1. *One-time carbon paper.* The carbon paper is interleaved in the form, used once, and thrown away. It is the most economical method for multipart forms.

2. *Carbon-backed paper.* The carbon surface is painted on the back of each copy so it transfers to the next lower copy. This method might be avoided because it is messy when handling the copies from the form.

3. *Chemical-coated paper.* An invisible chemical dye allows an image to be transferred to the next lower copy.

If there is a need to prevent some of the data from printing on all of the copies, the carbon paper can be blocked out (see Figure 14-11). Another method is to use short carbons (see Figure 14-12). A third method involves disguising the data by printing a random design on the copy in the area where the data will appear. The random design renders the carbon impression unreadable.

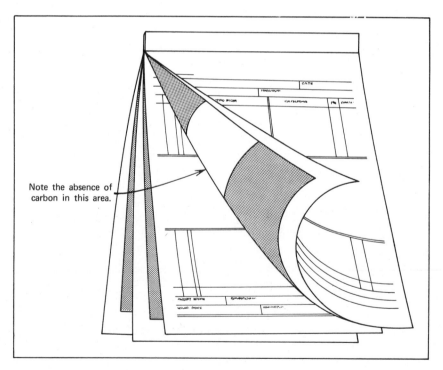

Note the absence of carbon in this area.

FIGURE 14-11   Blocked-out carbon paper.

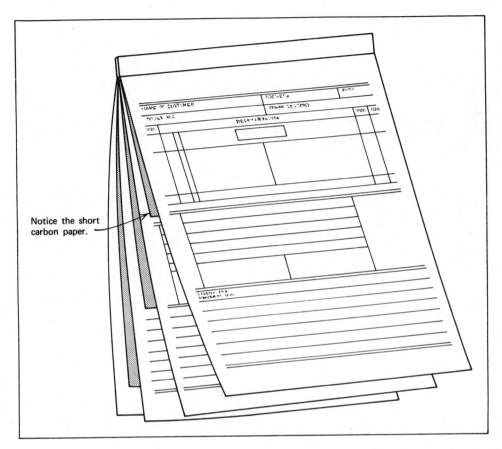

Notice the short
carbon paper.

FIGURE 14-12 Short carbon paper.

TYPES OF FORMS

Flat Forms

Flat forms are single-copy forms, that is, an original only. If copies of the original are required, carbon paper must be inserted between copies prior to filling them out. Or a copy machine can be used. Often pads of flat forms are printed identically to the original (top) copy of a unit-set. In this way a salesman might write in the data and give it to a secretary for typing onto a unit-set and thus not use the more expensive unit-set for the rough handwritten data.

343

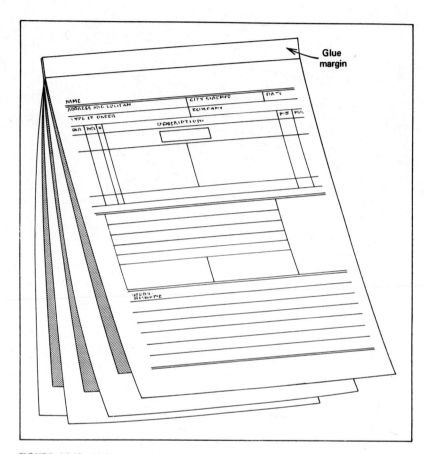

FIGURE 14-13   Unit-set/snapout forms.

Unit-Set/Snapout Forms

These forms have an original copy and several copies with carbon paper interleaved between each copy. The entire set is glued together into a unit set (see Figure 14-13). The unit-set should have the carbon paper cut $\frac{3}{8}$ inch shorter than the copies. The copies should be perforated at the glue margin and the carbon paper should not be perforated at all. This makes it possible to easily tear copies out and leave the carbon paper still attached to the glue margin for easy disposal. The easy detachment of the copies in this design resulted in the name "snapout."

Continuous Strip Forms

Continuous strip forms are joined together in a continuous strip with perforations between each form. They are delivered in a fanfold arrangement. Figure 14-14 shows an

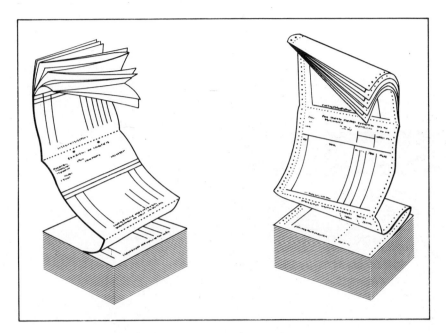

FIGURE 14-14 Fanfold continuous strip forms.

example. On the left are fanfold forms for a typewriter (no pin-feed holes) and on the right for a computer line printer (pin-feed holes). The analyst should know that, besides forms, the following items are available in a continuous strip for printing on a computer line printer.

1. Envelopes.
2. Fifty-one- or eighty-column tab cards.
3. Envelopes already stuffed with a preprinted form. As the envelope is addressed, the form also can have a minimum amount of data simultaneously printed on it by the computer line printer. Chemical-coated paper causes printing on the form *inside* the envelope because the character impact is transferred through the envelope.

Tab Card Form Sets

A tab card form set has a tab card as the last (bottom) copy and a couple of paper copies above with carbon paper interleaved between each copy (see Figure 14-15).

The advantage of a tab card form set is that certain equipment is available to emboss, much the same as with a credit card, and at the same time punch ten numeric digits.

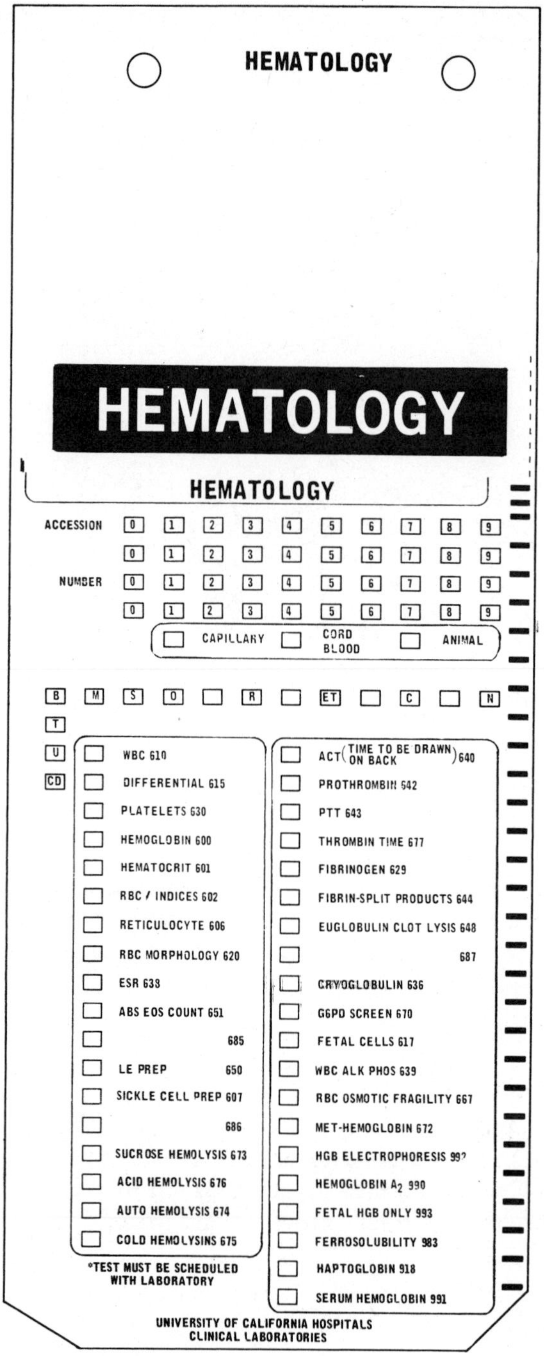
HEMATOLOGY

HEMATOLOGY

HEMATOLOGY

ACCESSION 0 1 2 3 4 5 6 7 8 9
0 1 2 3 4 5 6 7 8 9
NUMBER 0 1 2 3 4 5 6 7 8 9
0 1 2 3 4 5 6 7 8 9

CAPILLARY | CORD BLOOD | ANIMAL

B M S O R ET C N
T
U
CD

WBC 610
DIFFERENTIAL 615
PLATELETS 630
HEMOGLOBIN 600
HEMATOCRIT 601
RBC / INDICES 602
RETICULOCYTE 606
RBC MORPHOLOGY 620
ESR 638
ABS EOS COUNT 651
685
LE PREP 650
SICKLE CELL PREP 607
686
SUCROSE HEMOLYSIS 673
ACID HEMOLYSIS 676
AUTO HEMOLYSIS 674
COLD HEMOLYSINS 675

*TEST MUST BE SCHEDULED WITH LABORATORY

ACT (TIME TO BE DRAWN ON BACK) 640
PROTHROMBIN 642
PTT 643
THROMBIN TIME 677
FIBRINOGEN 629
FIBRIN-SPLIT PRODUCTS 644
EUGLOBULIN CLOT LYSIS 648
687
CRYOGLOBULIN 636
G6PD SCREEN 670
FETAL CELLS 617
WBC ALK PHOS 639
RBC OSMOTIC FRAGILITY 667
MET-HEMOGLOBIN 672
HGB ELECTROPHORESIS [illegible]
HEMOGLOBIN A_2 990
FETAL HGB ONLY 993
FERROSOLUBILITY 983
HAPTOGLOBIN 918
SERUM HEMOGLOBIN 991

UNIVERSITY OF CALIFORNIA HOSPITALS
CLINICAL LABORATORIES

FIGURE 14-15 Tab card form set.

There are tab card readers that will read both the forms punched digits and any mark-sensed data directly into a computer. In this way, as the form is filled out, it is being made machine readable at the same time.

SUCCESS IS A MISCELLANEOUS CHECKLIST

Probably the best advice that the analyst could follow is to call in an experienced forms salesman to assist in the design of any major new form, especially if there will be a high usage of the form. Over 500 a month would be considered high usage.

The analyst should be aware of a process called crash imprinting. Most printers have blank unit-sets of forms, already glued together, with different colored copies. You can get a quick unit-set of forms from this blank by printing with inked metal type and crashing the metal type down on the blank unit-set so hard that a carbon impression is transferred to all copies below the original. Crash imprinting costs less than regular printed unit-sets and offers a quicker delivery time. The quality usually is quite good. Another use of crash imprinting is to correct an omission made by the forms designer, or to revise an existing stock of forms with some additional details.

The analyst should always work through the purchasing department. The purchasing agent should arrange any meetings with the salesmen from various forms vendors. Quantities of the form used in the past should be checked to see if the same quantities are still being used. Try to obtain quantity price breaks. Also try to obtain reasonable delivery data promises. It usually takes 60 to 120 days from the time the new form order is placed until the new forms are delivered. Before placing the order have three or four co-workers check the final proofs before approving them for printing. Be sure that you are satisfied that the form will do its job before ordering it in large quantities.

TYPES OF PRINT

Legibility and simplicity are the primary consideration in the choice of typefaces. Stick with one basic design such as sans serif. The following example shows various sans serif styles.

UNIVERS — 11 Pt. Light

ABCDEFGHIJKLMNOPQRSTUVWXYZ
abcdefghijklmnopqrstuvwxyz
1234567890!†+$%/&*()—@¼ ½ ¾:",.;-=?[]

UNIVERS — 11 Pt. Medium

ABCDEFGHIJKLMNOPQRSTUVWXYZ
abcdefghijklmnopqrstuvwxyz
1234567890!†+$%/&*()—@¼½¾:",.;-=?[]

UNIVERS – 11 Pt. Medium Italic

ABCDEFGHIJKLMNOPQRSTUVWXYZ
abcdefghijklmnopqrstuvwxyz
1234567890!†+$%/&()–@¼½¾:",.;-=?[]*

UNIVERS — 11 Pt. Bold

ABCDEFGHIJKLMNOPQRSTUVWXYZ
abcdefghijklmnopqrstuvwxyz
1234567890!†+$%/&*()—@¼½¾:",.;-=?[]

UNIVERS — 11 Pt. Bold Condensed

ABCDEFGHIJKLMNOPQRSTUVWXYZ
abcdefghijklmnopqrstuvwxyz
1234567890!†+$%/&*()—@¼½¾:",.;-=?[]

UNIVERS — 10 Pt. Medium

ABCDEFGHIJKLMNOPQRSTUVWXYZ
abcdefghijklmnopqrstuvwxyz
1234567890!†+$%/&*()—@¼½¾:",.;-=?[]

UNIVERS — 8 Pt. Medium

ABCDEFGHIJKLMNOPQRSTUVWXYZ
abcdefghijklmnopqrstuvwxyz
1234567890!†+$%/&*()—@¼½¾:",.;-=?[]

UNIVERS — 7 Pt. Medium

ABCDEFGHIJKLMNOPQRSTUVWXYZ
abcdefghijklmnopqrstuvwxyz
1234567890!†+$%/&*()—@¼½¾:",.;-=?[]

Character counts are based on the standard unit of printer's measure which is a pica (no relationship to pica type on a typewriter). There are approximately 6 printer's picas to an inch and 12 points to a pica; thus there are 72 points to an inch, that is, one point is $\frac{1}{72}$ inch. Type sizes smaller than 6 points are difficult to read.

The following are combinations of ruled lines available to the analyst.

Hairline ————————————————————

½ Point Rules————————————————————

1 Point Rules————————————————————

1½ Point Rules————————————————————

3 Point Rules▬▬▬▬▬▬▬▬▬▬▬▬▬▬▬

4½ Point Rules▬▬▬▬▬▬▬▬▬▬▬▬▬

6 Point Rules▬▬▬▬▬▬▬▬▬▬▬▬

Parallel ═══════════════════════

In summary, it is evident that to be a successful form designer one must have knowledge of certain basic information in order to design a form that will be usable. This information includes knowledge of why forms control is desirable, types of machines used to fill out the form, types of paper and carbon that can be used, the types of forms that are available, and the techniques involved in the forms design itself.

SELECTIONS FOR FURTHER STUDY

Since forms have always been an integral part of any well-run business, most basic books on office administration will include something on the subject. For example, the *Dartnell Office Administration Handbook* contains a complete chapter on "Forms Control and Design" including standards for common forms design.

The systems analyst will be particularly interested in Prentice-Hall's *Handbook of Successful Operating Systems and Procedures, with Forms.* In it the forms and reports through which a system works are illustrated and explained. Emphasis is heavy on accounting; but many other areas are also included such as records for personnel, insurance, and production control.

1. ASSOCIATION FOR SYSTEMS MANAGEMENT. *Forms Design and Control.* Cleveland, Ohio: The Association for Systems Management, 1971.

2. CARLSEN, ROBERT D., and JAMES F. MCHUGH. *Handbook of Production Management Forms and Formats.* Englewood Cliffs, N.J.: Prentice-Hall, Inc., 1978.

3. CARLSEN, ROBERT D., and JAMES F. MCHUGH. *Handbook of Research and Development Forms and Formats.* Englewood Cliffs, N.J.: Prentice-Hall, Inc., 1978.

4. CARLSEN, ROBERT D., and JAMES F. MCHUGH. *Handbook of Sales and Marketing Forms and Formats.* Englewood Cliffs, N.J.: Prentice-Hall, Inc., 1978.

5. DYKE, FREEMAN H., JR. *Practical Approach to Information and Data Retrieval.* Boston: Industrial Education Institute, 1968.

6. ELFENBEIN, JULIUS. *Handbook of Business Forms and Letters.* New York: Monarch Press, 1971.

7. FETRIDGE, CLARK, and ROBERT MINOR. *Office Administration Handbook.* Chicago: The Dartnell Corporation, 1975.

8. GOTTERER, MALCOLM H. *KWIC Index; a Bibliography of Computer Management.* Princeton, New Jersey: Brandon/Systems Press, Inc., 1970.

9. HARPOOL, JACK D. *Business Data Systems: A Practical Guide.* Dubuque, Iowa: William C. Brown Company, 1978.

10. JENSEN, RANDALL W., and CHARLES C. TONIES. *Software Engineering.* Englewood Cliffs, N.J.: Prentice-Hall, Inc., 1979.

11. KISH, JOSEPH L. *Business Forms: Design and Control.* New York: Ronald Press, 1971.

12. KNOX, FRANK M. *Knox Standard Guide to Design and Control of Business Forms.* New York: McGraw-Hill Book Company, Inc., 1965.

13. OSTEEN, CARL E. *Forms Analysis: A Management Tool for Design and Control.* Stamford, Conn.: Office Publications, 1969.

14. PRENTICE-HALL EDITORIAL STAFF. *Handbook of Forms and Reports for Forty-Eight Representative Accounting Systems.* Englewood Cliffs, N.J.: Prentice-Hall, Inc., 1964.

15. PRENTICE-HALL EDITORIAL STAFF. *Handbook of Successful Operating Systems and Procedures, with Forms.* Englewood Cliffs, N.J.: Prentice-Hall, Inc., 1964.

16. SANDERS, DONALD H. *Computers in Business.* New York: McGraw-Hill Book Company, 1975.

17. WILLOUGHBY, THEODORE C., and JAMES A. SENN. *Business Systems.* Cleveland, Ohio: Association for Systems Management, 1975.

QUESTIONS

1. Name five objectives of centralized forms control.
2. What is the major cost associated with forms?
3. How many spaces per inch horizontally does an elite typewriter provide?
4. In designing forms for use with a typewriter, what are the vertical spacing considerations?

5. What is the first thing to do when designing a form?
6. Where is the best placement for form number and revision date?
7. Why should the forms designer avoid horizontal spacing lines as much as possible?
8. What is the minimum vertical space that should be allowed for writing on a handwritten form?
9. Name the three basic types of carbon paper.
10. Name four types of forms.

SITUATION CASES

CASE 14-1

A junior analyst, Brian Smith, had completed the major portion of a significant interdepartmental system study and design. He was now prepared to begin the design of a new form which would become part of the new system proposal.

The senior analyst provided Mr. Smith with a copy of the System Department manual on forms design and commented that he would check back with Smith, especially in the area of cost considerations. The manual proved to be very useful since it included an extensive step-by-step outline of procedures, tips, and pitfalls for all aspects of forms design.

Mr. Smith wrote down the purpose of the form as suggested, taking great pains to consult personally the offices and departments involved. He prepared written summaries of the problem definition and the new system requirements. He verified that the form was set up to be prepared by a pica typewriter in Department A with an original and three copies. (Distribution is shown in Figure 14-16.) The present system of using four original flat forms from a tablet and interleaving carbon paper seemed grossly inefficient. Considerable time was wasted in preparing the carbons and forms and straightening them in the typewriter. In addition, it was often hard to tell where copies of the form were going, since all copies were the same color and destinations were written by hand in the margin after the form was typed and removed from the typewriter.

Satisfied that no other machines were involved, Mr. Smith proceeded with the layout of the form. He collected all the pertinent information from the old form and again consulted his written documentation on problem definition, understanding the existing system, and defining new system requirements. Armed with the required contents of the form, he could now list all the items of information in the most logical order and compute a maximum number of alpha and/or numeric characters to be

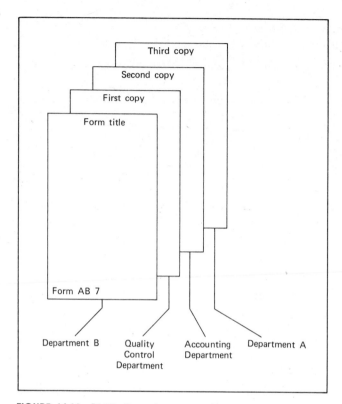

FIGURE 14-16   Distribution scheme.

allotted for each one. In sequencing the items, he provided for continuous flow (left-to-right and top-to-bottom) by the person who would be filling out the form. The analyst selected boxed design over caption and line, and used ballot boxes wherever possible. He decided to remain with standard lengths and widths of paper, concluding that this would facilitate filing and routing procedures and perhaps result in a cost savings as well.

The type of form was Mr. Smith's next consideration. He ultimately decided on a unit-set/snapout form to alleviate the interleaving problem of the old form.

When the design of the new form was near completion, Mr. Smith consulted the company's reproduction manager as to his part in the process. The type of paper and size of print, predetermined by the analyst, presented no problems that the manager could foresee. The Reproduction Department had access to equipment for making the unit-set/snapout type of form. Mr. Smith and the reproduction manager went over all other details, including a cost estimate.

To satisfy a suggestion noted in the System Department manual for high-use forms, Mr. Smith consulted an experienced forms salesman for advice. The salesman was carefully prescreened by the Purchasing Department on the basis of its past experience. The forms salesman they selected praised Mr. Smith's form, but he could not resist showing him a copy of Form AB 7 which his firm handled. This was a somewhat standardized form developed for other similar departmental operations. It was a unit-set form with multiple colored copies that had exactly the required file and destination information on them. Form AB 7 was designed for pica typewriters. Mr. Smith was impressed with the introduction of carbon-backed paper over the interleaved carbon. In addition to covering all the information incorporated in his form, Form AB 7 also included some extra blocks and spaces for information Mr. Smith felt could be used. Completion instructions were preprinted conveniently on the last page.

When Mr. Smith heard the price he was very impressed that purchasing the form could be so much less expensive than reproducing it in the company's Reproduction Department. He and the salesman then took a copy of Form AB 7 to the preparing and using locations. Form AB 7 was reviewed and discussed in all offices and departments affected by the form. Managers and clerks alike were as impressed as Mr. Smith with the salesman's form. The preparing secretaries were especially delighted to see a form that broke with traditional black print and used blue ink instead.

The new system, including Mr. Smith's recommendation for use of Form AB 7, was accepted by top management. Shortly after implementation of the new system, the senior analyst left a note on Smith's desk which said in part, ". . . it seems to me that perhaps you dwelled too much on the comparative acquisition cost of the vendor's form over one we could produce internally. In addition, the following problems were observed: (1) Form AB 7 does not include space for important information used by Department B and the Quality Control Department. (2) The carbon-backed paper is extremely messy, causing each receiving location to make a photocopy immediately upon receipt before the copy gets too blurred to read. (3) The additional information blocks create confusion on the part of users. A few users sent their copies back to Department A for redesign; others simply had a more difficult time finding space for needed information."

QUESTIONS

1. Why did the senior analyst comment that Mr. Smith "dwelled too much" on the comparative acquisition costs? What is the most important cost?

2. What can be said of the assertion that Department B and Quality Control Department should have spoken sooner if the form they had reviewed and discussed was deficient in content from what they really needed?

3. What basic forms design objective did Mr. Smith ignore in deference to the

salesman's argument? What would you have done under the same circumstances?

The forms control center of a large wholesaler requested one of the firm's systems analysts to design a more efficient sales form for use by the sales staff. The analyst was told by the forms control center that too much time was being wasted by both salespeople and typists in filling out the existing sales order form, number 2013. It was filled out in pen by the salespeople at the customer's place of business. At the end of each day, the form was sent to the typing pool to be typed in triplicate. The original accompanied the merchandise to the customer, the second copy was sent to the Accounting Department for filing upon payment, and the third copy was sent to the Shipping Department.

The analyst, Elizabeth Rutledge, began the project by writing, as precisely as possible, the purpose of the form. She stated that the purpose of the form was to facilitate the placing of customer orders to be filled by the company. In addition, the form was to be used by Accounting to bill the customer and by the Shipping Department to fill the order.

With some research, Ms. Rutledge learned that the sales staff needed only one copy of the form. As a result, she decided that when the revised forms were printed, they should come in two types, a three-copy "snapout" and also a pad of draft forms for the salespeople. The analyst discussed the contents of the revised form with the sales staff, typists, Shipping Department, and the Accounting Department.

After checking and double checking on the form contents, all users of form 2013 agreed that it should contain date, customer number, name and address, terms used, quantity, item number, wholesale price, total, authorized by, grand total, less a discount, and a net total to be used in that order. Next Ms. Rutledge outlined a rough draft of the form (see Figure 14-17) and reviewed it with all users to obtain a consensus of opinion regarding the revisions.

Ms. Rutledge then did some research and learned that the form could be printed to hold a maximum number of handwritten or typed characters. She also assured herself that the form could be typed by either the pica or elite typewriter since the company had both.

Instructions on how to fill out the form would be printed at the top of the form (the glue margin). The original copy was labeled at the bottom "Original," the second and third copy had a "2" and "3," respectively, at the bottom.

1. How could Ms. Rutledge have improved the design of the form?

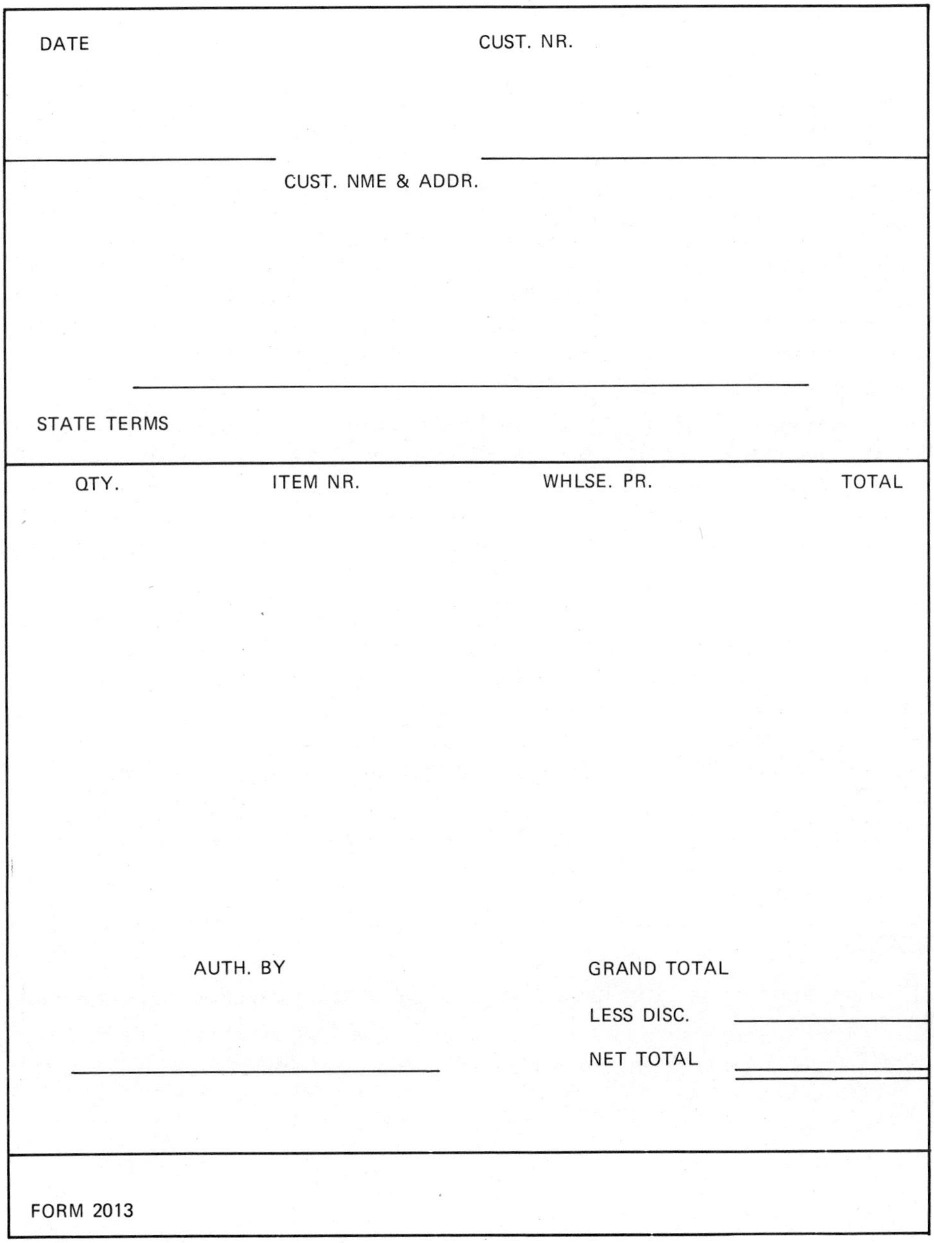

DATE		CUST. NR.	

CUST. NME & ADDR.

STATE TERMS

QTY.	ITEM NR.	WHLSE. PR.	TOTAL

AUTH. BY

GRAND TOTAL

LESS DISC.

NET TOTAL

FORM 2013

FIGURE 14-17 Rough draft of form.

CASE 14-3

Because of the recent growth of Supra Company and a general rise in interest rates, Supra's credit application form had become outdated. As a result, the credit manager met with the systems manager and requested a revision of the form to include provisions for new finance charges, a more complete description of the applicant's work history, and a large area for office use. The systems manager assigned an analyst to the job.

The first thing the analyst did was examine the existing credit application form. Next, he made a list of all the information that had to be included on the revised form. He then reviewed the list carefully, made some decisions about space allocation, and decided to limit the form size to 8½ by 11 inches to facilitate filing in the existing letter-size file cabinets.

The analyst decided that a name change for the form was not required, but a revision date was added to the new form. Because of company policy, he placed the date on the bottom right-hand corner of the form. The analyst had difficulty in trying to decide whether to design the form to facilitate using the standard company elite typewriter, or to make it more convenient to read and gather information. He finally decided on the latter.

The analyst now felt ready to outline a rough draft of the new form. He decided to use the line and caption design over the boxed design. He designed a three-part form with a continuous writing flow from left to right, top to bottom. Filing information was placed near the top of the form so it could be seen readily in the files. The analyst chose white paper and black ink for the original and three copies because it looked neater. Similar information was grouped together on the form as was done on other company forms.

After the analyst completed the draft, he drew a border around it and reviewed it with the credit department manager.

QUESTIONS

1. What could the analyst have done to be assured an effective and complete contents list?

2. What general layout principles has the analyst overlooked or decided against in revising the form?

CHAPTER 15

RECORDS RETENTION

Records can be considered as a vital extension of knowledge and memory. This is true also of the business firm since a firm begins to build records as soon as it enters the marketplace. Once records are created, it is inevitable that a cost-effective means of handling and storing the records will be sought. The systems analyst is charged with the responsibility for devising secure and economical methods of their retention. This chapter discusses records and how they should be handled.

PURPOSE OF A RECORDS RETENTION PROGRAM

Records retention is a systems technique which provides control over records. It provides for more efficient use of current office space, more efficient use of the space used to store old records, and reductions in paper usage. A formalized records retention program provides more effective access to both active and inactive business records. Usually it also must fit into the firm's information network.

The ultimate goal of a records retention program is to control the records from their source, through their use, on to their inactive storage, and finally, to their destruction. The single most important function of records retention is removing unneeded records from the current office space and moving them to an out-of-the-way storage area. In this way, space is made available for newly created records. The second most important function is to ensure that old records are kept for the proper length of time. Some are needed for longer periods than others. For example, a firm may elect to dispose of all production records on a job as soon as the job is finished. But it might elect to retain a copy of the job contract for ten years. The records retention system prescribes when the records are to be moved from the current work areas to the storage area, how they will be filed in the storage area, and when they will be destroyed.

ACTIVE . . . STORAGE . . . LEGAL LIFE

Active records are those records that are used in the day-to-day operation of the firm. Active records are located in the current office space and they are used frequently. Their total retention time might be from two or three days to twenty years.

Storage life (inactive) begins when the active life ends. Any record that is not referred to more than four times a year should be considered inactive and should go into its storage life. Schedules for retention and disposal provide timetables for the movement of each type of record from active records storage, located in the current office space, to inactive storage. Usually, when records are transferred from active to inactive storage, they should be analyzed to determine whether any of them can be immediately destroyed according to the retention period scheduled for the particular record.

The *legal life* of a record is the retention period that is required by law, or regulatory agencies which have the force of law. While the task of identifying which laws apply to any individual firm is large, it is not impossible. The *Federal Register,* published daily by the National Archives, provides proposed and new regulations that may affect business practices and records retention requirements. Further, the National Records Management Council (NAREMCO) maintains an index to federal records retention requirements.

RECORD RETENTION SCHEDULE

A record retention schedule is a timetable governing the retirement and destruction of all company records. The schedule shows how long the record should be kept in the active file located in the current office space, and it tells how long the record should be kept in storage. The overall life requirement of the record (both active and storage) may be governed by law (its legal life).

Many state and federal requirements for records retention times are vague. The firm often must try to learn the intent of the regulations and set its retention schedules accordingly. Never assume that the six- or seven-year statute of limitations is sufficient life for all records. There are many exceptions imposed for certain records by customers and by the state and federal governments. The firm itself should consider the value of retaining certain records for extended periods of time, or permanently.

Some of the laws or acts that affect records retention are the Fair Labor Standards Act, Code of Federal Regulations, Armed Services Procurement Regulation, Industrial Security Manual of the Defense Department, insurance company regulations, and the National Records Management Council (New York City).

As we have discussed, the usual route of most records is to go from active to inactive to disposal. As stated, the general rule that can be used to initiate the movement from active to inactive storage is a referral frequency of less than four times per year. Some records may go directly from active use to disposal, of course. Figure 15-1 shows an example of a general retention schedule and some of the codes often found in records control systems.

RECORDS INVENTORY

A records inventory is the starting point in initiating a records control program. The systems analyst should be given the authority of a project manager. The analyst must be authorized to go into the involved departments to examine their records. Of course, in government work the analyst also may be required to have a security clearance at or above the level of classification of the records being examined. In private industry, the authority given the analyst must be sufficient to overcome reluctance on the part of departments not wishing to allow access to their records.

The analyst should begin by making an inventory of the active records being maintained in each department. This should be accomplished in person. Later, if the job gets too large, a questionnaire can be designed to complete the records inventory; but it is important that the analyst get the feel of the job before designing the questionnaire. Questionnaires are fast, but an inventory personally performed by the analyst will usually be more accurate and realistic. During the physical inventory,

**LEGEND FOR
AUTHORITY TO DISPOSE**

AD—Administrative Decision
ASPR—Armed Services
    Procurement Regulation
CFR—Code of Federal Regulations
FLSA—Fair Labor Standards Act
ICC—Interstate Commerce
    Commission
INS—Insurance Company
    Regulation
ISM—Industrial Security Manual,
    Attachment to DD Form 441
* After Disposed   ** Normally

**LEGEND FOR
RETENTION PERIOD**

AC—Dispose After Completion
    of Job or Contract
AE—Dispose After Expiration
AF—After End of Fiscal Year
AM—After Moving
AS—After Settlement
AT—Dispose After Termination
ATR—After Trip
OBS—Dispose When Obsolete
P—Permanent
SUP—Dispose When Superseded
+ Govt. R&D Contracts

TYPE OF RECORD	RETENTION PERIOD YEARS	AUTHORITY
ACCOUNTING & FISCAL		
Accounts Payable Invoices	3	ASPR-STATE, FLSA
Accounts Payable Ledger	P	AD
Accounts Receivable Ledgers	5	AD
Authorizations for Accounting	SUP	AD
Balance Sheets	P	AD
Bank Deposits	3	AD
Bank Statements	3	AD
Bonds	P	AD
Budgets	3	AD
Capital Asset Record	3*	AD
Cash Receipt Records	7	AD
Check Register	P	AD
Checks, Dividend	6	
Checks, Payroll	2	FLSA, STATE
Checks, Voucher	3	FLSA, STATE
Cost Accounting Records	5	AD
Earnings Register	3	FLSA, STATE
Entertainment Gifts & Gratuities	3	AD
Estimates, Projections	7	AD
Expense Reports	3	AD
Financial Statements, Certified	P	AD
Financial Statements, Periodic	2	AD
General Ledger Records	P	CFR
Labor Cost Records	3	ASPR, CFR
Magnetic Tape and Tab Cards	1**	
Note Register	P	AD
Payroll Registers	3	FLSA, STATE
Petty Cash Records	3	AD
P & L Statements	P	AD
Salesman Commission Reports	3	AD
Travel Expense Reports	3	AD
Work Papers, Rough	2	AD
ADMINISTRATIVE RECORDS		
Audit Reports	10	AD
Audit Work Papers	3	AD
Classified Documents: Inventories, Reports, Receipts	10	AD
Correspondence, Executive	P	AD
Correspondence, General	5	AD
Directives from Officers	P	AD
Forms Used, File Copies	P	AD
Systems and Procedures Records	P	AD
Work Papers, Management Projects	P	AD

Revised and printed by Electric Wastebasket Corp.® 1974.

FIGURE 15-1   Records retention schedule.

TYPE OF RECORD	RETENTION PERIOD YEARS	AUTHORITY
COMMUNICATIONS		
Bulletins Explaining Communications	P	AD
Messenger Records	1	AD
Phone Directories	SUP	AD
Phone Installation Records	1	AD
Postage Reports, Stamp Requisitions	1 AF	AD
Postal Records, Registered Mail & Insured Mail Logs & Meter Records	1 AF	AD, CFR
Telecommunications Copies	1	AD
CONTRACT ADMINISTRATION		
Contracts, Negotiated. Bailments, Changes, Specifications, Procedures, Correspondence	P	CFR
Customer Reports	P	AD
Materials Relating to Distribution Revisions, Forms, and Format of Reports	P	AD
Work Papers	OBS	AD
CORPORATE		
Annual Reports	P	AD
Authority to Issue Securities	P	AD
Bonds, Surety	3 AE	AD
Capital Stock Ledger	P	AD
Charters, Constitutions, Bylaws	P	AD
Contracts	20 AT	AD
Corporate Election Records	P	AD
Incorporation Records	P	AD
Licenses - Federal, State, Local	AT	AD
Stock Transfer & Stockholder	P	AD
LEGAL		
Claims and Litigation Concerning Torts and Breach of Contracts	P	AD
Law Records - Federal, State, Local	SUP	AD
Patents and Related Material	P	AD
Trademark & Copyrights	P	AD
LIBRARY, COMPANY		
Accession Lists	P	AD
Copies of Requests for Materials	6 mos.	AD
Meeting Calendars	P	AD
Research Papers, Abstracts, Bibliographies	SUP, 6 mos. AC	AD
MANUFACTURING		
Bills of Material	2	AD, ASPR
Drafting Records	P	AD†
Drawings	2	AD, ASPR
Inspection Records	2	AD
Lab Test Reports	P	AD
Memos, Production	AC	AD
Product, Tooling, Design, Enginneering Research, Experiment & Specs Records	20	STATUE LIMITATIONS
Production Reports	3	AD
Quality Reports	1 AC	AD
Reliability Records	P	AD
Stock Issuing Records	3 AT	AD, ASPR
Tool Control	3 AT	AD, ASPR
Work Orders	3	AD
Work Status Reports	AC	AD

FIGURE 15-1 Continued.

TYPE OF RECORD	RETENTION PERIOD YEARS	AUTHORITY
OFFICE SUPPLIES & SERVICES		
Inventories	1 AF	AD
Office Equipment Records	6 AF	AD
Requests for: Services	1 AF	AD
Requisitions for Supplies	1 AF	AD
PERSONNEL		
Accident Reports, Injury Claims, Settlements	30 AS	CFR, INS, STATE
Applications, Changes & Terminations	5	AD, ASPR, CFR
Attendance Records	7	AD
Employee Activity Files	2 or SUP	AD
Employee Contracts	6 AT	AD
Fidelity Bonds	3 AT	AD
Garnishments	5	AD
Health & Safety Bulletins	P	AD
Injury Frequency Charts	P	CFR
Insurance Records, Employees	11 AT	INS
Job Descriptions	2 or SUP	CFR
Rating Cards	2 or SUP	CFR
Time Cards	3	AD
Training Manuals	P	AD
Union Agreements	3	WALSH-HEALEY ACT
PLANT & PROPERTY RECORDS		
Depreciation Schedules	P	AD
Inventory Records	P	AD
Maintenance & Repair, Building	10	AD
Maintenance & Repair, Machinery	5	AD
Plant Account Cards, Equipment	P	CFR, AD
Property Deeds	P	AD
Purchase or Lease Records of Plant Facility	P	AD
Space Allocation Records	1 AT	AD
PRINTING & DUPLICATING		
Copies Produced, Tech. Pubs., Charts	1 or OBS	AD
Film Reports	5	AD
Negatives	5	AD
Photographs	1	AD
Production Records	1 AC	AD
PROCUREMENT, PURCHASING		
Acknowledgements	AC	AD
Bids, Awards	3 AT	CFR
Contracts	3 AT	AD
Exception Notices (GAO)	6	AD
Price Lists	OBS	AD
Purchase Orders, Requisitions	3 AT	CFR
Quotations	1	AD
PRODUCTS, SERVICES, MARKETING		
Correspondence	3	AD
Credit Ratings & Classifications	7	AD
Development Studies	P	AD
Presentations & Proposals	P	AD
Price Lists, Catalogs	OBS	AD
Prospect Lines	OBS	AD
Register of Sales Order	NO VALUE	AD
Surveys	P	AD
Work Papers, Pertaining to Projects	NO VALUE	AD

FIGURE 15-1 Continued.

TYPE OF RECORD	RETENTION PERIOD YEARS	AUTHORITY
PUBLIC RELATIONS & ADVERTISING		
Advertising Activity Reports	5	AD
Community Affairs Records	P	AD
Contracts for Advertising	3 AT	AD
Employee Activities & Presentations	P	AD
Exhibits, Releases, Handouts	2 - 4	AD
Internal Publications	P (1 copy)	AD
Layouts	1	AD
Manuscripts	1	AD
Photos	1	AD
Public Information Activity	7	AD
Research Presentations	P	AD
Tear-Sheets	2	AD
SECURITY		
Classified Material Violations	P	AD
Courier Authorizations	1 mo. ATR	AD
Employee Clearance Lists	SUP	ISM
Employee Case Files	5	ISM
Fire Prevention Program	P	AD
Protection - Guards, Badge	5	AD
Lists, Protective Devices		
Subcontractor Clearances	2 AT	AD
Visitor Clearance	2	ISM
TAXATION		
Annuity or Deferred Payment Plan	P	CFR
Depreciation Schedules	P	CFR
Dividend Register	P	CFR
Employee Withholding	4	CFR
Excise Exemption Certificates	4	CFR
Excise Reports (Manufacturing)	4	CFR
Excise Reports (Retail)	4	CFR
Inventory Reports	P	CFR
Tax Bills and Statements	P	AD
Tax Returns	P	AD
TRAFFIC & TRANSPORTATION		
Aircraft Operating & Maintenance	P	CFR
Bills of Lading, Waybills	2	ICC, FLSA
Employee Travel	1 AF	AD
Freight Bills	3	ICC
Freight Claims	2	ICC
Household Moves	3 AM	AD
Motor Operating & Maintenance	2	AD
Rates and Tariffs	SUP	AD
Receiving Documents	2 - 10	AD, CFR
Shipping & Related Documents	2 - 10	AD, CFR

FIGURE 15-1 Continued.

using a worksheet such as the one shown in Figure 15-2, the analyst should obtain the following.

1. The current filing system of each area. (Do not get involved in improving the current filing methods except as a separate project from the records inventory.)

2. Title of each record (form number if applicable).

RECORDS INVENTORY WORKSHEET

Department	Analyst
Title of Record	Form Number
Type of Copy (original, carbon, etc.)	Size of Record in Inches
Inclusive Dates of Accumulation in Department	Quantity of Record

Number of Times This Record is Used Per Year

Current Filing System (Active Life Area)

Storage filing System (Storage Life Area)

Active Life + Storage Life = Total Life
(Then Destruction)

Active Life	Storage Life	Method of Destruction
		☐ Burn ☐ Shred ☐ Local Trash

Remarks (Relationship to Other Records)

FIGURE 15-2 Records inventory worksheet.

3. Type of copy, such as original or carbon copy or whatever.

4. Size of the record, for example, $8\frac{1}{2} \times 11$ inches.

5. Quantity of each record (approximate).

6. Inclusive dates of records accumulation.

The analyst then should research the retention requirements imposed by the government, customers, other regulatory agencies, and the firm itself, assigning appropriate retention times to each type of record.

The schedule shown in Figure 15-1 may be used as a guide to the types of records which should be retained. The analyst, however, should always keep in mind that since laws change, any list such as in Figure 15-1 may become obsolete very quickly. This is true especially of personnel records which are being retained for increasingly long periods of time. The only *sure* way of knowing how long to keep a record is to check the law governing that record. The corporate lawyer can assist in this, or if the firm is quite small, the nearest municipal or county law library should be visited. Remember, too, that records not covered by a federal law may vary from state to state.

After all record types have been assigned tentative retention times for both active and inactive phases, the analyst should seek the concurrence of the most knowledgeable people in each affected department. The Contracts Department is usually an excellent source of information regarding exception retention requirements imposed by customers. The Quality Assurance Department is another source of such information.

When the retention times for a department's records are finalized, the analyst should obtain a signature of concurrence from the department manager. When the schedule is finalized for all departments, and signed by each department manager, a concurring signature should be obtained from an executive at general manager level or higher.

TYPES OF FILING SYSTEMS

There are many types of filing systems: some good, some bad. When working on a records retention project, it is easy to get sidetracked in the current records area if it has a poorly designed filing system. The analyst should not get involved in the filing system used in the current work area (active life) except as a separate project. The goal is to develop a records retention schedule, get records moved from active areas to storage areas, and to develop a good retention system in the storage area. The storage area filing system must allow for the retrieval of the records whenever they are requested, and when their destruction date arrives.

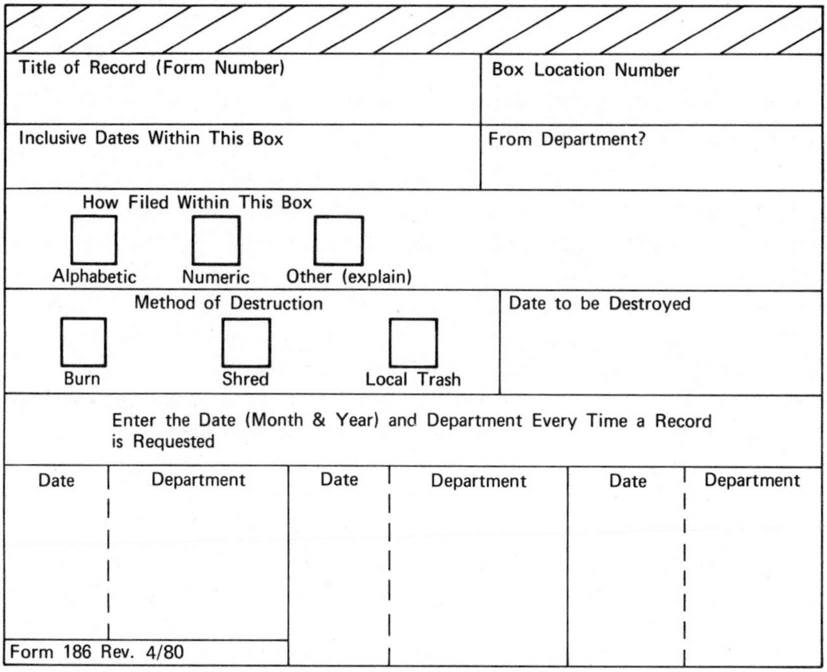

Title of Record (Form Number)		Box Location Number
Inclusive Dates Within This Box		From Department?

How Filed Within This Box

☐ Alphabetic   ☐ Numeric   ☐ Other (explain)

Method of Destruction	Date to be Destroyed
☐ Burn ☐ Shred ☐ Local Trash	

Enter the Date (Month & Year) and Department Every Time a Record is Requested

Date	Department	Date	Department	Date	Department

Form 186 Rev. 4/80

FIGURE 15-3 Storage box label form.

The most economical and quickest filing system for the storage area is to file the records by the title of the record. Each record is listed in alphabetical order using the Storage Box Label Form (Figure 15-3). The records are stored in boxes and then marked with the appropriate information. The label in Figure 15-3 is actually a three-part form. The top copy is a gummed label, the second copy is kept by the sending department (active life area), and the third copy is for the storage area files. The gummed label is put on the box of records before transfer to the storage area. The storage area files its copy, by the title of the record, in alphabetical sequence.

The records within the box are kept in the same filing sequence as they were sent to the storage area from the active life area. This eliminates any refiling in the storage area. It does, however, result in storing multiple copies of the same record because all records from all departments are stored without trying to consolidate. This has the minor disadvantage of losing some storage space, but has the positive advantage of safety in numbers of copies stored. In addition, the stored copies are in the same filing sequence that was used in the active area. This method, filing by title of the record, allows for quick retrieval because the storage area can quickly send a specific box of records to the requesting department. The requesting department can then locate the needed record since they will be familiar with the filing order within the box.

Other methods of filing are subject classification, record classification, and numeric classification. All three of these methods require the storage area to refile the records. In subject classification, each record is classified and arranged by subject in alphabetical order. In record classification, only one copy of each record is arranged in alphabetical order; this eliminates the storage of multiple copies of the same record. In numeric classification, all records are arranged in a unique redefined numeric sequence.

Unless the firm is willing to spend the extra money for the refiling of records by storage area personnel, it is recommended that the records be filed by title of the record and be kept in the same filing sequence as they were in the active life area, that is, the department from which they came. The cost of storage space for storing multiple copies of the same record from different departments is far less than the cost of refiling to eliminate the duplicate copies.

The personnel in the work area where the records have their active life are responsible for the boxing of the records for transfer to the storage area. When the records are received at the storage area, they should be checked to make sure they are properly identified and sequenced. A unique *box location number* is stamped on the box and also entered in the upper right corner of the filed copy of the storage box label form (see Figure 15-3). The object of the box location number is to enable the storage area to store boxes in the most efficient manner. The boxes are stored in a numeric sequence starting with 1 and going as high as required. With this index there is really no need for worrying about whether records with the same title are all in the same area on the storage shelves. To find a record, one looks up the filed copy of the storage box label form to find the unique box location number and retrieves the box.

Before placing the box in its designated shelf area, one other file should be utilized by the storage area. A file box of 5 × 8 inch cards, each with the month and year in consecutive order, must be developed. A fifty-year file will require only 600 cards of the kind shown in Figure 15-4. The personnel in the storage area enter the unique box location number on the card that contains the month and year in which the box of records is supposed to be destroyed. A "tickler file" of this type ensures that boxes are retrieved and destroyed. Prior to destruction, the supervisor of the storage area should check the dates of retrieval at the bottom of the storage box label (see Figure 15-3). If these records have been accessed numerous times in the recent past, then their destruction schedule should be reevaluated. The system should ensure that records still being used are not destroyed.

During the storage life of the records, whenever a box of records is returned to a specific department, the storage area should mark their file copy of the storage box label with the name of the department that has the box of records. When the box of records ultimately is returned, the storage area should then cross out the department listed at the bottom of Figure 15-3.

```
┌─────────────────────────────────────────────────────────┐
│  1981, JANUARY                                            │
├═══════════════════════════════════════════════════════════┤
│                                                           │
│    Unique box location number:  ·                         │
│                                                           │
│                                                           │
│                                                           │
│                                                           │
│                                                           │
│                                                           │
│                                                           │
│                                                           │
│                                                           │
│                                                           │
│                                                           │
│                                                           │
└─────────────────────────────────────────────────────────┘
```

FIGURE 15-4 Destruction cards.

SECURITY OF VITAL RECORDS

The loss of records may or may not hurt the firm. For this discussion, some records may be classified as "vital." Vital records are those records which would enable the firm to reconstruct its operations after a disaster such as a major fire. Vital records might be stockholder records, articles of incorporation, contracts and agreements, accounts receivable and payable, patents, production drawings, customer name and address files, client project files, and so on.

Copies of such vital records may be destroyed by fire, floods, terrorists, or nuclear destruction. Because of the importance of these records, many firms store them outside the plant in a safe area. The analyst must examine how the files might be destroyed and seek to find a location that will be protected from such events. Some firms store vital records in underground vaults that are available for rent.

Not only must the analyst evaluate where the vital records are to be stored, but a set of priorities also must be established. Generally speaking, those records that are necessary to resume operations after a disaster are given the highest priority (computer tapes should be stored relatively near the computer site in a different building). The priorities assume less importance as they go down the hierarchy to such items as product and price information.

MICROFORMS/MICROGRAPHICS

Microforms, or micrographics, have been with us since the 1920s; but they did not come into widespread use until the 1960s. Microform proponents have loudly touted their advantages, and rightly so. These include

- Microforms provide savings in storage space of 98%.
- They are relatively permanent, rarely lost, and seldom mutilated.
- They provide copies of rare items inexpensively (of concern primarily to libraries).
- They preserve rare or vital documents that could be destroyed by frequent use.
- Copies can be made quickly and inexpensively.
- Distribution of microform copies is simple and much less expensive than mailing the same thing in hard copy.
- Microforms generally cost less initially.
- For libraries in particular, binding costs are eliminated.
- Microforms can be used by most professions, including accounting, purchasing, education, personnel, hospitals, banking, insurance, engineering, retailing, and airlines, to name a few.
- Updating and retrieval are easy.
- Microforms meeting government specifications may be used for vital records.
- If color is needed, it is now available in some formats.
- Microforms have the added ecological advantage of preserving our forests since less paper is needed.

In short, microfilm has been referred to as the "transistorization of the printed page."[1]
 There are a number of disadvantages, too, which must be considered and which are often glossed over by the advocates of microforms. The disadvantages include

- There is some user reluctance, which varies from claims of inconvenience to claims of illness from watching a moving image on the screen; there is reluctance to depend on a machine for reading.
- Users cannot browse easily through microforms and usually need to be directed by an index or other secondary reference.
- File integrity, particularly with some microforms, is not well understood by some systems designers. It may be vital that only people knowledgeable in the system perform manual refiling.

- The user organization must provide a sufficient number of readers.
- Machines *do* break down, which can cause major problems.
- Continuous microfilm means that reel travel time must be experienced in order to arrive at the needed "page."
- Purchasers of microforms have to choose from more than 200 companies engaged in producing microfilm equipment and services.
- Although there have been many National Microfilm Conventions, there is a definite lack of standardization in the industry. For a company concerned only with one application and one format, this may not be a problem. But for companies purchasing microforms from both civilian and government agencies, different equipment may be needed for the various microforms since equipment is often incompatible and not adaptable.
- Microfilm may have many advantages in many situations, but it is not always an economic solution to a records storage problem. (On the other hand, economics may be less important than security for vital records.)
- Viewers must be used to read microforms. Since it may not be convenient or economically feasible to locate readers in all work areas, there may be an increase in employee lost time due to travel to and from readers.
- Each film must be inspected carefully to verify that it is completely readable; faint copies and some tissue copies do not photograph well.

Designing a microform information system of any kind is complicated by both the advantages and disadvantages. In addition, a major dimension is added to the system when you interject microforms because humans are habitual animals. For the human to communicate with a machine, the human must learn new thinking habits, and some humans cannot adapt. This human-to-machine interaction often is referred to as the man-machine interface.

The systems analyst follows the same procedures when designing a system of this type; but it is generally more complicated in that one must learn not only what routines are needed, but the microforms and their equipment capabilities and drawbacks as well. It is further complicated by the fact that costs are often miscalculated; and since one piece of equipment often leads to another, the *true* costs of the project are often not fully anticipated. Technology also is advancing so quickly that various types of microforms can now be used within the same system. In effect, many new systems are experimental, each one building upon previous ones. Robert Bodkin of Microfilm Service Corporation pointed this out when he said,

Generally, you are asking a customer to change his entire method of record-keeping and people just don't jump at something like that. . . .

When we first started selling microfilm we made something of a mistake by talking about it in general terms. Just about everyone knows you can record documents on microfilm, that you will save space by doing so and that it gives an added measure of security. But those aren't the advantages that sell microfilm. Microfilm has to be sold with a systems approach toward specific applications. It has to be sold as a system that does something better, not necessarily cheaper. If microfilm were being sold as something to make storage easier, it would be next to a nothing market for us.[2]

Fortunately, a whole new generation of office workers has grown up in an age in which high-reduction microforms and sophisticated micropublishing techniques have made microfilm less expensive than paper. The psychological barriers to the new media are disappearing as the paperless office emerges. Micronet, Inc. opened the "world's first paperless office" in Washington on May 2, 1979. The paperless office is intended as a working demonstration of how advanced information management concepts can be applied to practical business situations.

Microfilm is the oldest and best known of the microforms, and the term is used often to describe other microforms. It has been the principal medium for records storage since the 1920s. Microfilm, by reducing storage space by 95%, could eliminate the entire inactive storage area because all microfilmed records could be kept in the current office space (active life) and then go directly to destruction when that time arrived.

Microfilm, as used here, is a specific type of photo-optical reduction of a document onto a continuous strip of film that usually is placed on a reel (see Figure 15-5). Historically this film has been 35-mm (millimeter), which meant an optical reduction of anywhere from 10× to 40× (× means times). This size film had sufficient definition that copies could be made in the original size without appreciable loss of detail.

Since the 1950s, 16-mm has become increasingly popular (see Figure 15-6). For the many companies subscribing to periodicals microfilmed exclusively in 35-mm it causes a major equipment problem if a new type of film is introduced. If their equipment cannot utilize the new 16-mm film, they must purchase additional equipment. The systems analyst should endeavor to account for such unforeseen developments when designing for such an expensive long-term commitment.

There are currently two types of 35-mm microfilm available: *positive* and *negative*. Initially positive microfilm was the most widely used because only microfilm readers were of concern. Positive film provided the user a black print on a white background. The rapid rise of reader-printers has caused a trend toward negative film. Positive film prints look very much like photostats, that is, they have white letters on a black background. Negative film, on the other hand, provides a copy with black print on a white background. Equipment manufacturers have been working on printers for

FIGURE 15-5   A variety of microforms. Reading left to right, top row: negative micro-fiche (NMA standard), Recordax Micro-Thin jacket (COSATI standard), aperture card; middle row: 35-mm microfilm roll, Recordax 16-mm thread-easy microfilm magazine, another style of 16-mm magazine; bottom row: Recordak Microstrip holder. (Photograph graciously provided by Eastman Kodak Company, Rochester, N.Y.)

positive film to provide a black print on a white background. Color film, although beautiful in viewing, is still experimental in terms of reproduction.

There is a distinct possibility that standardization of continuous strip film will come with 16-mm microfilm. Its advantages include greatly improved film resolution, storage not on reels but on easy-to-use cartridges (see Figures 15-5 and 15-6), and reader-printers that need no manual threading and have push-button access to the needed page. The latter is made possible by coding each "page" or frame enabling retrieval through odometer readings or visually (see Figure 15-7). Copies made just by pushing a button save time, money, and errors in copying. Manufacturers claim that the high cost of purchasing this new equipment is inconsequential compared with the time saved in retrieval and printout by customers or by highly paid research people.

Microfilm rolls and cartridges generally are considered best for inactive files that no longer need updating. Updating of continuous roll film can be performed by only two methods. The first is to replace the entire roll or cartridge. This replacement often leads to the need for a new index if the cartridge/frame index is used for retrieval. The second method is to add on the end of individual rolls or at the end of the set of rolls. In this case either the current index or the historical index may have to be revised.

Another storage method is the *aperture card* (see Figure 15-5). This is another use of 35-mm or 16-mm film. An aperture card is simply an 80-column tab card upon which

FIGURE 15-6 A variety of microforms. Reading clockwise from bottom center: Recordax Microstrip holder, 16-mm microfilm roll, Recordax Micro-Thin jacket, Recordax 16-mm thread-easy magazine, another style of 16-mm magazine, and a positive microfiche. (Photograph graciously provided by Eastman Kodak Company, Rochester, N.Y.)

one or more frames of film have been mounted. The aperture card has been used successfully for the storage of voluminous materials such as engineering drawings. In 1966 the U.S. Patent Office contracted to put each patent granted since 1970 on aperture cards. At that time the Commissioner of Patents, Edward J. Brenner, estimated the cards would save $500,000 per year in filling the demand for 25,000 copies of patents every day.[3]

To confuse matters further, there are *opaque microcards*. These cards vary quite a bit in size, including 3×5, 5×8, 4×6, and 6×9 inches with equally varying reduction ratios. They, too, are available in both positive or negative form. The cards are opaque as the name implies and have the appearance of a black and white photograph.

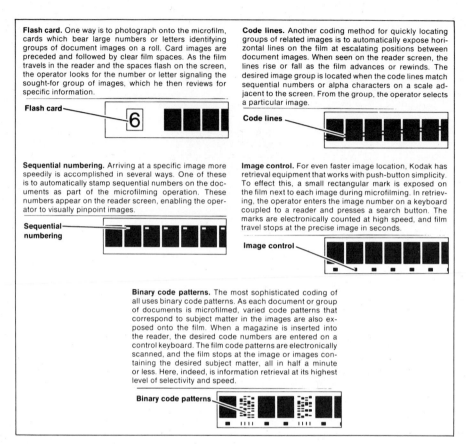

Flash card. One way is to photograph onto the microfilm, cards which bear large numbers or letters identifying groups of document images on a roll. Card images are preceded and followed by clear film spaces. As the film travels in the reader and the spaces flash on the screen, the operator looks for the number or letter signaling the sought-for group of images, which he then reviews for specific information.

Code lines. Another coding method for quickly locating groups of related images is to automatically expose horizontal lines on the film at escalating positions between document images. When seen on the reader screen, the lines rise or fall as the film advances or rewinds. The desired image group is located when the code lines match sequential numbers or alpha characters on a scale adjacent to the screen. From the group, the operator selects a particular image.

Sequential numbering. Arriving at a specific image more speedily is accomplished in several ways. One of these is to automatically stamp sequential numbers on the documents as part of the microfilming operation. These numbers appear on the reader screen, enabling the operator to visually pinpoint images.

Image control. For even faster image location, Kodak has retrieval equipment that works with push-button simplicity. To effect this, a small rectangular mark is exposed on the film next to each image during microfilming. In retrieving, the operator enters the image number on a keyboard coupled to a reader and presses a search button. The marks are electronically counted at high speed, and film travel stops at the precise image in seconds.

Binary code patterns. The most sophisticated coding of all uses binary code patterns. As each document or group of documents is microfilmed, varied code patterns that correspond to subject matter in the images are also exposed onto the film. When a magazine is inserted into the reader, the desired code numbers are entered on a control keyboard. The film code patterns are electronically scanned, and the film stops at the image or images containing the desired subject matter, all in half a minute or less. Here, indeed, is information retrieval at its highest level of selectivity and speed.

FIGURE 15-7   Automated microfilm retrieval methods. (Information and pictures graciously provided by Eastman Kodak Company, Rochester, N.Y.)

Transparent microforms are generally considered superior to opaque types because they cost less, they are more readable because the image is better, equipment is considerably better, it is easier to use higher reduction ratios thus using still less space, and copies cost less to make and are of superior quality.

A much more popular microform is *microfiche*, the latter part of the word probably derived from the French word "fiche" (pronounced fēēsh) meaning card. Microfiche is sometimes referred to as a unitized microform. It is generally a flat sheet of film, 4×6 inches in size, with a number of images (anywhere from sixty to ninety-eight) arranged in rows (see Figures 15-5 and 15-6). In reality, the "4×6 inches" size is known internationally as the ISO A6 size, or as 105 mm by 148 mm. This size is the only one that has been standardized by the International Organization for Standardization (ISO), the National Micrographics Association (NMA), the American National Standards Institute (ANSI), the Department of Defense (DoD), and the Committee on Scientific and Technical Information (COSATI). These organizations have the responsibility of recommending standardized methods for processing and handling microforms. The U.S. organizations must work with the international organizations so that technology can flow easily across national boundaries.

As with the other microforms, microfiche may be either positive or negative. If positive, it appears on the viewer as black print on a white background (see Figure 15-6). Positive "fiche" are good when large numbers of photographs are used because it is easier to interpret positive images. Negative microfiche, or white print on a black background (see Figure 15-5), are used when a paper copy is of major importance. Equipment is improving quickly, however, so this emphasis may soon lose its importance.

Microfiche received its greatest boost when various government agencies began disseminating their technical reports and specifications by this method. Today many engineers have desk-top viewers and store all the documents they need in one small drawer of their desk.

Microfiche advantages include the relatively low cost of the readers, the fact that full-size copies can be made quickly, easily, and inexpensively, and that wide distribution can be handled with little cost. This form can be used also for active storage files since it can be updated just by adding or pulling a microfiche (assuming an open-ended filing system). In addition, microfiche lends itself to a wide variety of applications including manufacturers parts catalog, in-house reproduction of research and development reports, maintenance information for complex equipment installations, efficient filing of customer invoices, and patient medical records.

Disadvantages include a lack of standardization in reduction ratios and formats with corresponding confusion in equipment. Cost of input equipment is high; one method of producing microfiche is by a step-and-repeat camera which costs more than cameras used for rolls, cartridges, or aperture cards. File integrity is also of major importance. Unless automated equipment is available, a good records administrator must oversee refiling procedures. Few people realize that each misfiled microfiche generally means other microfiche also will be misfiled in a building block fashion. Twenty percent of the file can, for all intents and purposes, be considered unusable when file integrity is lost. On the other hand, users tend to react negatively toward locks, keys, security, and steel cases. On the opposite side, however, is automated retrieval equipment for which no coding systems have been standardized. Another storage problem is that microfiche curls when improperly stored. If equipment does not have provision for keeping the microfiche packed flat and tightly together, users will soon be pulling out not one but several of the curled and bent microfiche.

Jackets are similar to microfiche. This technique uses roll microfilm which is cut and placed in a specific sequence within a "jacket." The jacket is usually of mylar film having good optical properties (see Figure 15-6). The advantages and disadvantages of jackets are similar to those of microfiche with one outstanding exception: the capability of adding, deleting, or making frequent changes is good with jackets. Once a jacket is filled and is ready to become a permanent record, it can be filmed and put into regular microfiche format. Two of the most frequent applications of jackets are in medical records and personnel records. Jackets are available in a variety of sizes.

Ultrafiche is the newest of the microforms, and it is surrounded by controversy. The process used by the companies producing ultrafiche is to first film the original on 35-mm and then further reduce it. Since it is still not very well known, reduction ratios are reported by varying sources to be from 60× to 150× and from 100× to 400× (× means times). One manufacturer claims 3200 pages to a single 4×6 inch transparency, while another claims 10,000 pages, also on 4×6, while still another puts 1000 pages on 3×6 using reduction ratios varying from 55× to 90×. Generally speaking ultrafiche images are reduced by at least 90 times.

The advantages of ultrafiche are obvious: they have greatly increased storage capacity, the cost of reducing the whole sheet is very inexpensive once the master is made, they are as easy to handle as microfiche, and manufacturers claim their equipment is inexpensive (always a relative term).

The disadvantages of ultrafiche include the alarming need for still another generation of unstandardized viewing and printing equipment. If a uniform system is to be developed, the existing commonly used reductions would have to be refilmed, or else the equipment would have to be considerably more sophisticated than it is now to accommodate all reductions. Cost of the original master is extremely high since it requires two steps in the process, under "clean-room" conditions. Additionally, the equipment itself is high priced, and the materials are metallic oxides requiring monochromatic light sources. Speculation concerning the tie-in of ultrafiche to data communication/computer facilities has yet to be proven.

No discussion of microform records would be complete without including COM, or *Computer Output Microfilm*. COM can be called the marriage of the computer to microfilm. It is a sophisticated new technology which has come about because of the computer's ability to overwhelm its users with paper and because of the limited speed of computer line printers. With COM, unwieldy paper is eliminated entirely, if desired. Instead the information may be put directly onto microfilm, or the information may be put onto magnetic tape to be put onto microfilm later at the convenience of the user.

The COM system design will take into account the accessibility of records, flexibility both in adding and deleting data in future applications, costs of microfilm preparation, recorders and readers, a cost/performance evaluation, and timeliness. It may become necessary to make the trade-off, or to sacrifice one item for another.

Advances in COM technology are taking place rapidly since it is obvious that to print a page and then microfilm the page is expensive and unnecessary in many cases.

Although high-speed impact printers will no doubt be in use for many years, the EDP manager may need to select a COM unit when computer usage changes from a high-speed reproduction unit to a true information tool. When evaluating a COM system some of the following advantages and disadvantages may need to be considered.

1. Adequate supplies of paper are getting more difficult to find. As a result, paper costs are rising and a shortage of business forms is on the horizon.
2. COM printers are one hundred times faster than impact printers.
3. CPU (Computer Processing Unit) time is significantly reduced over that of other methods.
4. New COM equipment is easier to use.
5. New COM equipment utilizes both dry and wet processing techniques. Film handling is eliminated in some.
6. It has been estimated that COM can reduce the cost of computer output as much as 80%.[4]
7. Labor costs are lower with COM since paper does not need to be decollated, burst, or bound.
8. COM can be updated easily.
9. COM is flexible and can be indexed easily for efficient retrieval.
10. COM has many applications.
11. COM equipment is expensive to purchase, with basic hardware and software costing anywhere from $75,000 to $150,000.[5]
12. Entry of a firm into COM usage is made easy by the use of service bureaus.
13. Storage space is reduced since voluminous paper records are no longer needed.
14. Shipping costs are reduced for reports that must be mailed.
15. COM has the ability to produce graphics.

Selection of the correct COM unit should take many factors into account and the systems analyst should play a major role in the decision-making process. The analyst will want to determine the functions for which the system is designed and for which it will be used. Some of the following steps should be undertaken before selection of a COM device.

1. *Check the current system of using microfilm to determine if a COM device is needed; that is, if a COM device will enhance the system.*

2. *Determine if only alphanumeric or graphic output, or both, are required. (Graphic units produce both kinds of output but are, naturally, more expensive.)*

3. *Determine if an on-line or off-line operation is required.*

4. *Check the output that the computer is capable of producing against the input acceptable to COM devices.*

5. *Check the form of output of COM devices in relation to the form required by the current microfilm usage.*

6. *Determine the needs for overlays and indexing.*

7. *Determine the expected quality.*

8. *Establish speed requirements.*

9. *Determine the need for special features such as on-line processing and color recording.*

10. *Check the price range that makes the acquisition of a COM device feasible. Also consider the operating cost.*[6]

When COM is selected as the processing medium, the firm must decide whether to implement an in-house COM processing facility or to use an outside service bureau. *Computing Canada,* assisted by Datacrown Limited, has listed the following criteria that should be used to determine whether a service bureau should be used.

1. Turnaround. The standard may be 24 hours, but some provide 4 hours.

2. Quality. Either method should adhere to standards set by the National Micrographics Association and the Canadian Micrographics Society. Quality control tests should be conducted at the end of each run.

3. Hours of service.

4. State of the art. Does the service bureau use a minicomputer to do processing that frees time from the host computer?

5. Is professional assistance available for selection of the proper applications and integration with other media for an effective system?

6. Is there a backup facility for emergencies?

7. Are appropriate cost accounting methods used?

8. Cost. Is the cost quotation best in terms of what you get, based on the above factors?

In summary, microforms are quickly becoming the records retention medium of the future. While the terms "video file," "remote terminals," and "optical encoding" are still relatively new, they will be commonplace concepts to the systems analyst in the year 2000.

SELECTIONS FOR FURTHER STUDY

Because improper records retention practices can put considerable strain of a firm's pursestrings and security, chapters on the subject will be found in most office management and systems texts. Also, since records retention is often treated as an integral part of forms control, the subject is frequently covered in forms control texts.

Kish and Morris' *Microfilm in Business* describes for records managers the various types of microforms, how they may be utilized, indexing systems for retrieval, cameras, readers, printers, and storage. This book has two important appendices: the first makes note of the Uniform Photographic Copies of Business Records as Evidence Act and discusses the legal aspects of microfilm for records retention, while the second discusses specifications for microfilm as required by the Department of Defense.

As noted, there are many laws which specify lengths of time during which a firm may be required to produce various records. The U.S. National Archives and Records Service publishes a *Guide to Record Retention Requirements*. This publication tells users (1) what federal records must be kept, (2) who must keep them, and (3) how long they must be kept. It is arranged by Departments (e.g., Labor, Postal Service, Transportation, etc.) and contains a thorough subject index. Since each item has its own individual number, finding entries is simplified.

Because records retention involves high-technology areas, the analyst's best reading may be in current periodical articles.

1. "An Aid for Effectively Designing and Documenting Micrographic Systems," *Microfilm Techniques*, volume 2, number 6, November 1973, p. 10.

2. AUTOMATED EDUCATION CENTER. *Records Management Handbook*. St. Clair Shores, Mich.: Managment Information Services, nd (looseleaf).

3. AVEDON, DON M. "Terms and Definitions Used in Standardization Work," *Microfilm Techniques*, volume 5, number 1, January 1976, p. 20.

4. AVEDON, DON M. "The User's Guide to Standard Microfiche Formats," *Microfilm Techniques*, volume 6, number 6, November-December 1977, p. 14+.

5. BJORK, G.V. "A Drawing Control System at Sandia Corporation: and the Engineering Paperwork Necessary To Do It," *Microfilm Techniques*, volume 3, number 2, March 1974, p. 6+.

6. CANNING, RICHARD G. "Long Term Retention of Data," *EDP Analyzer*, volume 11, number 7, July 1973.

7. CARDELLO, JOHN. "Honeywell Information Procedure for Microfiche," *Microfilm Techniques*, volume 3, number 6, November 1974, pp. 14–19.

8. CONNORS, RICHARD J. "Microfilm: Past, Present, and Future," *Infosystems*, volume 20, number 3, March 1973, pp. 39–41.

9. *Corporate Records Retention.* New York: Controllership Institute Research Foundation, 1958.

10. DORFMAN, HAROLD H. "Where Do I Start?" *Microfilm Techniques*, volume 2, number 3, May 1973, p. 14+.

11. EXELBERT, RODD, and BOB SAMPLE. "Micronet: World's First Paperless Office Opens," *Information & Records Management*, volume 13, number 6, June 1979, pp. 54–55.

12. FULLMAN, FRANCIS A., JR. "Legal Aspects of Corporate Records Problems," *Information & Records Management*, volume 7, number 9, October 1973.

13. GILDENBERG, ROBERT F. *Computer-Output-Microfilm Systems.* New York: John Wiley and Sons, Inc., 1974.

14. HARPOOL, JACK D. *Business Data Systems: A Practical Guide.* Dubuque, Iowa: William C. Brown Company, 1978.

15. HEDLIN, EDIE. *Business Archives: An Introduction.* Chicago: Society of American Archivists, 1978.

16. "How To Cope With Three Million Documents: And Put a House Full of Records in Ten Square Feet," *Microfilm Techniques*, volume 2, number 4, July 1973, p. 6+.

17. "Important Criteria on When To Use COM Processing," *Computing Canada*, April 2, 1979, p. 18.

18. KISH, JOSEPH L., JR. "Active Systems To Stalk a Paper Tiger," *Business Week*, April 24, 1978, pp. 12–13.

19. KISH, JOSEPH L., JR. "Effective COM Evaluation," *Computerworld*, volume 7, number 3, January 1973.

20. KISH, JOSEPH L., JR., and JAMES MORRIS. *Microfilm in Business.* New York: Ronald Press Company, 1966.

21. KISH, JOSEPH L., JR. "Satisfying You: And the Government," *Business Week*, April 24, 1978, pp. 17–18.

22. MAEDKE, WILMER O., et al. *Information and Records Management.* Encino, Calif.: Glencoe Publishing Company, Inc., 1974.

23. MATERAZZI, ALBERT T. "Archival Stability of Microfilm: A Technical Review," *Microfilm Techniques.* Issued in three parts, November-December 1978, January-February 1979, March-April 1979.

24. "Microfilm Takes Off at Sikorsky Aircraft," *Microfilm Techniques,* volume 2, number 5, September 1973, pp. 9–11.

25. "Microfilming Records at the National Level," *Microfilm Techniques,* volume 7, number 1, January-February 1978, pp. 15–25.

26. MOORE, RUSSELL F., ed. *AMA Management Handbook.* New York: American Management Association, 1970.

27. MOSKERINTZ, MICHAEL. "Converting to Micrographics," *Microfilm Techniques,* volume 7, number 5, September-October 1978, pp. 10–12.

28. PLACE, IRENE M., and E.L. POPHAM. *Filing and Records Management.* Englewood Cliffs, N.J.: Prentice-Hall, Inc., 1966.

29. *Records Management.* Cleveland, Ohio: Association for Systems Management, 1976.

30. *Reducing Made Easy: The Elements of Microfilm.* Durham, N.C.: Moore Publishing Company, 1976.

31. SAA Business Archives Committee. *SAA Directory of Business Archives in the United States and Canada.* Chicago: Society of American Archivists, 1975.

32. SMITH, CHARLES. *Micrographics Handbook.* Dedham, Mass.: Artech House, 1978.

33. SNYDER, PAUL D. "Computer Output Microfilm," *Journal of Systems Management,* volume 25, number 3, March 1974, pp. 8–13.

34. TAVERNIER, GERARD. *Basic Office Systems and Records.* Brooklyn Heights, New York: Beekman Publishers, Inc., 1971.

35. U.S. National Archives and Records Service. "Guide to Record Retention Requirements," *Federal Register,* volume 36, number 39, part II, February 26, 1971, pp. 3701–3792.

QUESTIONS

1. What is the single most important function of records retention?
2. Define active records.
3. What does the record retention schedule show?
4. Name six points of information the analyst should obtain during the physical inventory of records.
5. What is the most economical and quickest filing system for the storage area?
6. Where should copies of vital records be stored?

7. Name three advantages of using microfiche.

8. For what does COM stand?

9. Discuss some of the advantages and disadvantages of microfilm.

10. Discuss important features you should consider in selecting a COM unit.

11. What should be considered when choosing between an in-house or an outside service bureau that specializes in COM?

SITUATION CASES

CASE 15-1

The files of a company that specializes in installing burglar alarms were so numerous that the active files were difficult to locate among the inactive files. As a result, the company procured the assistance of a systems analyst to set up a records retention schedule. A recent college graduate, Doris Fuller, was hired to perform the analysis.

Approaching the job with much enthusiasm, Ms. Fuller immediately started to take a personal inventory of the records. She had been given the authority to go into the various departments and examine their files. For each department, she made a list of the type of filing system in each department, the titles of the files, the size, quantity and type of copy of the records contained in each file, and the inclusive dates of records accumulation. She used a worksheet, such as the one in Figure 15-8, to obtain this information.

During examination of the files, Ms. Fuller discovered that several departments did not have consistent filing systems. Thinking that a consistent filing system was important for a records retention program, she decided to reorganize the filing systems of several departments. Her involvement with the filing systems caused her to get behind in the analysis.

After reorganizing the departmental files, Ms. Fuller proceeded with the analysis by obtaining the retention requirements of the organization's insurance company, and then she assigned appropriate retention times to each type of record. Next, she obtained the signed concurrence of each department head.

In setting up the retention schedule, Ms. Fuller suggested that all inactive records (those used less than four times a year) that were needed to satisfy retention requirements be stored in the supply room in boxes. The records were to be filed by title. Each record was to be listed in alphabetical order using the storage box label form (Figure 15-9) which was to be a three-part form. The top copy, a gummed label, would be put on the box of records before being transferred to the supply room. The second copy was to be kept by the sending department which was the active life area. The third copy

RECORDS INVENTORY WORK SHEET	
Department	Analyst
Title of Record	Form Number
Type of Copy (original, carbon, etc.)	Size of Record in Inches
Inclusive Dates of Accumulation in Department	Quantity of Record

Number of Times This Record Is Used Per Year

Current Filing System (Active Life Area)

Storage Filing System (Storage Life Area)

Active Life + Storage Life = Total Life (Then Destruction)

Active Life	Storage Life	Method of Destruction
		☐ ☐ ☐
		Burn Shred Trash

Remarks (Relationship to Other Records, Legal Requirements)

FIGURE 15-8   Records inventory work sheet.

Title of Record (Form Number)	Box Location Number

Inclusive Dates Within This Box	From Department?

How Filed Within This Box

☐ Alphabetic   ☐ Numeric   ☐ Other (Explain)

Method of Destruction   **Date To Be Destroyed**

☐ Burn   ☐ Shred   ☐ Local Trash

Enter the Date (Month & Year) and Department Every Time a
Record Is Requested

Date	Department	Date	Department	Date	Department

FIGURE 15-9 Storage box label form.

was for the supply room files where it would be filed alphabetically by the title of the record.

The records within the box were to be kept in the same filing sequence as they were when sent from the active life area. When the supply room received a box of records, it was to be checked to ensure the records were properly identified and sequenced. Then, a box location number was to be stamped on the box and also entered in the upper right corner of the storage box label form. The numbering was to help the supply room store the files in the most efficient manner.

Before a box could be placed on the shelf, the reference file would be annotated to facilitate later retrieval or destruction of the records. The reference file contained cards with the destruction month and year; they were filed in consecutive order. The supply clerk would enter the box location number on the card that contained the month and year in which the box of records was to be destroyed. Each month the scheduled boxes would be located, pulled, and destroyed. Before destroying the records, however, the supply room supervisor would check to be sure that none of the records were still needed. This would be done by checking how many times the records had been accessed in the recent past. Each time a record in storage was accessed, this information would be written on the botton of the storage box label.

The analyst suggested that all of these records, including vital records such as the articles of incorporation, agreements, and contracts, be stored in the basement supply room so they would be out of the way, yet accessible when needed.

Ms. Fuller submitted her retention program to the manager of the company for approval. The manager endorsed the plan after commenting that the study took a long time. The retention schedule was initiated and Ms. Fuller was given a new assignment.

QUESTIONS

1. Did the analyst check into the retention time requirements thoroughly? Where else could she have looked?

2. During this study, did Ms. Fuller stay within the scope of the problem?

3. With regard to storing the records, what mistake did the analyst make?

CASE 15-2

The Accounting Department of Slendra Manufacturing had experienced some difficulty in handling and storing old records. The Accounting Office retained old records in filing cabinets located in a small corner of the office. Over the years, this area had become the location of all active and inactive financial records (i.e., contracts and vouchers). Available space had been exhausted, and frequent use of certain old records by Internal Audit had left the area in unsightly disarray.

The Accounting Department manager obtained the services of a systems analyst to develop a records retention program for the department. The analyst was given complete authority to examine the records of each section and interview the individual supervisors.

The analyst began the assignment by conducting an inventory of all active records. During this examination, he decided that all current year records should be classified as active; certain other types of records, with ages ranging from one to three years, should be classified semiactive. This was done to satisfy federal government requirements for accurate prior year accounting records for Slendra contracts. Some of these other records were used also by Internal Audit to satisfy their requirements. None of the other records were used by the Accounting Department.

The analyst became familiar with the filing system which was basically numeric. That is, each document was recorded and filed by a number. A contract could be filed with a pay voucher in the active files. This was done to create an efficient cross-reference system.

The analyst decided that records that could be moved to the warehouse should be filed by the title of the document in order to expedite retrieval, even though this would require refiling in the storage area. He could not get involved with the filing system of the Accounting Department office, so Slendra would have to live with the increase in cost and effort to refile the records.

The next goal of the analyst was to develop a retention schedule for both the active records in the Accounting Department and the records that would be moved to the storage area. He decided to keep the records in the Accounting Department for a period of two fiscal years. The extra space required for a third prior year was not warranted since the demand for them was small. This schedule was approved by all concerned with the active records.

The retention program in the warehouse was left to the analyst. He decided on a system that would identify easily the types of records in the boxes. To facilitate access, each box was numbered. A list of the boxes, and the records in each, would be filed in the warehouse and the Accounting Department. The system also would identify which boxes were accessed for particular record or report. The destruction schedule required burning boxes that were left untouched for a period of one year. The analyst thought this was adequate because records over three years old were never used by the Accounting Department. The warehouse foreman reluctantly approved the program but he did not want the extra responsibility of caring for the boxes of records.

QUESTIONS

1. Was the analyst correct in the decision to have the records refiled in the storage area?

2. Was the retention program in the warehouse satisfactory? Discuss why or why not.

FOOTNOTES

1. Franklin D. Crawford, *The Microfilm Technology Primer on Scholarly Journals* (Princeton, N.J.: Princeton Microfilm Corporation of New Jersey, 1969), p. 5.

2. "Out of the Desert, Into the Green." *Office Products,* Vol. 135 (1), January 1972, p. 28.

3. "Wallet Libraries," by Lee Berton. *Wall Street Journal,* CLXVIII, No. 62, September 28, 1966, p. 1.

4. "The New Freedom in Computer Output: Part III, Computer Output Microfilm (COM)," by Dale Rhodabarger. *Computer Decisions,* Vol. 2. (8), August 1979, p. 48.

5. Ibid., p. 50.

6. "Selecting the Right COM Unit," by George H. Harmon. *Datamation,* December 1969, pp. 102–106.

CHAPTER 16

REPORT ANALYSIS

As organizations grow in complexity, the interacting functions become more difficult to coordinate. Today's systems analyst is very involved in the effort to provide this needed coordination through systems of communication and action. A technique called Report Analysis has been developed to aid in this process.

REVIEWING MANAGEMENT REPORTING SYSTEMS

Because of the rapidly changing business environment, many management reporting systems have not kept pace with management's needs. In many cases the traditional paper reports (manual or computer generated) could be replaced by, or supplemented with, computer terminal inquiry and display capabilities. As a result there is a need for periodic reviews to keep pace with the changing environment. In addition, there are special circumstances that arise which require a revision in the reporting scheme. For example, a reporting system revision would be warranted by changes in

1. The size of the organization.
2. The structure of the organization.
3. Market conditions.
4. Management style or approach to decision making.
5. Management personnel.
6. Operational technology.

Most organizations should plan to review their reporting systems on a scheduled basis, with the time interval between reviews being relatively short. An effective review of an on-going reporting system would be composed of three fundamental phases.

1. Analysis of the information supply and the related demand.
2. Review of the utility of current reports and data collection techniques.
3. Preparation of a report on suggestions for improvement.

After all the reports have been reviewed, the data collection system should be analyzed to determine how it can be utilized best to meet the information reporting needs. Some of the questions that must be considered are

1. Does the system facilitate the summarization of data as it is collected?
2. Is all the data that is collected ultimately used in one or more reports?
3. Can the data collection system itself be utilized to meet some reporting requirements?
4. What steps can be taken to reduce errors?
5. Are controls being utilized to provide for the timeliness and accuracy of resulting reports?
6. Is there any way to reduce the manual effort in the reporting system?
7. Is the system properly documented?
8. Are all the reports being used or should some be eliminated?

After the review has been completed, a report presenting the results of the survey should be prepared for top management. The report should comment on the strengths and weaknesses of the existing reporting system, and it should contain specific recommendations for improvement. In order to identify the weaknesses of the existing reporting system, a thorough report analysis must be conducted. The remainder of the chapter is devoted to this subject.

WHAT IS REPORT ANALYSIS?

The objective of report analysis is the elimination of time and money spent on the preparation and distribution of unnecessary or redundant reports. In small firms, managers can personally gather the information they need for decision making by way of direct observation and contact with operating personnel. As firms grow, however, the report becomes a necessary medium for transmitting information to management. Report quality then becomes a vital factor in the firm's operations. Consistent, Accurate, Timely, Economically feasible, and Relevant (CATER) reports are necessary to assist management in its guidance of the firm. It is vitally important that reports supply information in the most useful form to the people who need it. Reports must be tailored to meet the specific needs of the management team.

Report analysis has become even more important since the advent of the computer because the computer can print reports at thousands of lines per minute. Too many reports can be worse than too few reports, especially when the information needed by management is spread too widely through the reports for convenient reading.

Usually, no single group or person is assigned the responsibility for developing a report-analysis function until the firm finds itself in trouble with too many reports. The systems analysis department should handle this function from a central base, similar to the approach recommended for forms control. The report analysis group or analyst performs a coordinating function rather than one of an authoritarian nature. The analyst should obtain a copy of every report, and any instructions that go with it. A report-number file, a functional file, and a cross-reference file should be developed as explained in the section "Objectives of Forms Control" in Chapter 14.

Reports can be classified into the following four general categories: *action reports,* which initiate or control a necessary procedure or operation; *informational reports,* which provide data or information for further analysis and control; *reference reports,* which keep the manager informed on operations; *feeder reports,* which consist of bits of data to be used later in conjunction with other data, another report, or for accumulating data for a decision.

Within each of the above four general categories of reports, a specific report might be presented using one of the following formats. First is the *scheduled report.* A scheduled report is one that is prepared and distributed at a fixed time, such as once a week, once a month, or once a year. The person who receives a scheduled report expects the copy at a prescheduled time. Second is the *on-demand report*. An on-demand report is one that is generated upon the user's request. The report may be generated instantly or it may take a few hours or even overnight to obtain the report. The point is that when the user demands a copy of the report, the report is generated and a copy is delivered to the user. Third is the *exception report*. An exception report is a report that is generated only when certain parameters are out of line with what is expected of these parameters. When the exception report is set up, the analyst decides what will be normal, within a given set of parameters. The computer-based system automatically checks the input data against these parameters and a report is prepared only if the data are outside the limits of the parameters. If the data are within the limits that were originally specified when the report was set up, then no report is generated. One of the advantages of a computer-based reporting system is that exception reports are generated easily. An exception report saves the time of management because no report is generated unless the input data are outside of some predetermined parameters. Because of its importance, exception reporting is again discussed later in this chapter.

In summary, scheduled reports appear on a manager's desk at fixed time intervals, on-demand reports are generated only when management asks for a report, and exception reports are generated only when the input data is outside of some preset limits that were developed originally by the analyst and the management of the company.

ANALYZING REPORTS

The analyst should proceed to analyze the reports to determine whether any can be eliminated, combined, rearranged, or simplified, or whether new reports are required. The first step is to collect one copy of each report. Next the frequency of each report should be determined, that is, how often it is prepared per week or per month. The distribution of each report (who gets it and why), the total number of copies being prepared, and any other relevant information about the reports have to be evaluated. As mentioned, if the *report analysis* is to be an on-going program, the analyst should handle it like *forms control* (see Chapter 14). In analyzing each report, the analyst should consider the following points.

1. The amount and level of detail.
2. The effect, or lack thereof, that each piece of information will have on management's decision-making capability.
3. The extent or completeness of the information provided. In some cases, two ineffective reports can be combined to produce one that is effective.
4. The degree of accuracy.

5. The effectiveness of the presentation. A good report presents information in a manner that will be an aid in the decision-making process.

6. The clarity of the report format. Proper layout can make a report much easier to read and understand, thus also making it more effective.

7. Key information should be highlighted. For example, the use of comparative data helps pinpoint critical information.

8. The report formats should be consistent from one period to another. When report formats are stable, it is much easier to review data from period to period.

9. The report "package" should be organized logically. Summary information should be on the first page. It should provide a clear index to each section of detail that supports the summary.

10. The report should be structured to provide answers to the questions that usually are asked when the report is reviewed.

After each report is studied, the analyst still has to learn the value of each report to the ultimate users. This is accomplished best through interviewing the persons to whom the report is distributed. Questionnaires do not work well because people are especially reluctant to criticize or order the cancellation of a report in writing. The same people, however, may be quite candid in a discussion of the report. The analyst must be certain to interview every person on the report's distribution list. The following is a list of questions which the analyst might ask to determine the importance of the report being studied.

1. How is this report essential to the work of your department?

2. How often do you use your copy of this report?

3. How many data or pieces of information on this report are *not* used?

4. How many people use the report?

5. Are the data and information on this report necessary for
   a. making day-to-day action decisions?
   b. establishing control over your operations?
   c. checking the accuracy of specific procedures?
   d. keeping informed on general conditions?

6. What would be the effect on your work if you
   a. did not receive this report?
   b. received this report less frequently?
   c. received less data or information than at the present time?
   d. received more data or information than at the present time?

7. What other reports, records, or forms are prepared from the data contained on this report?

8. Can the data or information on this report be located in any other sources?

9. Is this report easy to understand and easy to use?

10. How long do you keep your copy of this report?

11. How and where do you file your copy of this report?

12. How often do you refer to this report after its original distribution and use?

13. The cost of preparing this report has been estimated at $——. Do you consider your use of the report worth this expense?

The analyst now studies all answers to the questions for the report being evaluated. For example, in questions 1 and 2 the analyst is trying to learn the user's ranking of importance for the report. In question 5 the analyst is cross-checking the user's ranking of importance because any report used to make day-to-day action decisions is much more important than a report that just keeps the user informed on general conditions. Questions 3 and 8 show what parts of the report may be eliminated, from that specific user's viewpoint. Questions 4 and 7 might uncover heretofore unknown, bootleg, or unauthorized uses of the report. Questions 6, 9, 10, 11, and 12 are used to get general background on the feelings of the user about the real importance of the report. And finally, question 13 reveals the economic importance of the report. If *all* the users say "It is not worth the expense for me but think of the other users," then perhaps the report should be eliminated.

After careful study the analyst should decide upon some definite recommendations that conclude whether any reports can be eliminated, combined, rearranged, or simplified, or whether new reports should be developed. The analyst should prepare a summary of the report-analysis survey. The summary should mention whether interviews and/or questionnaires were used. Also note any discrepancies between actual utilization of a report and the claimed utilization of a report. Make graphical comparisons between the current and any proposed reports, including detailed instructions on how the new proposed reports will be prepared. Show the recommended distribution of all reports even if it is the same as before the report-analysis survey. Make recommendations and cost comparisons between the old and any recommended new reports. Remember, economy is usually the best method of justifying change in a profit-oriented firm.

CRITERIA OF A GOOD REPORTING SYSTEM

A good reporting system generates just the right amount of reports, at the right time, with the right amount of information. All too often a reporting system either buries the manager with too many reports or details, or it does not offer enough *usable* information.

A good reporting system conforms to the company's formal communications structure so that accountability for results is clarified in the organization. For each separate department or functional area the reporting system reports on all the important controllable elements of performance.

A good reporting system represents a plan of control. The information furnished by each report should tie in with the other reports for good coordination and standardization of concepts. As reports go up through the higher levels of the organization, they should become more condensed. The detail belongs in the lower-level reports.

A good reporting system is under continuous surveillance so it can be modified to meet changing needs. It is very important that an easy method exist for managers to delete unnecessary reports or to criticize the reports that are not doing their job.

As mentioned, a good reporting system puts out reports that are consistent, accurate, timely, economically feasible, and relevant; and the reports are distributed only to those persons responsible for the area being reported upon.

CHARACTERISTICS OF A GOOD REPORT

A good report covers an element of performance that has a significant bearing on the goals of the area receiving the report. It measures performance by comparing actual results with the planned or forecasted results. It reports only the essentials so the manager can quickly learn the whole story.

Managers are at the mercy of their reporting system. When they are not given consistent information, decision making is undertaken at reduced confidence levels and intuitive judgment is the order of the day. Managers often get

1. Too much information and information that is not relevant.
2. Too little information and too few specific facts.
3. No information because it is suppressed. This is caused by reluctance to impart information that may reflect unfavorably upon some department.
4. Information that is too late.
5. Information that is unverified. Information of questionable validity may be used, but with the highest risk.

A good report is aimed at controllable items. It segregates controllable and noncontrollable items so the report is easy to understand and easy to use. It focuses attention on out-of-line performance by accenting significant trends.

A good report appears in a format that is easy to understand, and it is expressed in the language of the report user. For example, if the report user wants the figures in person-hours per week, report it in person-hours per week—so long as all managers are able to make conversions to each other's preferred terms and ratios. Or, better yet, let the computer make the conversion.

WEAKNESSES OF REPORTS

We know that reports are essential to the operation of an organization; however, they generally have their weaknesses. Unfortunately, reports are not usually designed to forecast or promote action; they are designed to report historical actions, such as accomplishments. So, for example, why not publish a "cost increase report" rather than a "cost reduction report"? (Because costs are increasing more often than they are being decreased.) This focuses attention on the costs that are increasing the fastest.

EXCEPTION REPORTING

Exception reporting is a convenient tool for management control. Although it is not a new technique, it is being used more successfully than ever now that the computer is here to process the huge quantities of data generated by today's complex organization.

Basically, exception reports spotlight only the unusual situations. All situations not reported can be assumed to be normal. Exception reporting facilitates the management-by-exception principle, wherein

1. Management plans are made for operating a program.
2. Milestones and standards are agreed upon for measuring program performance.
3. Valid methods of measuring *actual* performance are devised.
4. Report formats are designed to spotlight only the significant deviations from planned performance.
5. The computer usually is used to process the data and print the reports, so that the information is available in time for meaningful action.
6. The reports are distributed to the personnel responsible for keeping plans on the right track. If a report shows a detrimental deviation from standard, either corrective action is taken to put things back on course, or the plan or standard itself is adjusted.

Exception reporting generally is used as a tool for the control of a function. For example, the *quality* control of a firm's products is becoming an increasingly important focus. Product quality is a vital weapon of business competition, and must be preserved at some economic standard. Anything that deviates from this standard is reported.

A firm may decide to install a system of reports which spotlight unusual situations or trends in the following.

1. Customer complaint rate.
2. Warranty claims.
3. Product returns.
4. Gain or loss of customers because of quality.

For each of these items, some rate must be agreed upon as standard, either in economic terms or in whatever terms are appropriate to the product or company. For example, in a given company it may be decided that product returns for quality deficiencies will be considered normal if they are no higher than 2% of sales. Management may decide to apply this standard to several of their products.

Next, a valid means of obtaining the data needed to measure the dollar value of returns at each of six outlets must be found. They may solve this formidable problem with a central computer receiving the data from terminals in the outlet stores. Even though enough data may be collected to present a detailed quality control report for each product, the final report may contain only a handful of problem situations (Figure 16-1) because the final report only reports exceptions to the 2%, as stated above.

The report also could contain analysis data related to the problem tools, such as last month's figures, department responsible, actual defects reported, and so on. All of these things relate to the exceptional cases listed on the report. The report should not, however, list the figures for tools that are not causing any problems. Instead the report

---

Tools with return rate over 2% of all tools sales due to quality deficiencies

For the month of October

Problem tools	Return rate (% of sales)	Outlets reporting problem
40162 Wrench	6.4	1 2 3 4 5 6
40671 Pliers	5.2	1 2 3
51611 Clamp	3.1	1 2 3
62361 Hammer	2.7	2 3 5
No other problems	- - -	- -

FIGURE 16-1 Exception report.

reads at a glance since only the problems are reported. All other products (there may be 1000 more) are assumed to be in control (less than 2% returns) and require no special attention.

Many problems exist in today's complex organizations, but the serious problems are few and far between. Exception reporting enables management to concentrate its action where it is needed most. And computerized exception reporting is enabling the organization to make a faster response to trouble than ever before in business history. Some other areas where it has been employed successfully include: (1) out-of-balance conditions in accounting systems; (2) inventory stock-outs and back-orders; (3) micro-electronic part failures; (4) project delays; (5) excess overtime worked; (6) excess and obsolete inventory reports; and (7) budget overruns.

Our examples above are only a few of the hundreds of exception reporting applications now in common use in business and industry.

SELECTIONS FOR FURTHER STUDY

Most of your knowledge on report analysis can be gained from a liberal amount of common sense applied to an excellent comprehension of the needs of the system users. Many of the books already mentioned may be of benefit, but it should be noted that information on this subject is very sparse. Report analysis became necessary with the advent of large-capacity computers and their high-speed printers. Computer-generated reports mushroomed and many managers were flooded with reports. Many of these reports were not needed, were little understood, and were not wanted. Ackoff commented[1] that most management information systems were designed on the assumption that managers suffered from a *lack of relevant information.* He felt managers suffered more from an *overabundance of irrelevant information.* As a result, analysis and refinement of these computer-generated reports is now an important function in computerized organizations.

1. ALBRECHT, LEON K. *Organization and Management of Information Processing Systems.* New York: Macmillan Publishing Company, 1973.

2. ALEXANDER, MILTON J. *Information Systems Analysis: Theory and Application.* Chicago: Science Research Associates, 1974.

3. BENWELL, N.J. *Data Preparation Techniques.* Washington, D.C.: Hemisphere Publishing Corporation, 1976.

4. BOUTELL, WAYNE S. *Computer-Oriented Business Systems, 2nd edition.* Englewood Cliffs, N.J.: Prentice-Hall, Inc., 1973.

5. ELLIOT, C. ORVILLE, and ROBERT S. WASLEY. *Business Information Processing Systems, 4th edition.* Homewood, Ill.: Richard D. Irwin, Inc., 1975.

6. ENGER, NORMAN. *Management Standards for Developing Information Systems.* New York: American Management Association, 1977.

7. KENNEDY, MILES. "Exception Reporting Systems," *Ideas for Management.* Cleveland, Ohio: Association for Systems Management, 1973.

8. KENNEY, J.W. "Financial Reporting, RVFQ, and the Computer," *Data Management,* volume 10, number 3, March 1972.

9. KING, DONALD W., and EDWARD C. BRYANT. *The Evaluation of Information Services and Products.* Washington, D.C.: Information Resources Press, 1971.

10. LANCASTER, F. WILFRED. *Information Retrieval On-Line.* New York: John Wiley and Sons, Inc., 1973.

11. LANCASTER, F. WILFRED. *Information Retrieval Systems: Characteristics, Testing and Evaluation.* New York: John Wiley and Sons, Inc., 1968.

12. LANCASTER, F. WILFRED. *Toward Paperless Information Systems.* New York: Academic Press, Inc., 1978.

13. MARTIN, MERLE P. "Making the Management Reports Useful," *Journal of Systems Management,* volume 24, number 5, May 1973.

14. MURDICK, ROBERT G., and JOEL E. ROSS. *Information Systems for Modern Management, 2nd edition.* Englewood Cliffs, N.J.: Prentice-Hall, Inc., 1975.

15. MYERS, CHARLES A. *Impact of Computers on Management.* Cambridge, Mass.: MIT Press, 1967.

16. ORLICKY, JOSEPH A. *Successful Computer System: A Management Guide.* New York: McGraw-Hill Book Company, 1969.

17. SANDERS, N. *The Corporate Computer: How To Live with an Ecological Intrusion.* New York: McGraw-Hill Book Company, 1973.

18. SHARRATT, J.R. *Data Control Guidelines.* Rochelle Park, N.J.: Hayden Book Company, Inc., 1976.

19. STEWART, ROSEMARY. *How Computers Affect Management.* Cambridge, Mass.: MIT Press, 1972.

20. WEAVER, BARBARA N., and WILEY L. BISHOP. *The Corporate Memory: A Profitable, Practical Approach to Information Management and Retention.* New York: John Wiley and Sons, Inc., 1974.

21. WITHINGTON, FREDERIC G. *The Use of Computers in Business Organizations, 2nd edition.* Reading, Mass.: Addison-Wesley Publishing Company, Inc., 1971.

22. WOOLDRIDGE, SUSAN. *Computer Output Design.* New York: Van Nostrand Reinhold Company, 1975.

QUESTIONS

1. With regard to report quality, to what does the acronym CATER refer?
2. What are "feeder reports"?
3. When is an exception report generated? How does an exception report save the time of management?
4. After each report has been studied, what does the analyst still have to learn? How is this best accomplished?
5. What does a good reporting system generate?
6. What does a good report cover and measure?

SITUATION CASES

CASE 16-1

After noting a seemingly unwarranted cost increase within the Production Division of a major manufacturing concern, the division's vice-president requested that the firm's systems analysis section study the division and submit recommendations to achieve cost reductions. As her part of the study, a junior analyst was assigned the task of analyzing all reports originating from, and received by, the Production Division. This portion of the overall study would enable management to determine if there was a correlation between the increasing costs of the division and the management reporting system.

Upon completing the initial steps of the analysis (problem definition, outline, etc.), the analyst visited the firm's data center to obtain copies of all reports issued or received by the Production Division. She then determined the frequency of each report as follows.

PERCENTAGE OF TOTAL REPORTS	
Issued daily	71%
Weekly	17%
Monthly	8%
Quarterly, semiannually, annually	3%
On request	1%

After determining the frequency of the reports, the analyst separated them into four major categories: action, information, reference and feeder. From this, she determined that the majority of the action reports were in the "on request" category.

At this point, the analyst felt it was necessary to obtain the users' viewpoints and feelings toward the reports. She prepared a questionnaire for distribution, which appears as Figure 16-2. By the return deadline, the analyst had received completed

From: Joan Gilbert, Systems Analyst
To: _____

Survey of Usage of: _____

1. As compared to other reports you receive, do you consider this report:
   a. More important?
   b. Of equal importance?
   c. Of less importance?
   d. Totally insignificant?

2. Do you use this report:
   a. On a day-to-day basis?
   b. For establishment of controls?
   c. For accuracy checks?
   d. For general informational purposes?
   e. Other: _____ .

3. How much data or information contained in this report is not used?

4. How many of your personnel use this report?

5. What other reports, records, or forms are you aware of that are totally or partially compiled from data contained in this report?

6. Can the data or information included in this report be located in any other sources also available to you?

7. Do you consider this report easy to understand and use? (Please qualify your answer.)

8. On the average, how long do you keep this report?

9. Where and how is this report filed?

10. How often do you refer to this report after its initial distribution and use?

11. Do you consider the data or information contained in this report essential to your work?

FIGURE 16-2 Questionnaire.

questionnaires totaling 87% of those sent. She then proceeded to evaluate the questionnaires as part of the final analysis.

After a careful review of the findings concerning frequency and usage, and the results of the survey, the analyst compiled some recommendations, by report, for combination, rearrangement, simplification, and if needed, elimination. The analyst felt that too many informational and/or reference reports were being distributed too frequently. These included separate reports on nearly every item being sent from the Production Division, and a separate report on the cost figures for each item. She therefore recommended that a summarization of these reports be prepared on a weekly, quarterly, semiannual, and annual basis, as needed. In addition, she suggested that a current report on a specific item or cost could be obtained on a request basis, thus saving the time and money previously wasted in searching through numerous reports for one item. The analyst estimated that this procedure would eliminate approximately 72% of the current reports concerning the Production Division. This, in turn, would result in a total savings of approximately 17% of the Production Division's total budget.

The analyst supported these findings in the analysis summary by showing an actual usage of current reports of about 18% as compared to a claimed usage of over 63%. She also mentioned this in the summary, and included a copy of the questionnaire. In addition she outlined procedures for preparation of the proposed reports, included new form layouts, procedures for improved data collection, a recommended distribution schedule for the proposed reports, and a graphic comparison of the current and proposed reports.

After a careful review of the report analysis and accompanying summary, the analyst presented them to her group leader for review and inclusion in the overall study of the Production Division.

QUESTIONS

1. What was the main fault in the method used by the analyst to survey users and originators of the Production Division reports?

2. Which, of the thirteen items listed in this chapter concerning report analysis, did the analyst omit from the survey?

3. The analyst's group leader noted an omission in the report analysis summary, which might cause rejection by management. Review the final two paragraphs of the case and state what was omitted by the analyst in her summary.

CASE 16-2

To ease the ever-increasing complexity of modern business operations, the president of a small, but rapidly growing, stereo components company, decided to acquire a computer. Not only would the corporate image brighten, but the coordination and processing of all paperwork would improve. All functional reports would be generated by the computer. Communications to distant points of distribution would be instantaneous. A terminal/computer tie-in system with the company's distributors was created. These decisions, like all decisions, were made by the president since he firmly believed that centralized control of all major decisions rested with him.

The president next established a quality control section and designated measures of performance for each of the product divisions. A weekly quality control report was to be generated by the computer and it would list the percentage of returned merchandise to total sales. The president felt that he could immediately correct any deviations from established quality control standards if there was a weekly scheduled report. The three departments also were told that frequent on-demand reports would be requested by the president so he could spot check quality control performance.

Since implementing the computerized reporting system, the president has been working 10 to 12 hours a day because of the expanding number of reports directed to his attention. Too many reports had lessened his ability to respond to problem areas and had made him unusually short-tempered.

Realizing there was a serious problem, the president requested that a consulting systems analyst conduct a detailed study of the reports. The analyst first studied the quality control report and then arranged interviews with all recipients. He questioned each person as to the need, availability, cost, and timeliness of the quality control report. All prior weekly quality control reports were analyzed. The analyst found that quality control for each department normally was within the established parameters.

After extensive study, the analyst summarized all the data and arrived at the following conclusion: the quality control report is essential in ensuring that actual system performance is in accordance with that desired by management.

In his recommendation to the president, the analyst stated: to avoid the distribution of sometimes redundant reports, the weekly control report should be generated monthly. Information formerly contained in a weekly quality control report should be stored in the computer until the monthly report is generated. On-demand reports are a good idea for spot performance measures and should be retained.

QUESTION

1. If you were the consulting analyst, what would be your recommendations to the president on the report study?

CASE 16-3

The Cost Control and Analysis Department of a large firm has among its tasks (1) the manual publication of certain cost reports, (2) analysis of costs that deviate from established standards and budgets, (3) "make or buy" analysis of parts or operations upon request, and (4) the forecasting of direct and indirect personnel for the production departments.

The department's workload has increased steadily over the last quarter, causing a large amount of overtime work. The department's forecast of future work shows that there will be no slackening of the workload in the future. The manager of the department is faced with a decision to either hire additional personnel or reduce the current workload.

The manager knows the department is over budget, which precludes adding personnel this year. Realizing this, the manager has requested that the Systems Analysis Department study the reports currently being published by the department and recommend which can be rearranged, simplified, combined, eliminated, or reproduced by the computer.

The analyst assigned to the project proceeded to define the problem and prepared an outline for her study. She then obtained from the Cost Control and Analysis Department one copy of every report they produce and any instructions that go with them. She developed a numerical file, a functional file, and a cross-reference file. Next she determined the frequency of each report and obtained a distribution list. After studying and evaluating the relevant information on each report, the analyst made plans for interviewing the report recipients. Since the composite of all the distribution lists was quite large, she randomly picked 25% of the people listed and made appointments to interview them.

In the interviews, the analyst followed the original outline and asked only previously formulated questions. To learn how the users ranked the importance of each report, she asked them if the report was essential, how often it was used, and if the report was necessary for day-to-day action decisions, for controlling their operations, for checking accuracy of their operations, or for keeping informed on general conditions. Then she asked how much data or information on this report was not used and if the information could be obtained from any other source. This enabled her to see what parts of the reports could be eliminated. To learn of unknown, bootlegged, or unauthorized uses of the report, she asked how many persons use each copy and what other reports, records, or forms are prepared from the data contained in the report. For general background on the users' feelings as to the real importance of the report, the analyst asked these questions: (1) What would be the effect on your work if you: did not receive this report? received this report less frequently? received less data or information than at the present time? or received more data or information than at the present time? (2) How and where do you file your copy of the report? (3) How often do you refer to the report after its original distribution and use? She then finished the interview by asking if the user felt that his or her use of the report was worth the expense of preparation. After each interview, the analyst returned to her department to summarize her notes and store the information in files she had set up.

After conducting the last interview, the analyst carefully studied all the information obtained. Then she prepared a final summary of the report-analysis survey. In the report, she recommended which reports could be eliminated, rearranged, simplified, or combined. After the report was typed and approved by her supervisor, the analyst furnished a copy to the requesting manager.

QUESTIONS

1. Has the analyst obtained enough information to make a sound recommendation regarding the reports?
2. What could the analyst have done to be more thorough?
3. What other questions could the analyst have asked during the interviews?

FOOTNOTE

1. Russell A. Ackoff, "Management Misinformation Systems," *Management Science*, XIV, No. 4 (December, 1967), pp. B147–B156.

CHAPTER 17

PROCEDURE WRITING

The written instruction enables us to retain and utilize good methods that were devised by others. The systems analyst provides the firm with such capability by researching and writing detailed procedures for the firm's operations. This chapter delineates why and how the analyst does this.

WHY WRITE PROCEDURES . . . FOUR BASIC REASONS

Before discussing the four basic reasons for writing procedures, the reader should be familiar with the following definitions.

SUBJECT: The topic or central theme of the procedure.

SCOPE: The area or range that the procedure will encompass.

REFERENCES: The titles of any documents that have a governing or otherwise vital bearing upon the procedure. For example, if the procedure is necessary for compliance with a military specification, that specification title should be referenced.

GOALS: What the firm is trying to accomplish with the procedure.

POLICY: Management guidelines for regulating progress toward the firm's goals. They set rational limitations to manager's actions. Policies are behavioral guides that may be originated by management, appealed to superiors from subordinates to resolve particular problems, or imposed by external forces that demand compliance. Policies set objectives and usually are given as general statements.

PROCEDURE: These are guides to action; they are more specific than policies. Procedures seek to avoid disorganized activity by directing, coordinating, and articulating operations. They are a series of step-by-step instructions that explain how to carry out the policies. Procedures explain what is to be done, who will do it, and how it will be done.

SYSTEM: A network of interrelated procedures that are joined together in order to perform an activity.

Procedures are the road map of a system. The procedures explain, usually in minute detail, how the system should be operated. Four basic reasons for writing procedures are

1. To record and preserve the company's methods of operation and previous experiences. They record historically what has proven to be good practice and what has failed. They elicit economy in operations by enabling management to avoid the cost of recurrent investigations. By imposing consistency across the organization and through time, procedures help direct all activities toward common goals. The company's methods of operation must be preserved because employees forget the details, or purposes, or technical considerations in-

volved, and so on. Hopefully, recorded experiences ensure that mistakes of the past are not repeated.

2. To facilitate the training of new employees and acquaint experienced employees with new jobs or new systems. Written procedures standardize the job, and they ensure that the employee gets all the details of the job.

3. To establish a basis of control. Procedures serve to delegate authority to subordinates to make decisions within the framework of the policies devised by management. The written procedure gives a standardized basis from which to regulate and evaluate employee performance.

4. To force an examination and evaluation of the procedure or of the system itself. Written procedures help establish a bench mark to compare with past or future operating methods. Written procedures help both management and employees resolve questions about how the job should be performed.

STYLES OF PROCEDURE WRITING

Basically, procedures are written in one of three styles, but it is quite permissible to write a procedure in whatever manner makes it clear and easy to understand. The three basic styles of writing procedures are narrative, step-by-step outline, and play-script.

The *narrative procedure* is composed of words that make sentences and sentences that make paragraphs. The overall objective is to write a story that tells what is to be done, who will do it, when it will be done, and how it will be done. The narrative should include everything that is important to the procedure, including charts and graphs that simplify things for the reader. Narrative format is the hardest to write because of the smooth transition required between the various steps of the procedure.

The *step-by-step outline procedure* walks the reader through the process. Item-by-item the reader sees what each step is. References to various parts of the procedure are made easily because of the roman numeral, letter, or number identification of each step. An example of good outline format appears below.

I. Refund desk.
   A. Pay all refunds by check.
      1. Do not make a direct cash payment.
      2. If registrant wants cash, send person to cashier to cash the check.
   B. Do not pay refunds on
      1. Meal tickets unless surrendered one day preceding the event.
      2. Seminar tickets unless surrendered one day preceding the event.

II. Registrant.

    A. Sign name, member number, and name of chapter on

        1. Back of seminar ticket.

        2. Blue refund card (form 71-162).

    B. Member number may be omitted.

III. Refund desk.

    A. Use refund journal (form 78-180).

        1. Enter cancellation.

        2. Enter refund paid.

    B. Prepare refund check (form 80-10) and give refund check to registrant.

    C. Send blue refund card to the head cashier's office.

A step-by-step outline of a procedure is the easiest style of procedure to read because each phase has a roman numeral (I, II, III, etc.). The divisions within these phases (A, B, 1, 2, a, b) spell out the what, who, when, how, and other information required to explain the procedure.

The *playscript method* of procedure writing is also a what, who, when, and how type of procedure. The playscript style uses sequence numbers, actors, action verbs, and a straight chronological sequence of who does what in the procedure. The sequence numbers (1, 2, 3, etc.) list the sequence of steps in their chronological order. The actors are the employees. They are listed by their job function or job title. Action verbs are present-tense verbs like those used in flowcharting. An example of the playscript method is given below.

1. Refund desk   Pay all refunds by check. Do not make direct cash payment. If registrant wants cash, send person to the cashier to cash the check. Do not pay refunds on meal tickets or seminar tickets unless surrendered one day preceding the event.

2. Registrant   Sign name, member number, and the name of chapter on the back of the seminar ticket and also on the blue refund card (form 71-162). Member number may be omitted.

3. Refund desk   Enter cancellation and refund paid on refund journal (form 78-180). Prepare refund check (form 80-10) and give refund check to registrant. Send blue refund card to the head cashier's office.

TECHNIQUES OF WRITING PROCEDURES

Written procedures do not spell out the *why* of each step in the procedure. Because of this, they are considerably more simplified and condensed. Written procedures should specify clearly the actions required by the employee. Written procedures may be easier to write and may give the user a much clearer concept of the course of action if the following techniques are adopted by the procedure writer.

Devise a logical outline before writing the procedure. This outline should be a rough draft of the sequence of steps within the procedure. Avoid writing in circles. Instead, follow the step-by-step sequence. Use charts, graphs, and examples of forms whenever it makes the procedure clearer and simpler. Even decision tables or other matrix methods may be used, for example, in an administrative routine. Flowcharts are sometimes used, but the analyst should keep in mind that they should be used only for people who thoroughly understand them. Obviously, a person who does not know one flowchart symbol from another could not locate needed information through the use of a flowchart.

Main headings, divisional headings, and subdivisional headings may be used to break the procedure into separate understandable steps. Write short sentences (ten words or less) or short paragraphs (one idea per paragraph). For simplicity in making necessary changes, it should be structured in an "open-ended" format. For example, a decimal arrangement allows for additions and deletions. The use of boldface print for new subjects helps the user spot what is needed quickly.

There are three essentials to consider in procedure writing. Layout is the physical placement of the items on the page. Style is the wording that is used. It may be imperative telling who, what, how, when, and so forth; or it may be declarative in the manner of a policy statement. The third consideration is the format. This is the logical sequence of events mentioned above.

Indexing is necessary for a usable procedure manual, for it is the key to information retrieval. A table of contents is helpful for a general approach to a manual; but an index using the vocabulary of the users makes it a unique usable tool. The most frequently cited reason for nonuse of manuals is that "no one can ever find anything." There are usually two main reasons for this lack of utility; either the correct (i.e., needed) information is not included; or if it is included, indexing is inadequate so it cannot be found.

When a form is introduced, include its identification number. Then, in the index to the procedure, list all the forms used in that procedure. List the forms both in "form-name" sequence and in "form-number" sequence. Also include a filled-out facsimile of each form in the appendix to the procedure.

Explain the filing systems used for records, reports, forms, or any other paper used in the procedure. Describe the type of filing equipment used and the method of indexing. Give examples if it will make the indexing method more clear to the reader.

Procedures should be direct and to the point. Use present tense verbs and other action words in a straightforward style of writing that moves the reader through the operations smoothly.

There should be a method of disseminating and updating the procedures. One method is to bind the procedures manual like a book and reprint the entire manual every two or three months. Only *critical* procedure changes are distributed, by memo, between reprintings of the book. With this method, a large, complex organization can feel reasonably sure that its procedure manuals are always current (within three months) and that old manuals are disposed of completely (since all pages are bound together).

An alternative method is to distribute the procedures manual in a loose-leaf notebook. With this method, new or revised procedures are distributed on loose-leaf pages. Users update their own procedure manuals by removing obsolete pages and inserting the updated pages. (If this method is used, all revision pages should be dated.) It is a good policy to have the department secretary keep one copy of a procedure manual updated for the department even though various people in the department have personal copies. This can be the master copy for department use.

TYPES OF WRITTEN DOCUMENTATION

Written procedures are one kind of documentation. Let us take this opportunity to list several forms of documentation which are usually found in medium-size or larger firms.

1. Procedure manuals contain detailed step-by-step information about how a particular operation or activity is to be carried out.

2. Policy manuals contain information on management's attitudes concerning how various phases of the business should be conducted. Policies normally state general guidelines and imply what course of action should be followed.

3. Organization manuals contain information concerning the structure of a business, such as corporate objectives, organization charts, lines of authority, extent of centralization or decentralization, management job descriptions, and so forth.

4. Systems studies contain a review of current systems, system requirements, and new system recommendations.

5. Programming documentation contains program flowcharts, in-line documentation, narrative descriptions, input/output format descriptions, file descriptions, and so forth.

6. Program run descriptions contain information on how a computer program is to be run. They include disk or tape instructions, restart procedures, and checkpoint indicators.

7. Computer library procedures specify magnetic tape and disk storage methods. Also included are computer hardware and software manuals.

8. Standard operating procedure manual "for the data processing area" contains the standard procedures for systems analysis, programming, and computer operations. This manual also may contain the department's organization charts and job descriptions.

PROCEDURE WRITING IN RETROSPECT

Procedures are unlike policies in that they are more specific and provide detailed instructions for operating activities. Policies imply a general course of action, not a specific set of steps to carry out that action.

When writing or evaluating written procedures there are many general considerations that the analyst must observe in order to get things right the first time. First, the analyst should be clear on the problems and objectives involved. The analyst should be satisfied that sufficient investigation has been performed to ensure that the procedure will be realistic and adequate. A hasty approach usually will result in an unusable procedure. The analyst must be sure that any important relevant contractual requirements have been covered, and that management planning and other systems in progress are compatible with the procedure being devised.

The analyst should stay alert to the possibility of becoming negatively involved in *departmental interactions* (Chapter 5). Sometimes a manager who seeks control of additional activities in the firm will request a procedure that gives him or her control in that area. When this occurs, the analyst usually is given a one-sided view of the proposal and goes to work on changes that may have considerable impact on another manager's department. The analyst should be on the lookout for this type of situation by always studying the potential effects of the changes on other departments. Suspicions should be aroused if no clearly objective need for the new or modified procedure can be established.

It is very important, also, to assess the effect the changes will have on other procedures. Sometimes hasty changes are written up and implemented before it is learned that they contradict a dozen other important procedures. This kind of backfire is especially bad for the analyst's image and reputation in the firm since it attracts a lot of negative attention. It is wise to remember that people resist change. If they can prove that the changes are detrimental to efficient or effective operations they often will attack the analyst's competence. The analyst *cannot* afford this very often because when the other employees have lost respect for the analyst, it is difficult to do any systems work around the firm. The analyst, therefore, should be very cautious in sizing up the potential hazards that exist in any project.

Analysts sometimes think that going on record (by memo to their supervisor or other authority) and predicting a fiasco will protect them from any ill effects that result. However, if the analyst is identified with a poor procedure in the eyes of the rank and file, the damage is already done. What can one do? It is hard to say, but we recommend that the analyst scrutinize such a situation and then try to convince management that the changes will do more harm than good. Unfortunately, a systems analyst can expect to be involved in many such difficulties.

There is a time and place for the fast, patchup procedure job. Whenever an urgent need for change occurs, the analyst should be able to assess the problem quickly and devise a temporary solution that will suffice until a better routine can be devised, or until the emergency is gone. For example, if the normal speed of paperwork in the firm is too slow to keep up with the schedule requirements of a certain customer's orders, the analyst must find a way to expedite the routine for that customer's paperwork. The analyst might shortcut some steps, or design a faster form to process that customer's transaction. In any event, fast action may be required since the customer may be a valuable one whose demands must be satisfied. Such emergencies are called "firefighting" by systems people, and they occur rather frequently. The important step is to follow up after the patch job is done to be sure that either the emergency has subsided, or that an adequate and economic procedure is developed to convert the emergency to a controlled routine. When dealing with firefighting, the analyst has to be careful to avoid a reputation-damaging sloppy job. But the criticism that will follow if one is too cautious and slow in taking action also must be avoided. Again the analyst walks a tightrope!

A procedure should be checked carefully to be sure that it will work on all shifts the firm runs. Usually there are fewer people on evening and graveyard shifts; and there are always fewer authorities and decision makers present then. If the analyst writes up a procedural step that calls for, say, an engineer's signature on a form, the procedure may not operate on a shift when there is no engineer available. There are many differences between the operations of different shifts. The analyst has to learn how they will affect the procedure being developed.

The analyst should ensure that the steps of the procedure are not unnecessarily fixed in serial sequence, that is, a rigid sequential order. It is best to design the steps so as many as possible can be performed simultaneously. For example, if certain extra data for future analysis are picked up by a procedural step somewhere along the line, do not allow this step to delay the more vital operations by placing it in series order with them. Instead, route a copy to the area that provides the data while the main document continues in the processing stream. Or find some other way of obtaining the data. The point is that the vital line functions of a procedure should be identified and the steps designed to get them done with minimum delay. Always explore the possibilities of simultaneous or optional sequences of procedural operations.

The analyst should keep track of the procedure's operating time during the design. There should be a reasonable estimate of the acceptable maximum time. The most effective procedure in the world is useless to the firm if it is too slow.

The analyst should evaluate the operations of the procedure to be sure that none are excessively rigid in their requirements. For example, an analyst might have the impression that a certain checking operation is vitally important to the management of the department. For this reason the procedural step might be written as one that requires the time and signature of some authority in the department. This might cause bottlenecks whenever the person is busy elsewhere. The analyst should always question operations that are liable to be slowed by such circumstances, and the procedure should be designed to run at the lowest possible level of authority in the area.

Briefly, the following checkpoints, if observed with those previously discussed, will bring the analyst and the procedure into the clear in most cases.

1. Are the procedural steps in the *best* order?
2. Can any steps be eliminated?
3. Will the procedure accommodate present and future work volumes as well as management-imposed requirements?
4. Are there enough copies of each form? Too many?
5. Can mechanization be used economically?
6. Will the procedure accommodate unusual transactions?
7. Are any of the steps too complicated for the abilities of operating personnel?
8. Has the procedure been rigorously scanned for potential bottlenecks?
9. Can statistics or sampling be used to shortcut any operations?
10. Are the steps designed for operation at the lowest level of authority possible?

In closing this section we might caution the analyst that all too often systems people insist upon being given a free hand to create what they feel will be a good system or procedure for the firm. This is fine; but the wise analyst will seek advice and concurrence on ideas before getting married to them. And one should stand ready at all times to change personal ideas to whatever form better suits the objectives of the company or the area under study.

SELECTIONS FOR FURTHER STUDY

The subject of procedures usually is touched upon in most of the office management, systems analysis, or technical writing texts. Turnbull and Baird's *Graphics of Communication* can aid in the technical phases of typography, layout, and design. The systems analyst should note that the first chapter of Bohl's *Flowcharting Techniques* discusses flowcharting a procedure in the context of programming.

1. BOHL, MARILYN. *Flowcharting Techniques.* Chicago: Science Research Associates, Inc., 1971.

2. BROGAN, JOHN A. *Clear Technical Writing.* New York: McGraw-Hill Book Company, 1973.

3. COUGHLIN, CLIFFORD W. "The Need for Good Procedures," *Journal of Systems Management,* volume 24, number 6, June 1973.

4. DAGHER, JOSEPH P. *Writing a Practical Guide.* Boston, Mass.: Houghton Mifflin Company, 1975.

5. ENGER, NORMAN L. *Documentation Standards for Computer Systems.* Fairfax, Va.: Technology Press, Inc., 1976.

6. GILDERSLEEVE, THOMAS R. *Organizing and Documenting Data Processing Information.* Rochelle Park, N.J.: Hayden Book Company, Inc., 1977.

7. GOULD, JAY R. *Directions in Technical Writing and Communications.* Farmingdale, N.Y.: Baywood Publishing Company, Inc., 1978.

8. HAGA, CLIFFORD. "Procedure Manuals," *Ideas for Management.* Cleveland, Ohio: Association for Systems Management, 1968.

9. *Handbook of Successful Operating Systems and Procedures, with Forms.* Englewood Cliffs, N.J.: Prentice-Hall, Inc., 1964.

10. HARPER, WILLIAM. *Data Processing Documentation: Standards, Procedures, and Applications.* Englewood Cliffs, N.J.: Prentice-Hall, Inc., 1973.

11. JACKSON, CLYDE W. "Documentation Is Spelled Communicating," *Journal of Systems Management,* volume 24, number 6, June 1973.

12. JENNINGS, LUCY. *Secretarial and Administrative Procedures.* Englewood Cliffs, N.J.: Prentice-Hall, Inc., 1978.

13. KENDALL, RAYMOND H. "A Manual for Systems Users," *Journal of Systems Management,* volume 23, number 10, October 1972.

14. LONDON, KEITH. *Documentation Standards, 2nd edition.* New York: Van Nostrand Reinhold Company, 1974.

15. MATTHIES, LESLIE H. *The Playscript Procedure.* Stamford, Conn.: Office Publications, Inc., 1961.

16. National Computing Centre, Ltd. *A System Documented: In Accordance with the Standards in the Systems Documentation Manual.* New York: International Publications Service, 1971.

17. National Computing Centre, Ltd. *Documenting Systems (The User's View).* New York: International Publications Service, 1972.

18. ROSS, H. JOHN. *How To Make a Procedure Manual, 4th ed.* Miami, Fla.: Office Research Institute, 1970.

19. TURNBULL, ARTHUR T., and RUSSELL N. BAIRD. *Graphics of Communication: Typography, Layout and Design, 3rd edition.* New York: Holt, Rinehart and Winston, Inc., 1975.

20. VAN DUYN, J. *Documentation Manual.* New York: Van Nostrand Reinhold Company, 1972.

21. WALTON, THOMAS F. *Technical Manual Writing and Administration.* New York: McGraw-Hill Book Company, 1968.

22. WENDER, RUTH W. "The Procedure Manual," *Special Libraries,* volume 68, number 11, November 1977, pp. 407–410.

23. WHITEHOUSE, FRANK. *Systems Documentation: Techniques of Persuasion in Large Organizations.* Brooklyn Heights, N.Y.: Beekman Publishers, Inc., 1973.

QUESTIONS

1. With regard to writing procedures, define "References."
2. What are the four things a procedure explains?
3. What is the overall objective of the "narrative procedure"?
4. Which of the three basic procedure styles is easiest to read?
5. In what sequence is the playscript procedure?
6. Define policy manual.

SITUATION CASES

CASE 17-1

A large corporation, which owns twenty hospitals in the United States, decided to change its billing system from posting machines to a computer service bureau. On June 1, all hospitals were to go on-line to the computer.

On March 10, a systems analyst arrived at one of the corporation's 180-bed facilities to prepare the procedure manual for admitting clerks. It was the analyst's responsibility to rewrite the manual to incorporate procedures needed for using the new system. This manual would be used to learn the new procedures involved in admitting a patient, and it would be the primary source used to answer questions regarding the new system.

Although the analyst was quite knowledgeable and experienced with the service bureau system, she was not familiar with the admitting office procedures.

By carefully examining the current admitting manual and by closely observing the day and night shift clerks' methods of admitting patients, the analyst gained a good understanding of patient admitting.

The analyst felt that procedures were the basis for performing any job well. In her opinion, a procedure manual should give, in detail, the precise series of step-by-step instructions that explain what is to be done, who will do it, when it will be done, and how it will be done. Although she had never actually written a procedure manual before, the analyst realized the basic reasons for a manual were (1) to record and preserve the company's methods of operation and previous experiences; (2) to facilitate the training of new employees, and (3) to acquaint experienced employees with new jobs or new systems. In this case, the analyst felt it would be effective in acquainting all employees with the new system, including changes in the operation of other departments that would be affected by the new system. She also knew that written procedures would establish a basis of control and force examination and evaluation of the procedures.

With this basic knowledge, the analyst began writing the procedure manual. By observing the clerks' work patterns, she was able to identify some procedures that appeared to hinder efficiency. Whenever a discrepency arose between what the manual stated and what one of the clerks was doing, the analyst would question the clerk about the difference. She would listen to reasons and jot down a few notes. Then all instructions that were no longer applicable to the new system were deleted and steps that were necessary to use the new system were added.

After compiling a list of what seemed to be every step necessary to efficiently perform the new admitting procedure, the manual writing was started. Its style was a step-by-step outline, since the analyst felt it would be the easiest style to understand. Each step was identified by numeral, letter, or number for easy reference. Sentences and paragraphs were short, direct, and to the point. Present tense verbs and other action words were used to help the reader move smoothly through the material. The analyst used main, divisional, and subdivisional headings to break the procedures into separate understandable steps.

Many forms were used in the procedures, all of which were identified by the "form-name" and "form-number" in the appendix. Although the analyst inserted a filled-in facsimile of each form in the appendix, no explanation of the new method of filing the forms was provided.

Pleased that the revision was completed ahead of the May 1st deadline, the analyst submitted the revised manual to the office manager for evaluation, criticism, and any needed changes. The manual was to be printed and distributed to each admitting clerk in a loose-leaf notebook. An open-ended system would allow the employees to update their own manuals by removing obsolete pages and inserting updated pages. A method for evaluation of the manual was to be developed after the procedures had been reviewed and modified.

Two days later the office manager brought to the analyst's attention certain areas in the procedure manual that needed more explanation. Since the analyst had never worked as an admitting clerk, all knowledge of the admitting operation had been gained through observation and reading of the manual. She did not know what to do next.

1. What important step did the analyst omit before writing the procedure manual?

2. What did the analyst exclude with regard to the new forms?

3. The analyst thought the observations had gone well and that reading the procedure manual had been quite helpful. What else could the analyst have done to get a better understanding of the admitting procedures?

CASE 17-2

The Xebac Newspaper Publishing Company has been experiencing some financial difficulties for the last three years. They have been due, at least in part, to the loss of subscribers at an increasing rate. Management became very concerned about this situation and requested that an analyst be assigned immediately to identify the problem and develop a reasonable solution.

The analyst began to work feverishly to define the problem. He identified a number of situations of concern including: (1) duplication of orders for supplies; (2) overlapping duties among the employees; (3) a lack of employee pride throughout the company; (4) a general sense of disorganization; and (5) an absence of effective management control.

The analyst decided, after much serious thought, that the first step toward resolving the problems should be the establishment of a sound framework of operating procedures. With the approval of top management, he began the task of reviewing the existing procedures. In order to determine how far out of date they were, the analyst requested key employees in problem areas to develop an informal procedure manual describing current duties. The analyst reviewed each of the informal procedures and compared them with any formal procedures. By using this technique, he hoped to identify what to add, change, or delete for each current procedure.

The first informal procedure the analyst reviewed was prepared by the business manager who was a "would-be procedure writer." The procedure covered the processing of classified advertisements and the ordering of supplies. It included the following steps.

I. Classified advertisements.

    A. Proofread Form 1A to ensure that all essential information has been recorded in the appropriate boxes.

        1. Record the date the advertisement is to run.

        2. Post the total cost of the advertisement to the receivables ledger.

        3. Give the advertising copy to the Pasteup Department so they can layout the ad properly.

    B. Prepare the customer bill (see Billing Procedure).

II. Ordering supplies.

    A. Use Form 658-A for all supply orders.

        1. Complete the form by entering the name, quantity, description, and unit cost of the item.

        2. Mail the order to the supplier eight weeks prior to stock-out because it takes approximately six weeks to process and ship the order.

        3. The white, yellow and green copies go to the supplier.

        4. The blue copy goes to the requesting department so they will know to expect a supply shipment.

        5. The pink copy is retained for the accounts payable ledger.

    B. After the receipt of the order has been verified by return of the blue copy from the department, update the inventory records.

In reviewing and rewriting the procedures, the analyst asked the following questions: (1) Was enough investigation performed prior to writing the new procedures? (2) Do the changes in the procedures affect other procedures in other departments? (3) Are the new job procedures too rigid? (4) Are the procedural steps in the proper order?

Upon completion of the draft procedures manual, the analyst sent copies to key employees and management personnel. Although the "usual" amount of questions and concern were expected, he was surprised to find a high degree of dissatisfaction among both the managers and their subordinates.

After discussing their concerns, the analyst decided to follow-up on the suggestion of one of the most powerful of the disgruntled managers because he felt this person's support would help gain approval of the manual. As a result, he studied the possibility of realigning some staff duties into other departments, subsequently incorporating the changes into the new procedures manual. In completing other revisions the analyst felt were necessary, he decided to incorporate the manager's suggestion in preference to the suggestions of the other employees.

When the analyst presented the final draft of the new procedures manual to the senior manager's committee, there was a barrage of criticism. The analyst left the meeting in a confused state and wondering what to do next.

QUESTIONS

1. What style of procedure writing did the business manager use?
2. How could the business manager's draft procedure have been improved?
3. In reviewing and rewriting the procedures, what key question should the analyst have asked?
4. What was the result of the analyst's efforts to gain political support for the new manual? How would you handle a similar situation?

CHAPTER 18

TECHNIQUES FOR THE SYSTEMS MANAGER

The Systems Department, as an advocate of the best organization and method techniques for the firm, must set a good example in its own house. And it must maintain the best diplomatic relations with all other areas of the firm. This calls for skillful performance by the systems manager. It has been said that managers who stop growing are dangerous not only to themselves but to the firm as well. Nowhere is this more true than in the Systems Department. To help the manager grow technically, there are introductory sections on information systems, data communications, data base, how to purchase software, and the use of consultants.

PROJECT CONTROL

Most systems and programming departments organize their work on a project assignment basis. Analysts are assigned various projects, such as a problem definition project or a full system study or perhaps a feasibility study. In fact, an analyst might have eight or ten projects underway simultaneously. Each project has its scheduled starting date, its scheduled completion date, its own start-to-finish elapsed time, and its own chargeable time. As discussed in Chapter 13, a Gantt chart is the easiest and most economical method of seeing the whole picture of work progress. An example of a Gantt chart in use in systems project control is shown in Figure 18-1.

All projects should be assigned in writing, including small projects. The larger or more vital projects should be described in greater detail than small projects. A Project Assignment Sheet like the one shown in Figure 18-2 can be used to make the written job assignments. It is filled out as follows. (Follow the numbers on Figure 18-2.)

1. The *subject* is the title, topic, or central theme of the project.
2. The *scope* is the area or range that the study will encompass. (Keep the scope within the boundaries of the subject.)
3. The *objectives* are the results that the analyst will be trying to accomplish.
4. The *project number* is a three-part number: the first box contains the analyst's initials; the second box contains a consecutive number that restarts on January 1 of each year; the third box contains the year.

PROJECT NUMBER	J.F.	01	80
	EMPLOYEE	NUMBER	YEAR

5. *Analyst* is the name of the analyst assigned to this project.
6. *Manager* is the name of the manager of the systems analyst.
7. *Date received* is the date that the project assignment sheet was prepared.
8. *Originating source of the project* is the name of the person who requested that this project be undertaken.
9. *Priority.* Check whichever box pertains to the priority assigned to this project, as follows.
   a. Emergencies . . . extreme hazards.
   b. Code violations . . . safety . . . labor relations.
   c. Urgent with high economic return—special surveys.
   d. Less urgent and smaller economic return.

PROJECT NAME	S or C	JAN 7	14	21	28	FEB 7	14	21	28	MAR 7	14	21	28	APRIL 7	14	21	28	MAY 7	14	21	28	JUNE 7	14	21	28	JULY 7	14	21	28	AUG 7	14	21	28	SEPT 7	14	21	28	OCT 7	14	21	28	NOV 7	14	21	28	DEC 7	14	21	28
Determine collection costs	S / C	x	x	x	x	x	x	x	x																																									
Relayout keypunch area	S / C						x	x	x	x	x	x																																						
Computer feasibility study	S / C						x	x	x	x	x	x	x	x	x	x	x	x	x	x	x	x	x	x																										
Production schedule slippage	S / C														x	x	x	x	x	x																														
Shipping dock study	S / C															x	x	x	x	x	x	x	x	x	x	x	x	x	x	x	x	x	x	x	x	x														
Redesign payroll form	S / C																						x	x	x	x																								
Draw work distribution chart	S / C																						x	x																										
New project control form	S / C																							x	x	x	x	x	x	x	x	x	x	x	x															

FIGURE 18-1 Project scheduling using a Gantt chart.

e. Routine projects.

f. Minor projects with low economic return.

10. *Project phases.* The analyst is expected to divide the project into phases. Each phase is a major task that the analyst completes on the way toward completion of the entire project. When all these tasks are completed, the project is completed.

11. *Project schedule.* Each phase of the project is now scheduled. Figure 18-3 shows phase 1 scheduled for the last week of April and the first week of May; phase 2 is scheduled for the last three weeks of May; and so on. It also shows that phase 1 is completed and only the first of the 3 weeks of phase 2 is completed. The overall time for this project is from the last week of April to and including the first week in July. The *project schedule* portion of the project assignment sheet can be used as a type of progress report showing work completed to date compare with the estimated completion schedule on a phase-by-phase basis.

12. *Follow-up notes.* This space is for any notes or remarks that the analyst wants to make.

13. *Project record.* Start date is the date that the project was started. Completion date is the date that the project was completed. Chargeable hours are the actual number of hours spent working on this project.

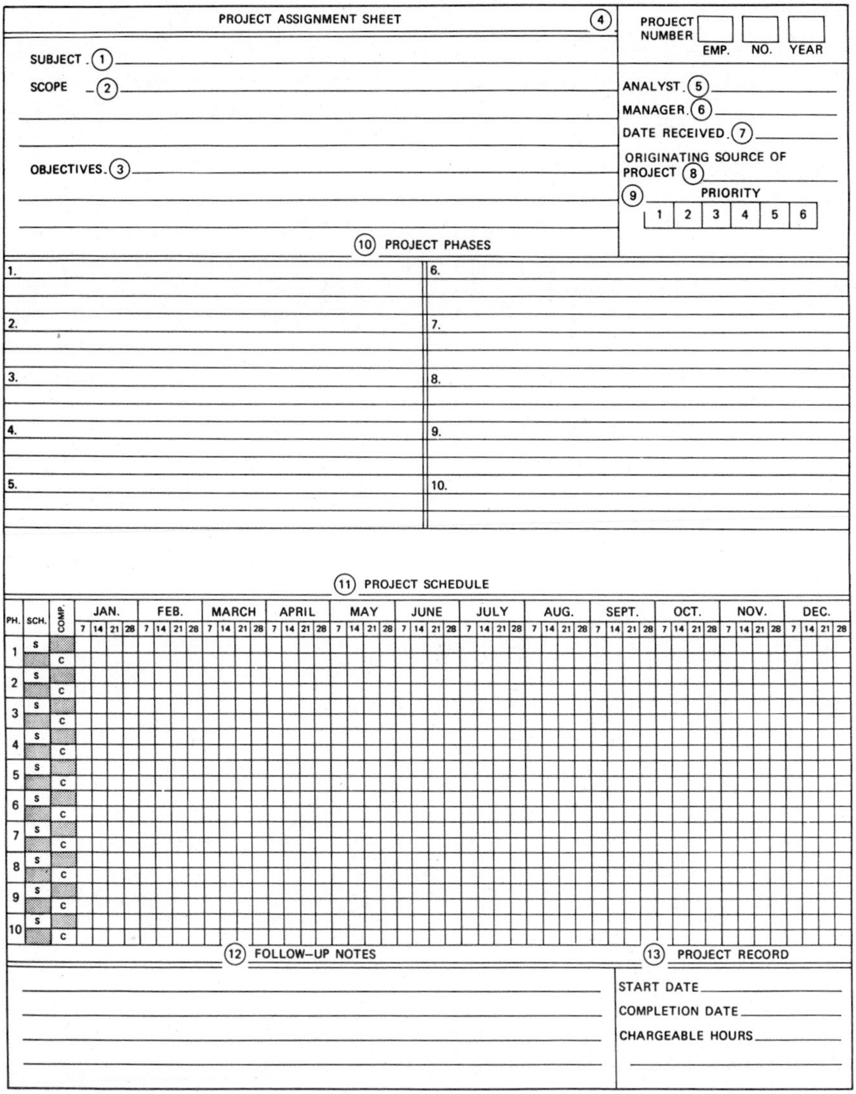
PROJECT ASSIGNMENT SHEET (4)

PROJECT NUMBER EMP. NO. YEAR

SUBJECT (1)

SCOPE (2)

OBJECTIVES (3)

ANALYST (5)

MANAGER (6)

DATE RECEIVED (7)

ORIGINATING SOURCE OF PROJECT (8)

(9) PRIORITY 1 2 3 4 5 6

(10) PROJECT PHASES

1. 2. 3. 4. 5. 6. 7. 8. 9. 10.

(11) PROJECT SCHEDULE

PH. SCH. COMP. JAN. FEB. MARCH APRIL MAY JUNE JULY AUG. SEPT. OCT. NOV. DEC.

7 14 21 28

1 2 3 4 5 6 7 8 9 10 S C

(12) FOLLOW-UP NOTES

(13) PROJECT RECORD

START DATE

COMPLETION DATE

CHARGEABLE HOURS

FIGURE 18-2 Project assignment sheet.

It is usually good practice to prepare three copies of the project assignment sheet. The analyst, the analyst's manager, and the person who requested the project should all receive a copy. This distribution assures that the three major parties in the project development are in agreement on how the project should proceed.

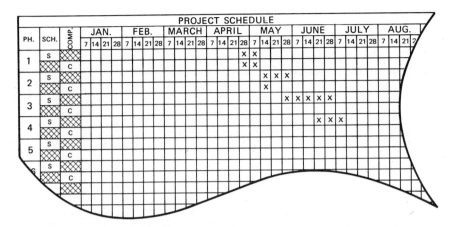

FIGURE 18-3 Project schedule.

PROGRESS REPORTS

In order to maintain control over the department, the manager of systems analysis may require periodic progress reports. The manager uses a progress report to monitor the progress of the various systems projects, whereas the analyst uses a progress report to report progress to date on a systems project.

Progress reports are usually concerned with work done during the interval since the last report. A typical progress report might contain a very short written description of the progress to date, approximate cost-to-date compared with estimated budgeted amount-to-date (Is the estimated budget being underspent or overspent?), and the work completed to date compared with the estimated completion schedule (Is the project ahead or behind schedule?). Some financial data, such as actual versus estimated budget, may be reported automatically via the firm's management information system. You should, however, report on *all* unexpected problems as soon as they are encountered.

The progress report is similar to a story told in serial fashion, where the reader is led from the previous progress reports to the current one. The progress reports should be tied together so there is an audit trail to trace back through all previous reports. For example, the first paragraph of the current progress report might point out the previous report's title, date, and accession number if it had one.

Progress reports may be written at irregular times, at the completion of major phases of the system development cycle, or they may be written at time intervals such as once a month. This depends on how often the manager of the systems department wants to see progress reports. In any case, progress reports should be concise and to the point so that neither the time of the reader nor that of the writer will be wasted.

Progress reports may be transmitted both verbally and in written format. Some managers want verbal progress reports, but usually written reports are requested. They are written in many forms: letters, memoranda, or special formats such as a preprinted form. The following example (Figure 18-4) is one style of a special format printed up as a formal company form.

PROGRESS REPORT		
PROJECT NAME (SUBJECT)		DATE
DATE OF PREVIOUS PROGRESS REPORT	ANALYST	PROJECT NUMBER
BUDGETED AMOUNT TO DATE	COST TO DATE	THIS PROJECT'S COSTS ARE ☐ UNDER ESTIMATE ☐ ON TARGET ☐ OVER ESTIMATE
ORIGINAL ESTIMATED COMPLETION DATE	NEW ESTIMATED COMPLETION DATE	THIS PROJECT WILL FINISH ☐ EARLY ☐ ON TIME ☐ LATE
NARRATIVE (SHORT WRITTEN DESCRIPTION OF PROGRESS SINCE LAST REPORT)		

FIGURE 18-4 Progress report form.

The previously mentioned project assignment sheet (Figure 18-2) can be used also as a type of progress report. If the analyst updates the Gantt chart on this sheet (let us say on a monthly basis), then the project assignment sheet also serves the purpose of showing work completed to date compared with the estimated completion schedule.

Progress reports, in one form or another, are a valuable aid to the systems manager for they relate in a concise manner the status of each project currently being completed under the auspices of the Systems Department. In addition, they provide a positive medium through which the manager can evaluate performance of individual systems analysts.

GENERAL FUNCTIONS OF THE MANAGER OF SYSTEMS ANALYSIS

Supervision and administration is the manager's primary duty. Good supervision requires a thorough knowledge of the technical skills used by subordinates. Proper administration results in good utilization of personnel and equipment. The five key tasks are *planning, organizing, directing, controlling,* and *staffing.* These are the five basic responsibilities of the manager of the Systems Department (or of any manager, for that matter).

A good system for the translation of technical progress and performance into easily understood reports is required by the systems manager. Project control can be achieved using progress reports that continually report on the progress of all jobs being worked on in the department.

As does the firm, the department should develop short-range plans (less than one year in the future) and long-range plans (more than one year in the future) as guidelines for its activities and direction. The systems manager should be a forward thinker. The systems manager also must ensure that high standards of performance are maintained. Work quality standards should be developed, qualified personnel have to be recruited, ongoing training programs for employees have to be maintained, job descriptions should be meaningful and up-to-date, and employees should get at least one annual performance review to keep them aware of their professional development and job performance.

The Systems Department must maintain good relations with other departments because departments that do not trust or respect the Systems Department will avoid the Systems Department, even when they need help. The systems manager should point out any department limitations before agreeing to do all the projects requested. Mature people can accept being turned down, and it is far better to decline a project than to accept it and fail because the department is overloaded with work. It is vital for the systems manager to know what the Systems Department's capacity is, to keep

427

track of how it is being used, and work to use it at maximum output. The systems manager has to know when to say yes or no to a project request.

SYMPTOMS OF ORGANIZATIONAL PROBLEMS

A number of symptoms may seem to indicate organizational problems, that is, seem to indicate that the organization structure itself is deficient. The presence of these symptoms may or may not warrant reorganization of the systems department, but each one should be corrected as soon as it is identified.

Since a key function of every manager is to actively receive and give information, communication failures or personality clashes may be at the root of some apparent organization problems. When an apparent organizational problem appears, it is wise to look first into the communications or personality areas for the causes. Internal communication problems may result from personal conflicts among employees or from strained relations among supervisory personnel. Careful observation of employee interactions is the key to uncovering the roots of such problems.

Deterioration of "diplomatic relationships" is another symptom of a possible need for reorganization. The systems analysis function is a service function to other departments in the firm. Its image, from the view of these departments, is crucial to its success. If other departments are unable to respect or trust the systems group, they may be reluctant to cooperate with it. One measurement technique that can be used to detect this symptom is for the manager of the Systems Department to interview the managers of user departments to learn just how they feel about the department's service function.

A high rate of turnover among the analysts is often a significant indicator of organizational and personnel management ineffectiveness. This symptom of organizational problems should be interpreted with caution because many factors can affect personnel turnover. The causes may be completely external to the department, or firm; however, the systems manager should always look out for the well-being of the analysts. Unfair and imbalanced workloads should be prevented, and the analysts should be informed about their status within the group. The potential damage that can be caused by an analyst's resignation in the middle of a complex project makes good human relations in the Systems Department essential (along with good documentation, of course). This is a major portion of the manager's duty to the firm.

Continuous scheduling difficulties, such as the failure to meet completion dates or failure to set realistic completion dates may be another indicator of organizational ineffectiveness. It may be that the analysts are not organized enough to monitor their own workload and progress accurately. Or it may be that they are allowing themselves to be pressured into unrealistic commitments and deadlines. Certainly every department that works to maximize its output will have some scheduling problems; but chronic slippage is a symptom that the manager should examine and correct.

JOB DESCRIPTIONS

The following job descriptions are for the functions typically found in a large systems group.

Manager of Systems Analysis

Plans, organizes, directs, controls, and staffs the activities of the Systems Analysis Department for the establishment and implementation of new or revised systems and procedures concerned with both manual operations and with electronic data processing. Usually considered as being in command of all systems analysis activities. Responsible for problem definition, feasibility studies, systems design, and implementation. Makes recommendations on the action to be taken. Assigns personnel to various projects, schedules projects, and directs analyst activities. Consults with and advises other departments on systems and procedures priorities. Coordinates department activities with the activities of other departments. Prepares progress reports regarding the activities of the entire Systems Analysis Department. Ensures that company policy with regard to system controls has been followed in all projects.

Lead Systems Analyst

Usually considered to be the assistant manager of systems analysis and has full technical knowledge of the activity comparable to a senior systems analyst. Has supervisory duties of instructing, directing, and controlling the work of the other systems analysts, including the senior systems analysts. Assists in planning, organizing, directing, controlling, and staffing the activities of the department. Assists in scheduling the work of the department and the assignment of personnel to various projects. May act as systems project manager in the absence of the manager. May coordinate the activities of the department with other departments. Able to perform all the duties of a senior systems analyst. Reviews and approves the definition of new system control requirements and the design for a framework of integrated controls to meet the needs of the new system.

Senior Systems Analyst

Under general direction, performs problem definition studies, feasibility studies, and full systems studies. Devises procedures for the solution of the problem through the use of manual systems and electronic data processing systems. Usually competent to work at the highest level of all phases of the study while working undirected most of the time. May give some direction and guidance to lower level personnel. Confers with all levels of personnel in order to define the problem. Prepares all types of charts, tables, and graphs to assist in analyzing problems. Utilizes various business and mathematical techniques. Devises logical procedures to solve problems by manual systems and electronic data processing, keeping in mind equipment capacity and limitations, operating time, and form of desired results. Analyzes existing system difficulties and revises the procedures involved as necessary. Prepares the formal definition of the new system control requirements by drawing together the work done by all analysts working on the project. Designs the framework of integrated controls to be utilized in meeting new system needs.

Systems Analyst I

Under general supervision, performs problem definition studies, feasibility studies, and full systems studies. Develops procedures to process information by means of electronic data processing equipment as well as manually. Usually competent enough in most phases of systems analysis to work undirected. Requires some general direction for the balance of the activities. Confers with organizational personnel to determine the problem and type of system to be designed. Analyzes problems in terms of systems requirements. Documents control points and procedures within the existing system. Defines specific control requirements and related control techniques. Proposes groups of related control techniques covering the requirements of major portions of the new system. Modifies the systems design to take maximum advantage of existing data processing equipment. Where necessary, recommends equipment modifications, manual systems, computer systems, or additions to enable efficient and effective systems applications. Defines the problem and makes recommendations for its solution. Confers with a programmer who prepares program flowcharts and machine instructions when the solution requires a computer.

Systems Analyst II

Under direct supervision, assists higher level classifications in problem definition studies, feasibility studies, and full systems studies. Usually sufficiently competent to work on several phases of systems analysis with only general direction, but still needs instruction and guidance for the other phases. Studies and analyzes existing systems as assigned. Prepares system flowcharts to describe existing and proposed operations. Documents control points and procedures within the existing system, either in flowchart or narrative form. Assists in defining specific control requirements of new systems and proposes specific control techniques.

Systems Analyst III

Under immediate supervision, carries out analyses of a less complex nature. Usually works only on a couple of activities under very close direction with the work being monitored. Prepares system flowcharts to describe existing and proposed operations. Documents control points and procedures within the existing system, either in flowchart or narrative form. Designs detailed record and form layouts. May assist in the preparation of program flowcharting. This classification is staffed usually by novices who have had sufficient educational background and experience to start in systems analysis.

INFORMATION STORAGE AND RETRIEVAL AS A TOOL

Information storage and retrieval systems are those systems that can store vast amounts of information, which can be retrieved when needed. Information retrieval is a field concerned with the structure, analysis, organization, storage, searching, and retrieval of information. The purpose of an information storage and retrieval system is to inform the user on the existence, or nonexistence, and location of information relating to a specific request.

Basically, information storage and retrieval systems are divided into three categories.

1. *Document retrieval systems.* A system that ultimately provides the user with the full text of the document that was requested.

2. *Reference retrieval systems.* A system that presents citations to document locations.

3. *Data retrieval systems.* A data retrieval system usually retrieves data displayed as words or numbers; it is not a true information retrieval system. Data retrieval systems are used for such things as airline seat reservation schedules or inventory control operations. This type of a retrieval system is commonly known as "lookup."

Information retrieval systems are needed in order to lessen the costs brought about by duplication of effort in storing data and to lower costs caused by duplication of error when storing the same data in numerous locations.

Management information systems require these information retrieval techniques in order to retrieve the information from a central data base. Better decisions can be made by the use of information retrieval systems that have timely and accurate information. Information retrieval systems can increase creativity, reduce paperwork, improve customer service, and improve internal and external communication. Problems created by manpower shortages in technical fields, along with the deeper specialization required, puts a great demand on information retrieval as an ideal solution toward increasing a technical person's creativity and productivity.

Information classification systems started with Aristotle. Aristotle classified the complete universe into ten basic types of being. Aristotle's system of classification was continued when the Dewey Decimal System was developed for libraries. The Dewey Decimal System also divides all knowledge into ten classifications, and is in use today as a retrieval device.

Some of the most common information storage and retrieval systems include edge-notched cards, uniterm cards, superimposable card systems, and high-speed digital computers. Edge-notched cards, uniterm cards, and superimposable card systems are examples of semiautomatic manual systems of storing information and retrieving it; a digital computer is an example of a fully automatic high-speed information storage and retrieval system.

The indexing of an information retrieval system is the most important single item of the system. The index is essentially a filter, and its purpose is to let through the wanted information and to hold back the unwanted information. There are two basic elements of an information retrieval system index. First are *the items* or the thing or things that the user is attempting to locate. When an information retrieval system is approached by the item, this is referred to as specific retrieval. For example, in a large university system, student records need to be filed in some sort of sequence. This sequence might be alphabetically by surname or numerically by a student identification number.

The second element is *the characteristic* or the key to the items. When an information retrieval system is approached by its characteristic or key, it is referred to as a class retrieval system. Using the same example of university records, the honor students could be readily identified if there was an index classifying student names or numbers by grade point ranking. In addition, each student could be classified within each grade point ranking by age.

A few ways of arranging the items or characteristics of an information retrieval system are as follows.

Conventional Grouping

In conventional grouping there is one record for each item. The characteristics or the codes are listed on each record. Every record has to be searched each time something is requested (search question). The advantages of conventional grouping are: its initial cost is lower, it provides direct access to the information, and it is easier to update. Depending on whether the system is manual or computerized, it may be the more time-consuming and more expensive method of information retrieval.

Inverted Method of Grouping

There is a record for each characteristic rather than for each item. Only the records pertinent to the characteristic or terms in the search question are searched in an inverted file. The inverted method has the advantage of producing output faster, it is easier to browse the file, and it is more flexible in vocabulary. To browse the file, a requestor puts in a query and receives some records. After looking at the results of the first query, the requestor may put in another request, perhaps requesting more specific information. This query/response procedure continues until the requestor is satisfied that no further information can be retrieved.

Permuted Indexes

Permuted indexes operate on titles. For example, through the use of computers every key word in the title of a document is allowed to be placed alphabetically in relation to the key words of other document titles. Any word in the title can be found in an alphabetic list along with the accession number of the document containing the title. Key Word in Context is a type of permuted index; it is commonly called a KWIC index.

In summary, information retrieval systems have grown primarily for three reasons. First, the volume of facts and other published information has expanded to the point where facts cannot be retrieved if they are not arranged in a logical sequence. Second, modern society demands rapid response time in filling an information request. And last, information retrieval equipment (whether manual or computerized) has become available at reasonable prices.

MANAGEMENT INFORMATION SYSTEMS

A management information system (MIS) is a system that provides historical information, information on the current status, and projected information, all appropriately summarized for those having an established need to know. The information must be provided in a time frame that will permit meaningful decision making at a nonprohibitive cost. It is a communication process in which data are recorded and processed for further operational uses. A MIS is a system that collects, processes, stores, and distributes information to aid in decision making for the managerial functions of planning, organizing, directing, controlling, and staffing a business organization. Management information systems include functions such as information storage and retrieval as well as all of the aspects of data communications. A management information system can be looked upon as the binding together of the entire organization into an effective integrated flow of information. MIS allows an information channel to serve as a means of improving the day-to-day operations and in future planning. Management information systems are built using

1. People, who are needed to operate the system.
2. Data processing, which provides the needed speed for information sorting and classifying.
3. Data communication, which is required in order to keep the information flowing between the different parts of the system and the people using the system.
4. Information storage and retrieval, which is required in order to store the information in its proper format and to make sure that the information can be retrieved when it is needed.
5. Systems planning, which is required in order to integrate the people, the data processing, the data communications, the information storage and retrieval, and the users of the system into an overall meaningful and well-organized management information system.

A management information system should provide information that is Consistent, Accurate, Timely, Economically feasible, and Relevant. Information with these characteristics will CATER to the needs of the user. A well-designed management information system should meet the needs of both the entire organization and each component part of the organization, and with a minimum of duplication of the stored data. It can provide reports to management that would not be feasible by other methods, and also provide them with more speed and more accuracy.

The attitudes at all the levels of management usually have to be reoriented to the fact that they do not "own" the information. For example, when information is put into a central data base, everyone who has a need to know in the organization will have access to that information. This eliminates individual pockets of information strategically located throughout the company. In preparing management for a new MIS, a constant fear on the part of many managers who feel they will be losing status or control may have to be overcome. As with any other system, involving the individual manager in the project is the best method of gaining the needed cooperation.

MIS is like any other system in that it, too, needs to be evaluated in terms of its effectiveness. Answers to the following questions can help the analyst determine the effectiveness of the MIS system. Does management still make profuse use of hunches even though there is a centralized data base and a management information system? Does the information from the MIS fit the plans or objectives of the organization? Does management seem to have confidence in the accuracy of the information disseminated by the MIS? Does the information meet the responsibilities of each individual manager? Are the data presented in a format that is easily understandable by its users?

The *data base* is the heart of a management information system. It is the centralized master file of basic information that is available to any authorized person within the firm. A data base should provide for rapid retrieval of accurate and relevant information. The terms data base and data bank (which are used interchangeably) are discussed in many different textbooks. They refer to a set of logically connected files that have a common method of access between them. They are the sum total of all the data that exist within the organization or within a specific department. It is essential that the data base satisfy the requirements of each organization that depends on it. If it does not, each organization will continue to maintain its own information system, thus defeating the purpose of the centralized data base/data bank. The key element in this concept of a management information system is that each organization utilizes the same data base in the satisfaction of its day-to-day information needs. The data within the data base must be retrieved easily. Computer programs must be written to enable quick retrieval of the requested data. As previously mentioned, a method of indexing the data must be chosen, such as conventional grouping, inverted grouping, or permuted indexes.

In summary, many organizations have computer-based systems for payroll, accounts receivable, inventory control, billing, and various other corporate activities. Very few of these organizations have fully integrated management information systems that have consolidated all these data from the independent computer-based systems into one data base available to all authorized personnel. For the analyst, the future shift is away from individual computer-based systems and into improved systems design for integrated management information systems that contain a computer-based data base for management use in the day-to-day business activities and for use in future planning.

DATA COMMUNICATIONS[1]

Data communications describes a part of the overall system which permits one or more users to access a remotely located computer or terminal. Some questions that should be considered when designing a data communication system follow. First, what is this communications system to do for the organization? For example, the information system might be used to link together the various users and the data base in a management information system, or it might be for data retrieval only such as in an airline reservation system.

Second, consider the number of points to be involved in the transmission of information. This is when the analyst decides where to place the various locations of the input/output terminals. These terminals may be located within one plant or between two or three plants or geographically separated by hundreds of miles.

Third, the volume or total amount of information that must be transmitted within a given period of time should be considered. The analyst determines the total traffic for each terminal by taking into account the type of message, how many messages are to be transmitted, and the average number of characters per message. When this information is broken down into the number of bits per hour, it can be decided which type of communications equipment to use. For example, communication lines (telephone lines) are rated in the number of bits per second that can be transmitted over them. Other equipment such as data sets and terminals also are rated as to the number of bits per second that can be transmitted over them. (This is a complicated technique and would require an entire book to do it justice.) It is with this information that the analyst decides what type of terminal is required, what type of data set is required, and what type of telephone lines are required. When deciding on the type of terminals, data sets, and telephone lines, the analyst should take into account the future company growth because it is almost always more desirable to buy or lease equipment that is able to handle next year's communication load rather than redesign the system next year.

A fourth consideration is urgency. How urgent will the messages be? Will high priority messages slow low priority messages? Will the system be able to handle peak loads such as a large number of requests between 9 A.M. and 10 A.M. if that is when a large number of requests is expected? Will the system be used more because it is both available and quick (this is called the turnpike effect)? Will errors in the transmission of data cause so much retransmission of the same data that it either slows or stops the flow of data from a centralized data bank to the user?

The fifth consideration is whether operators of the individual terminals will be able to learn the operational procedures easily and quickly. This is extremely important because if the managers who are supposed to use the system cannot quickly and easily interrogate that system through their terminal, the system will not be utilized to its fullest extent.

In summary, the data communication system is a link that connects a user to a centralized data base. The analyst must determine what the communication system is to be used for, the number of terminals or points involved in the system, and the volume of information that is to be transmitted over the system. All these factors are put together in order to determine the type of equipment to be used. This is an extremely important link in any management information system or information storage and retrieval system because whenever the data communication link breaks down no one has access to the centralized data files.

DATA BASE MANAGEMENT SYSTEMS

The growth and acceptance of Data Base Management Systems (DBMS) within the business community was a significant factor in the 1970s in promoting the timely and efficient processing of management information. As such, the systems manager must be familiar with basic DBMS concepts so the potential for their use in the organization can be evaluated. Information has a value to every organization; however, in many cases, the value of information to management decreases across time, that is, as it loses "currency." Traditionally, the function of data processing has been to turn data into information and disseminate it before its absence hinders normal operating activities. The advent of telecommunications has enhanced greatly the ability of data processing to provide a timely service to users..It is through the on-line data base systems that data processing can process and disseminate management information when it is at its peak decision-making value.

A data base can be viewed as a centralized collection of all data that relates to one or more applications. The technological advances in direct access hardware technology have provided capacities and speeds that allow the data for many applications to share the same physical storage device. As a result, the cost and complexity of data redundancy can be eliminated by having multiple applications use a common, single source of data.

Once the ability to integrate the data was available, it was necessary to be able to structure it in a manner that would meet the processing requirements of each user application. In order to accomplish this, it was recognized that the data and its description must be independent of any programming language. The concepts of data independence and language independence facilitate centralized data base maintenance, protection, and control over the physical aspects of the data base. The systems manager must view a data base as a generalized, common, integrated collection of company data which fulfills the individual data requirements of all the user applications. In addition, the manager must recognize that the data within the data base is structured to follow the material data relationships that exist in the company.

Conceptually, a data base can be viewed as having three elements, that is, (1) the physical structure, (2) the contents of the data base—data and control information, and (3) the logical relationships between the elements of data stored. Because of the differences in design among direct access storage devices, the physical structure of a data base can vary widely depending upon the manufacturers. Most data bases are maintained on one or more disk packs. Each disk pack is divided up into a fixed number of cylinders, tracks (per cylinder), and blocks (per track). A block is the physical unit of data transfer between the data base and the CPU, and is referred to as a *page*. Each page in a data base occupies a known location and has a unique identification number (address).

A data base usually is divided into specifically named sections of contiguous pages called *areas*. A *data item*, or *data element*, is the smallest unit of named data in a data base. (Before data base these were called fields.) A data element has three attributes, that is, (1) name, (2) type, and (3) length. A collection of one or more data elements is referred to as a record. To completely describe a record requires its name and the names and attributes of all data elements contained within it.

An *occurrence* of a specific record type is said to exist when a value for each data element within it exists within the data base (see Figure 18-5). Control over physical placement of records in the data base is established by identifying areas in which specific record occurrences can be stored.

The data base also contains system information which is used to: (1) provide an audit trail, (2) provide an inventory of available space on each page, and (3) control access to each page. In order to satisfy the complete data structure requirements of an integrated data base, provision is made for a flexible method of establishing logical relationships between record types. These relationships are established by a mechanism referred to as a *family*. A family may have only one *parent* and must have at least one record type

RECORD TYPE	VENDOR NUMBER	VENDOR NAME	VENDOR ADDRESS
Vendor record (occurrence #1)	23654	Jones Mfg.	Fresno, CA
Vendor record (occurrence #2)	27894	Williams Co.	Boston, MA
Vendor record (occurrence #3)	31468	Lane & Co.	Albany, NY

FIGURE 18-5 Illustrates three occurrences of the vendor record type.

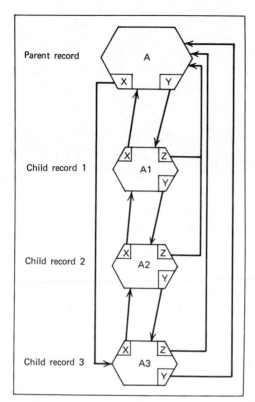

FIGURE 18-6 Illustrates a family with one parent record occurrence and three child record occurrences. X — pointer to previous record in family; Y — pointer to next record in family; Z — pointer to family parent.

that functions as a *child.* Figure 18-6 illustrates a family with one parent record occurrence and three child record occurrences. To facilitate the construction of complex data structures, three basic rules have been established for forming family relationships between record types.

1. Any record type may be a child in one or more families, and it also may be a parent in one or more other families.
2. Any record type may be the parent of one or more families.
3. A family may have only one record type as its parent, but may have one or more record types as children.

The description of the elements of a data base is not complete without a definition of the concepts of schema and subschema. The complete description of all the elements in a data base is referred to as a *schema*. A *subschema* is a consistent and logical subset of the schema from which it was obtained. A subschema names the areas, data elements, records, and families which are accessed by specific programs. Therefore, logical program or terminal user requests are subschemas; whereas the logical description of the entire data base is the schema.

One of the most difficult tasks facing the systems manager is the process of evaluating and selecting a data base package that best meets the needs of the organization. Data base packages are available principally from two sources, that is, hardware vendors and software package vendors. The number of packages available has grown rapidly in the past decade. The following list includes a few of the most popular packages from each type of source.

DATA BASE PACKAGE	HARDWARE VENDOR
IMS	IBM
DM4	HONEYWELL
IDS-2	HONEYWELL
DMS-90	UNIVAC
DMS-110	UNIVAC
DMS-2	BURROUGHS

DATA BASE PACKAGE	SOFTWARE VENDOR
MARK IV	Informatics
TOTAL	CINCOM Systems Inc.
IDMS	CULLINANE Corp.
SYSTEM 2000	MRI SYSTEMS Corp.
ADABAS	Software A6
RAMIS	Mathematica, Inc.
DS/2	System Development Corp.

The DBMS evaluation and selection process is well-suited to a team approach. The prime objective of the team effort is to develop and submit to management a report containing recommendations for use of a data base package. To achieve this objective, the team must study its organization to evaluate: (1) the need for a data base system, and (2) which of the data base packages is most suited to the organization and its information requirements. When forming a team for this effort, individuals with the following types of experience should be recruited.

1. Systems analysis and programming.
2. Systems programming.

3. Computer operations.

4. Budgetary planning and cost estimating for computer system projects.

5. User representatives.

When evaluating the need for a data base system, the team must consider the following factors.

1. Determine who will use the data base system.

2. Identify all types of services that should be provided to users.

3. Determine the frequency of use for each service.

4. Identify the significant data elements needed for each service.

5. Determine what functions will need to be performed on each data element.

Figure 18-7 illustrates the type of information that would be collected by the team for each type of service.

Service:	Verification of orders
Frequency:	720 orders/day
Significant Data Elements:	Inventory ID code
	Order number
	Part description
	Unit cost
	Volume
	Order date
	Shipping date
	Shipping destination
Functions:	1. Verify parts information.
	• Order number
	• Part description
	• Unit cost
	• Volume
	2. Verify shipping information.
	• Order number
	• Shipping date
	• Shipping destination

FIGURE 18-7 Illustrates the type of information that would be collected by a DBMS evaluation team to analyze what types of service should be provided by the system.

As a final step in evaluating the need for a data base system, the team must do a cost-benefit analysis. The personnel and computing requirements for implementation of the data base system must be estimated and viewed as a capital cost. The current costs of carrying out the various operations that will be incorporated into the data base system later must be estimated also. The current costs then should be compared to the estimated cost of operating the data base system on an ongoing basis.

The results of the team's study effort should be outlined in a report to management, along with a recommended course of action. Once the decision has been made to acquire a data base system, candidate packages should be compared on the basis of the five major elements of a data base system as well as other related criteria. The five major elements to be evaluated include: (1) data manipulation capabilities, (2) query capabilities, (3) application programming capabilities, (4) physical file organization and handling capabilities, and (5) data communication capabilities. Once the candidate packages have been evaluated in terms of the major elements of a data base system, the following additional points should be considered.

1. Cost of data base system.
   - Lease.
   - Purchase.
2. Installation fee.
3. Maintenance fees.
4. Performance guarantees.
5. Product deliverables.
   - Program listings.
   - Program libraries.
   - Program specification.
   - User manuals.
6. User training.
   - Requirements.
   - Availability.
   - Cost.
7. Language in which the system is written.
8. Mechanics of file conversion (what help is available to users).
9. Hardware and software environments supported.
10. Compatibility of system with long-range data processing plans.

The rapid growth in the use of data base systems has created a need for a new function to augment the functions of systems analysis and programming, that is, the data base administrator. The systems manager must be prepared to develop an effective data base administration function in order to ensure the successful implementation and operation of a data base system. The data base administration function requires some expertise in data base design and creation, operation of the data base management system, and the use of one or more data manipulation languages. Responsibilities of the data base administrator usually include such activities as

1. Specify the structure and content of the data base.
2. Establish appropriate operating recovery and rollback procedures to preserve the integrity of the data base in the event of software or hardware failure.
3. Coordinate data base restructuring to provide for changes in data structure, or simply to provide for additional physical space.
4. Monitor program performance and data base loading characteristics.
5. Assist programmers in the use of effective data manipulation language techniques.
6. Coordinate and control the addition of data items, record types, or set types to the data base.
7. Maintain documentation of the data base schema and subschemas.
8. Create subschemas as required by new applications.

IN-HOUSE SYSTEMS STAFF OR OUTSIDE CONSULTANTS?

As discovered earlier, an organization usually does not feel the need for any extensive system renovation until it has some serious growing pains. Then, as often happens, it discovers that the old, overloaded systems simply will not support the projected rate of growth or competitive pressure.

This is commonly the point at which someone suggests the installation of a Systems Department. In many instances this is the necessary solution to the firm's problems. But all organizations are not the same. Each is characterized by a unique set of operating details and problems. Each one also differs in its approach to objectives, style, and control; therefore each must have its own approach to systems management.

At this point we can discuss some of the pros and cons of a firm doing its own systems work as opposed to requesting help from outside sources. Should a firm maintain its own Systems Department? Or should it rely on periodic help from professional systems consultants? Arguments for both positions are outlined below.

The business manager should consider carefully each argument as it relates to the organization in question.

I. Arguments for having an in-house Systems Department are the following.

   A. Growth, present or future, is the reason given most often for wanting an in-house Systems Department.

      1. To tie the organization together across departmental boundaries.

      2. To develop and maintain up-to-date systems to control the most vital operations.

      3. To help manage the computer (if there is one).

      4. To plan ahead for change.

      5. To maintain competitive efficiency.

      6. To preserve economies of scale.

   B. Familiarity with the firm's intricacies is another argument in favor of in-house systems personnel.

      1. They know the firm and its management so they can more realistically evaluate the feasibility of management's proposals.

      2. They are always present and can keep the firm abreast of all opportunities that arise in systems methodology.

      3. Confidential company information is kept within the company, rather than carried away by consultants.

   C. Other reasons for an in-house systems department include

      1. It is more convenient because the personnel are always around when you need a systems problem fixed.

      2. If *someone* is not assigned the responsibility for systems maintenance, it does not get done.

      3. Since systems personnel are on the payroll, they have a stake in keeping up the good work.

      4. With proven staff members, taking chances on unknown outside consultants is eliminated.

II. Arguments for using outside consultants instead of a permanent in-house staff are the following.

   A. Consultants give an impartial outside viewpoint.

      1. They have no vested interest in the *status quo* and are thus free to recommend whatever is best for the firm.

2. They do not get stale and complacent as an in-house staff might if it is not well-organized and trained.

3. Members of the firm can be relieved of direct responsibility for unpopular recommendations such as
   a. Reorganization or eliminations.
   b. Personnel transfers or replacement.

4. They import good ideas from other firms.

B. They have access to any kind of specialist needed to assist with or set up a long-run system, such as

1. Up-to-date accounting systems.

2. Records management system.

3. EDP hardware and software security.

4. Data communications.

5. Distributed systems.

6. Data base.

C. You only pay them while you need them.

D. Use of consultants has proven to be sound practice in professions that were developed long before systems analysis, such as medicine, law, and architecture.

As mentioned, the business manager should consider each item as it relates to the specific firm. It also may pay to learn what the experiences of similar firms have been in this area; but again, since each organization is unique, comparisons should be made only with great care. The success or failure of a particular systems approach in one firm may have been due to circumstances that do not exist in another firm. As any seasoned manager knows, small differences can require very divergent approaches to the "same" problem. On the other hand, there are many general approaches that will fit almost all organizations.

Hopefully, the arguments we have presented will assist the manager in estimating the feasibility, risk, and cost of an in-house systems staff as opposed to the use of outside consultants. Both approaches have merit. Probably some blend of the two will fit most organizations.

SOFTWARE PACKAGE EVALUATION AND SELECTION

At the present, data processing organizations in the United States already are spending more for software than for hardware. Although in the past much of this effort came from within an organization, the emphasis on purchasing outside software is growing rapidly. In 1977, the software package market represented a $1.2 billion dollar industry. By 1980, it is expected to be a $4.0 billion dollar market.[2] IBM will soon be charging more for software than hardware.

When an organization is faced with the need for a new application system, the analyst must be prepared to assist with the make-or-buy software decision. Usually the purchase of a software package is appropriate where one or more of the following criteria apply.

- The estimated cost of developing a system in-house exceeds the purchase price of a software package.
- The organization does not have sufficient technical expertise to develop a system.
- The development of a system may require technical expertise in areas in which the data processing staff has little or no experience.
- The organization may need the use of a system much sooner than it could be developed in-house.

Prior to beginning the search for candidate software packages, it is assumed that the analyst has identified the system requirements, basic design objectives, and estimated costs to produce such a system in-house (see Chapters 3 through 7). In going through the evaluation and selection process the analyst may find it necessary to pursue the following three phases of activity.

- Research existing sources of software package information, including preparation of an RFP (Request for Proposal) when necessary.
- Evaluate the candidate software packages in matrix fashion and select those most appropriate to organization needs.
- Ensure that the contract for purchase or lease of the software contains appropriate considerations.

When an analyst begins the search for software packages, there are several sources of information that can be explored. The most widely known source of information on software packages is the *ICP Quarterly* which perhaps publishes the largest list of software package descriptions and vendor addresses. Figure 18-8 illustrates the type of information contained in this publication. It allows the analyst to identify vendors with products, and it facilitates preparation of a list of candidate software packages for later followup and evaluation. In addition, there are several organizations, such as Datapro, Business Automation Specification Reports, and Auerbach, which sell evalu-

ABC's ACCOUNTS RECEIVABLE SYSTEM
Number of Installations: 300
Narrative: The ABC Accounts Receivable System is available either in a batch or on-line mode.
System features include:

- Open item and balance forward processing.
- Invoice date (how old), "as of date," and past due (delinquent) aging.
- User-oriented report writer.
- Dunning letters.
- Paid history report covering any variable period of time.
- Various components of an invoice (various line items including freight and taxes) can be journalized for subsequent entry to either the user's or ABC's General Ledger System.

The System also provides the user with these capabilities:

- Setting up a company profile which enables each business entity to define its own processing options in regard to the application of cash, late payment finance charges, etc.
- Tracking the cumulative tolerance granted to each customer.
- Association of up to four different bank accounts with each account in order to simplify the automatic application of cash.
- Projecting and forecasting payments by day for the next 30 days.
- Extension of discount date or payment due date "grace days" at the account level.
- Accommodation of both anticipation and standard discounts.
- Application of cash on-line, using either a customer number, invoice number or bank account number.
- Simplification of correction correspondence by automatically printing unauthorized deduction and unearned discount invoices.
- Supplies data required for standard credit interchange services.
- Highlights accounts that are approaching their credit limit.
- Determines the appropriate cash algorithm to be used for each specific account when automatic cash is used.
- Maintains an on-line tickler file which automatically reminds credit persons, on the appropriate day, of accounts requiring follow-up and review.
- Allows indicative customer data to be updated on line.

One year's maintenance and warranty included in purchase price.
Operating Environment: IBM 360/370, DOS, OS, VS, CICS

FIGURE 18-8 Software package description.

ation services covering current software packages. As an example, Datapro provides evaluations based upon user surveys covering the following categories.

- Weighted averages of overall satisfaction, efficiency, ease of installation, ease of use, documentation, vendor technical support, and training (scale of 1–5, 5 being best rating).

- Advantages: flexible, inexpensive, saves human resources, saves system resources.
- Disadvantages: inflexible, costly, complex, slow, uses excessive resources, lacks key capabilities, compatibility problems.
- Performed as advertised.
- Required modification.

Figure 18-9 shows an example of how Datapro ratings can be placed in a matrix to assist in establishing a list of final candidate packages for evaluation. The ratings are based upon the user sample size shown on line one for each vendor package evaluated. The figures represent the number of users who responded positively in each category.

The size of a software package evaluation and selection project can vary widely. If the resources to be committed are substantial, the analyst may have to prepare a formal Request for Proposal (RFP) and send it to a final list of vendors whose software seems applicable. When the size of the project warrants the preparation of an RFP, the analyst should consider requesting the following information in the document: (1) organizational background, (2) description of currently operating systems, (3) new system requirements, (4) format and content of vendor proposals, (5) evaluation criteria for responses, and (6) appendix/exhibits.

The organizational background section of the RFP should provide prospective software package vendors with a concise picture of the company including

- Description of industry if useful to explain unique system requirements.
- Description of the purchasing organization's history, structure, and product lines.
- Description of how current management and administrative data is processed, that is, manual systems, accounting machines, in-house computer, or service bureau.

The description of the currently operating systems section of the RFP should be a narrative identifying pertinent aspects of each current manual or automated system which will be replaced, or affected, by the potential software packages. The analyst should include quantitative information on current activity levels and anticipated growth. Figure 18-10 illustrates the type of information that should be provided to vendors.

The new system requirements section of the RFP should identify specific design requirements and objectives that must be satisfied by the package selected. (See Chapter 7 and Chapter 8 for additional information on defining new system requirements and designing the new system.) If the package is to be installed in phases, this

		VENDOR A	VENDOR B	VENDOR C	VENDOR D	VENDOR E	VENDOR F
Weighted Averages	User sample size	36.0	36.0	9.0	20.0	3.0	5.0
	Overall satisfaction	2.8	2.8	3.0	2.9	3.7	2.0
	Throughput/efficiency	2.5	2.3	2.2	2.5	4.0	1.5
	Ease of installation	2.4	2.3	2.4	2.7	4.0	2.0
	Ease of use	2.9	2.4	2.7	2.8	3.7	2.0
	Documentation	2.9	2.8	2.9	2.9	3.7	2.6
	Vendor tech support	2.5	2.6	2.1	2.4	3.3	1.6
	Training	2.6	2.7	2.3	2.8	3.7	2.0
Advantages	Flexible	21.0	25.0	6.0	14.0	3.0	0.0
	Inexpensive	6.0	1.0	0.0	1.0	2.0	2.0
	Saves systems resources	2.0	3.0	2.0	4.0	1.0	0.0
	Saves human resources	18.0	14.0	4.0	10.0	3.0	2.0
Disadvantages	Inflexible	6.0	1.0	1.0	0.0	0.0	1.0
	Costly	8.0	6.0	0.0	4.0	0.0	1.0
	Complex	9.0	21.0	1.0	9.0	0.0	2.0
	Slow	8.0	11.0	4.0	5.0	0.0	1.0
	Uses excessive resources	12.0	18.0	2.0	7.0	0.0	1.0
	Lacks key capabilities	2.0	1.0	0.0	2.0	0.0	1.0
	Compatability problems	2.0	6.0	1.0	1.0	0.0	1.0
Performed as Advertised	Immediately	8.0	10.0	2.0	9.0	2.0	1.0
	Eventually	26.0	20.0	6.0	10.0	1.0	4.0
Required Modification	No	12.0	11.0	4.0	12.0	3.0	1.0
	Yes, by vendor	11.0	12.0	3.0	5.0	0.0	2.0
	Yes, by user	18.0	16.0	3.0	2.0	0.0	4.0

FIGURE 18-9   Software package ratings matrix.

should be spelled out in terms of the content and timing of each phase. The need for special system capabilities, such as data base management or report generator facility, should be specified.

The format and content of vendor proposals section of the RFP should enumerate the specific organization scheme and content that is to be provided in the vendor's proposal. The following is an example of how a typical proposal should be organized.

AREA	ITEM	AVE. VOLUMES	AVE. ANNUAL GROWTH FACTOR
Inventory	• # Raw material items	3000	30%
	• # Items—finished goods	150	20%
	• # Products in line	20	25%
	• # Master schedule items/run	15	—
	• # MRP planned orders (ave. # of P.O.'s and W.O.'s opened during master schedule period plus scheduled orders)	400	20%
	• # Purchase orders	250/mo.	20%
	• Ave. # line items per P.O.	4	—
	• # Open work orders	60	20%
Accounts receivable	• # Invoices	300/mo.	10–15%
	• Ave. # accts. per invoice	6	—
	• # Cash receipts (all)	330/mo.	10–15%
	• # Customers	1350	15–20%
Accounts payable	• # Vouchers	600/mo.	15%
	• # Checks:		
	• Hand checks	50/mo.	—
	• Computer checks	425/mo.	—
	• # Vendors	1100	35%
Property accounting	• # Asset items	1300	20%
Payroll	• Semimo. employees	100	20%
	• Bi-weekly employees	50	20%
Order entry	• # Customer orders	200/mo.	40–50%
	• (ave. 2 line items per order)		
General ledger	• # Accounts	800	—

FIGURE 18-10   Description of current system activity levels.

- Title page.
- Letter of introduction.
- Table of contents.
- Management summary.
- Benefits summary.
- Vendor credentials.
- Vendor experience with similar requirements.
- Hardware and firmware requirements.
- Mandatory software features.
- Other software features and related packages.
- System installation plan.
- Continuing installation technical support.
- Costs.
- Contract considerations.

The evaluation criteria of responses section of the RFP should describe the basis of award, the confidentiality of proposals, and the terms of evaluation. The basis of award is made usually on the recommendation of an in-house evaluation group or committee comprised of the analyst, users, and management representatives. This section should stress that all responses to the proposal become the property of the requestor and will be held confidential from parties other than the evaluation group and management. In general, the terms of evaluation are based upon the bidder's responses to the specific proposal requirements defined earlier. The analyst, however, should consider the following additional factors during vendor proposal evaluation.

- Specificity, clarity, and comprehension of the installation plan.
- The level of creativity and innovation demonstrated by the bidder's response to meet the needs of the firm in an efficient and effective manner.
- The ability and commitment of the bidder to complete the project within a reasonable time frame.
- Competency in, and understanding of, the proposed application as demonstrated by the proposed staffing and applications documentation.

The appendix/exhibits section of the RFP should contain additional detailed information that the analyst thinks prospective vendors will need to write a complete proposal. This section could include such things as samples of input documents, file layouts, reports, complex logical criteria, or computations.

Once the vendor proposals have been received, the final evaluation and selection process begins. The analyst must identify all the criteria that are of major importance to this project and evaluate how each of the packages meet the criteria. In Chapter 7, the section "Development of Evaluation Criteria for the New System" provides a basic framework of evaluation criteria that can be tailored to the specific needs of a software package selection project. Figure 18-11 illustrates how a detailed list of criteria can be used to evaluate software packages by placing the data in a matrix format. Once the analyst has prepared the final evaluation matrix, the evaluation committee usually meets to discuss the results, and to select a software package and vendor.

Once the decision has been made to purchase a specific software package, the analyst must ensure that all pertinent factors have been considered in contracting for the product. The analyst must recognize that

- Software is by no means as tangible or well-defined as hardware, and its operating characteristics are not as clear-cut as those of hardware.
- Software companies often are not as substantial as hardware companies.
- There are inherent risks in the purchase of software.

The analyst must understand that in any type of contractual relationship, whether for hardware or software, a few general rules apply. First, the contract seldom covers all eventualities; however, it should cover all major risks and clearly set forth the intent of both parties. A clear statement of intent by both parties will be of particular assistance if a situation arises which is not covered explicitly in the contract. Second, once the contract is written and executed, it should be filed. A good relationship between the parties generally will do more than a contract to ensure successful package purchase and installation. Third, an attorney should be used in contract formulation and negotiation when the value of the software package exceeds an amount the company cannot afford to lose, or when the risk of failure is high.

In arranging for the purchase of packaged software, the analyst must be cognizant of the following major risks.

- *Infringement:* The seller may not have clear title to the software package. As a result, a third party may assert rights unexpectedly against the purchaser. This risk is minimized when dealing with major software package vendors.
- *Vendor bankruptcy:* In the past, software companies often have been quite small in size, and under-capitalized. Their failure to perform sometimes resulted in litigation. If there are no significant assets, however, winning a case is meaningless.
- *Nonperformance:* A package may not function properly, or may take too much time, overhead, or resources.

VENDOR & PACKAGE NAME	VENDOR A G/L-SYSTEM	VENDOR B G/L-SYSTEM	VENDOR C G/L-SYSTEM	VENDOR D G/L-SYSTEM	VENDOR E G/L-SYSTEM	VENDOR F G/L-SYSTEM
1978 Revenue by vendor	$10M	$100M	$30M	$80M	$3.7M	Not applicable
Other financial software prod.	A/R, A/P, fixed assets	Payroll, A/P + P.O., A/R, fixed assets, financed forecast	Payroll, A/P, A/R, inventory	Fixed assets, A/P	A/P + P.O., construction in progress, A/R	G/L-system 3
Type of contract	License	Purchase	License	License	License	Purchase
Software warranties	1 year date of install, correct program defects	1 year date of install, correct program defects	1 year date of install, correct program defects	1 year date of install, correct program defects	Continuous warranty on performance	None
Approx. base price (OS-version)	$15–$52.5K	$40K	$33–$40K (+$5K on line)	$45K	$42.5K	$25K
Maintenance ($/year)	1st year free 0.10 × price	1st year free 0.10 × price	1st year free $3000	1st year free max. $5400	1st year free $4.2K	No maintenance. User developed enhancements will be distributed
Number of installations (1978)	700	700	300	200	52+	Over 100 (?)
Vendor training and installation support as part of purchase (onsite)	10	5	8	7	8	Supported by contract pgmrs. (@ $50/hr)
Vendor technical support cost	$300–$350/day	$425/day	$350/day	$500/day	$400/day	Supported by contract pgmrs. (@ $50/hr)
Documentation quality	Very good	Very good	Excellent	Good	Very good	Fair
First commercial install.	1970	1969 (?)	1969	1970	1977	1968 (?)
Adapted to govt. acctg.	1980	1975	No explicit capabilities	No explicit capabilities	No explicit capabilities	1969 (?)
Average of new releases issued by vendor	1 major release per year	1 major release every two years	1 major release every 1.5 years	1 major release every two years	None so far	New releases not supported
Restart recovery techniques	Update count in record	Generation data sets	External to system as defined by user	Internal integrity checks are used to check master files	No built-in recov.	None
Principal file access used	SCIF*-BDAM	Sequential/JSAM	Sequential/JSAM	DBS*-BDAM	JSAM/VSAM	Sequential
Record formats	Variable	Fixed	Fixed	Variable	Fixed	Fixed

	Package 1	Package 2	Package 3	Package 4	Package 5	Package 6
System throughput	Fair	Poor	Fair	Good	Very good	Poor
High volume maintenance capability	Yes	No	No	Yes	Yes	No
Source—input and edits	Good	Fair	Good	Good	Good	Fair
Single entry all transactions	Yes		Yes	No	No	Yes
User controlled edit tables	Yes		Yes	Preprocessor	Yes	No
No. of physical files in system	4	6	5	1	4 Standard 3 Additional	
I/O logic separated from processing logic	Yes	No	No	Yes	No	No
Peripherals needed: Min. / Opt.	1 disk, 2 tapes	1 disk, 2 tapes	1 disk, 2 tapes 2 disks, 4 tapes	1 disk, 2 tapes 2 disks, 4 tapes	1 disk, 2 tapes 2 disks, 2 tapes	4 tapes
Main memory used (OS-version)	(Segmentation) 150K	(Segmentation) 160K	(Segmentation) 150K	(Multimodule) 200K	(Single loads) 190K	(Extensive seg.) 50K
Number of programs	55	68	40	40 (includes budget and allocation prog.)	18	50+
Programming languages used	COBOL A/C-J/O	COBOL A/C	COBOL	COBOL (80%) FORTRAN IV A/C-I/O	COBOL	COBOL A/C
T.P. monitor (interface)	CICS TSO	TSO	TSO	CICS	CICS TAPS	TSO
Terminals supported	3270	3270	3270	3270	3270	
Generalized DBMS interface	IDMS IMS	Total IMS	None	IMS DBS*-UCC	IMS	None
Organizational reporting structure	Relationship file	Table	Table		Table	
Multiple chart of accounts supported	Yes	No	Yes	Yes	Yes	No
No. of standard reports produced	47	22	18	109	50	10
Ease of use by accounts user	Complexity based on relationship file—maintenance reqmts.	Complexity because of sensitivity to user errors	Good user interface	Good user interface	Good user interface	Poor
Management reporting	Excellent reporting capabilities based on flexibility of report writer		Supported by system report writer			

FIGURE 18-11 Software package evaluation matrix.

- *Progress payments:* Software package purchase and installation may require progress payments when nothing tangible has been delivered.
- *Product quality:* If internal standards for the development of software products are weak or nonexistent, it may be difficult to establish a framework in which to evaluate package software quality.
- *Excessive resource usage:* Most package installation contracts require that the purchaser supply some resources to the vendor. These resources could include machine time or keypunching. It is possible that the vendor may use far more of these than contemplated. For this reason some provision must be made to prevent this eventuality.

The analyst should recognize that a contract can be viewed as having both general and specific clauses. In terms of general clauses, there are a number that are applicable to all contracts such as

- *Arbitration:* If a dispute arises, arbitration settles things quickly. This prevents serious problems of long delays and disruption.
- *Force Majeure:* Some definition of force majeure (causes beyond the reasonable control of either party) should be included in the contract.
- *Confidentiality:* To protect the purchaser from breaches of confidence, the vendor and individual personnel should sign a nondisclosure agreement, or include such a clause in the contract.

In contracting specifically for the purchase or perpetual license of a software package, the analyst should consider the following specific clauses.

- *Specifications:* The best way to ensure that specifications are understood and met is to attach the package sales material to the contract. In addition, a set of specifications can be developed and appended to the contract as a schedule.
- *Documentation:* The documentation to be delivered with the package should be specified in detail.
- *Run time performance:* The best way to establish a standard here is to use a benchmark, or a level of volume which gives some operating time. This can be checked on the user's computer configuration, and rechecked after maintenance.
- *Fixes:* Corrections must be made quickly.
- *Source code availability:* The purchaser should have the right to obtain the

source code for possible modification, either after expiration of any maintenance agreement or in the event of vendor business termination.

- *Business termination rights:* If the vendor goes out of business, the purchaser
  should become an "owner" of the package.
- *Acceptance:* Acceptance of the package involves meeting the specifications,
  the run-time performance, documentation standards, and any other specified
  criteria.
- *Guarantee of ownership:* To prevent infringement, clear title to the software
  package should be warranted.
- *Right to rescind during a period of warranty:* As long as the vendor provides a
  warranty period, the purchaser should have the right to rescind the contract,
  paying only for use of the package on a lease basis.

The analyst has completed only the first phase of the project when the contract has
been signed. The next step is for the analyst to get involved in the implementation,
follow-up, and re-evaluation phases. This is discussed in more depth in Chapter 12.

SELECTIONS FOR FURTHER STUDY

While most systems analysis texts at least touch upon management of the systems
function, few go into the depth of Greenwood's *Managing of the Systems Analysis
Function.* Although a small volume in size, it contains practical information on the
selection and training of analysts, day-to-day systems management, the manager's
relationship with other departments and top management, the use of outside consultants, and obtaining support from computer manufacturers. Texts on the functions and
techniques of management are applicable to the Systems Department. These might
include Ecker's *Handbook for Supervisors,* Cleland's *System Analysis and Project
Management,* the National Computing Centre's *Systems Analyst Selection,* or
Taylor's *Successful Project Management.*

The tools of the system's manager are discussed in Faulkner's *Project Management
with CPM,* Delp's *Systems Tools for Project Planning,* the *Project Manager's Guide to
PERT,* Davis' *Introduction to Management Information Systems,* and FitzGerald's
Fundamentals of Data Communications.

1. BAUER, F.L. *Software Engineering.* New York: Springer-Verlag, Inc., 1977.
2. BLUMENTHAL, SHERMAN S. *Management Information Systems: A Framework
   for Planning and Development.* Englewood Cliffs, N.J.: Prentice-Hall, Inc.,
   1969.

3. BROWN, P.J. *Software Portability*. New York: Cambridge University Press, 1977.

4. CAMPBELL, BONITA J. *Understanding Information Systems: Foundation for Control*. Englewood Cliffs, N.J.: Winthrop Publishing Company, 1977.

5. CARDENAS. *Data Base Management Systems*. Boston: Allyn and Bacon, Inc., 1978.

6. CARLSEN, ROBERT D., and JAMES A. LEWIS. *Systems Analysis Workbook: A Complete Guide to Project Implementation and Control*. Englewood Cliffs, N.J.: Prentice-Hall, Inc., 1973.

7. CLELAND, DAVID I., and WILLIAM R. KING. *Management: A Systems Approach*. New York: McGraw-Hill Book Company, 1972.

8. CLELAND, DAVID I., and WILLIAM R. KING. *Systems Analysis and Project Management, 2nd edition*. New York: McGraw-Hill Book Company, 1975.

9. DAVIS, GORDON B. *Introduction to Management Information Systems: Conceptual Foundations, Structure and Development*. New York: McGraw-Hill Book Company, 1974.

10. DELP, PETER. *Systems Tools for Project Planning*. Bloomington, Ind.: International Development Institute, 1976.

11. ECKER, H. PAUL, VERNON OUELETTE, and JOHN MACRAE. *Handbook for Supervisors, 2nd edition*. Englewood Cliffs, N.J.: Prentice-Hall, Inc., 1970.

12. *Effective Project Management, 3 Units*. New York: Preston Publishing Company.

13. ENGER, NORMAN L. *Management Standards for Developing Information Systems*. New York: American Management Association, 1977.

14. FAULKNER, EDWARD E. *Project Management with CPM, 3rd edition*. Duxbury, Mass.: Means, Robert Snow Company, Inc., 1973.

15. FITZGERALD, JERRY, and THOMAS EASON. *Fundamentals of Data Communications*. New York: John Wiley and Sons, Inc., 1978.

16. GIDO, JACK. *An Introduction to Project Planning*. Schenectady, N.Y.: General Electric Training and Education Programs, 1974.

17. GREIBACH, S.A. *Theory of Program Structures: Schemes, Semantics, Verification*. New York: Springer-Verlag, Inc., 1976.

18. HARPOOL, JACK D. *Business Data Systems: A Practical Guide*. Dubuque, Iowa: William C. Brown Company, 1978.

19. HODGE, BARTOW, and ROBERT N. HODGSON. *Management and the Computer in Information and Control Systems*. New York: McGraw-Hill Book Company, 1969.

20. JENSEN, RANDALL W., and CHARLES C. TONIES. *Software Engineering.* Englewood Cliffs, N.J.: Prentice-Hall, Inc., 1979.
21. JONES, PAUL E., JR. *Data Base Design Methodology.* Wellesley, Mass.: QED Information Sciences, Inc., 1976.
22. KAIMANN, R.A. *Structured Information Files.* New York: John Wiley and Sons, Inc., 1973.
23. MARTIN, JAMES. *Systems Analysis for Data Transmission.* Englewood Cliffs, N.J.: Prentice-Hall, Inc., 1972.
24. MARTINO, R.L. *Project Management.* New York: Gordon & Breach Science Publishers, Inc., 1968.
25. MOORE, RUSSELL F., ed. *AMA Management Handbook.* New York: American Management Association, 1970.
26. NATIONAL COMPUTING CENTRE, LTD. *Systems Analyst Selection.* New York: International Publications Services, 1970.
27. *Project Manager's Guide to PERT.* Chelmsford, Mass.: New Division Publications, 1977.
28. ROSS, RONALD G. *Data Base Systems: Design, Implementation, and Management.* New York: American Management Association, 1978.
29. RYAN, ANTOINETTE T. *Systems Approach to Organization and Administration.* Danville, Ill.: Interstate, 1978.
30. SAMARAS, THOMAS T., and FRANK L. CZERWINSKI. *Fundamentals of Configuration Management.* New York: John Wiley and Sons, Inc., 1971.
31. SANDERS, DONALD H. *Computers in Business.* New York: McGraw-Hill Book Company, 1975.
32. SAYLES, LEONARD A. and MARGARET K. CHANDLER. *Managing Larger Systems: Organizations for the Future.* Scranton, Pa.: Harper & Row Publishers, Inc., 1971.
33. SCHNEIDERMAN, BEN. *Data Bases: Improving Usability and Effectiveness.* New York: Academic Press, Inc., 1978.
34. SIEMENS, NICOLAI. *Operations Research: Planning, Operating and Information Systems.* Riverside, N.J.: Free Press, 1973.
35. SILVERMAN, MELVIN. *Project Management: A Short Course for Professionals.* New York: John Wiley and Sons, Inc., 1976.
36. TAYLOR, W.J., and T.F. WATLING. *Successful Project Management.* Brooklyn Heights, N.Y.: Beekman Publishers, Inc., 1970.
37. WILLOUGHBY, THEODORE C., and JAMES A. SENN. *Business Systems.* Cleveland, Ohio: Association for Systems Management, 1975.

QUESTIONS

1. What type of chart is the most effective in portraying the whole picture of work progress?

2. Name six of the key data elements included on a project assignment sheet.

3. Why should progress reports be tied together?

4. What are the five basic responsibilities of the manager of the Systems Department?

5. When an organizational problem surfaces, where should one look first for causes?

6. Name the three categories of information storage and retrieval systems.

7. What does the Dewey Decimal System have in common with Aristotle's classification of the universe?

8. Name the three primary reasons information retrieval systems have grown.

9. What is a "data base"?

10. Name the three elements that a data base can be viewed as having.

11. Define area, page, and data element.

12. List the three basic rules for forming family relationships between record types.

13. Identify the three phases of activity an analyst may find it necessary to pursue in going through the software package evaluation and selection process.

14. Name three types of information that should be requested in an RFP.

15. List four major risks the analyst must be cognizant of in arranging for the purchase of packaged software.

SITUATION CASES

CASE 18-1

The president of XOF Cosmetic Company requested the Corporate Systems Group to develop a more effective timely system of historical, current, and projected planning information. The systems group manager discussed the situation in-depth with the president and suggested that perhaps a Management Information System (MIS) would be a viable solution. The president was not sure what the term MIS meant and asked the systems manager to explain. The manager explained that a company-wide MIS would collect, process, store, and distribute information in a format suitable for effective decision making in areas such as (1) planning, (2) organizing, (3) directing, (4) controlling, and (5) staffing. The president decided such a system was needed and approved the plan.

The systems group then set out to develop a company-wide MIS. As a part of the design phase, each department was surveyed to identify key information needed by the managers. All information was analyzed to identify data that demonstrated the flow of information within each department. The design effort took many months. As implementation neared, the systems manager assigned a junior analyst (systems analyst III) to meet with each department manager to explain the expected impact of the new system. The manager indicated that he would be available for counseling if the junior analyst needed help.

The junior analyst studied the new system plans and formulated an outline of what should be discussed with the department managers. A joint meeting with all the department managers was arranged and the following speech was given by the junior analyst.

The new system was designed by the systems group to provide more accurate and timely information for management. The system was planned so that key information would flow easily between the different departments and people who needed it. Computer terminals are to be placed in each department to facilitate data entry, validation, and on-line inquiries. The new system uses permuted indexing for information storage and retrieval.

The junior analyst spent some time in explaining the basics of the permuted indexing technique. In addition, copies were distributed of the input/output procedures to be followed, and the managers were told who to contact if there were any questions.

After the new system was implemented, there was a substantial amount of negative feedback. Upon further investigation, management discovered that the new system was not being utilized fully by the departments.

QUESTIONS

1. Did the systems manager use good judgment in selecting a junior analyst for this job? Why?
2. What could have been done to more effectively carry out this assignment?

CASE 18-2

CPU Inc., a high technology manufacturing and distribution company, was beginning to experience serious organizational problems in managing and controlling its growth effectively. Because of a combination of some recent legislation and unique marketing circumstances, CPU had an opportunity to double its effective size and market share. The Board of Directors met to discuss the opportunity, and decided growth should be guided by a team of professional system analysts. In order to proceed, the board recommended that the business manager either start an in-house systems group or engage a team of professional systems consultants from outside CPU.

The business manager was uncertain on how to proceed with this decision. Finally, a meeting was arranged with all the department heads to discuss the pros and cons of each alternative. At the meeting, the business manager asked each department head to discuss the two alternatives.

The Accounting Department head spoke first. "Our growth rate is such that an in-house systems group clearly is needed to ensure that the process is orderly. Such a group will be needed to support the use of the new computer being considered. Our growth may cause some departmental boundaries to change, and a systems group could assist by developing procedures to maintain a consistent flow of products. Vital operations must be controlled by up-to-date systems and any changes must be planned for well ahead of time. An in-house group could help us maintain a competitive efficiency and preserve the economies of scale."

The Purchasing Department head spoke next. "The use of outside consultants would allow us to select experienced specialists in each area to be covered, such as accounting, records management, and EDP hardware and software."

The Manufacturing Department head was next to speak. "Being familiar with our intricate operations is an important factor here. In-house systems personnel are bound to know the firm better, and can render more realistic opinions on our proposals. I also think that our confidential information should be kept here with trusted employees instead of being carried off by outside consultants."

In rebuttal, the Purchasing Department head stated, "No, an outside, impartial viewpoint is needed. Consultants wouldn't be involved directly with our operations so there could be no personal bias. Outside consultants wouldn't get stale or complacent and they could bring us good ideas from other firms. They also would eliminate in-house responsibility for unpopular recommendations like reorganization, personnel transfers, and replacement. Using consultants has proven to be a sound practice elsewhere, and costs about the same as an in-house systems group."

As a final comment, the Accounting Department head said, "We should not have to take chances on unknown outside consultants. Proven staff members, since they are a part of the company, have a much larger stake in keeping up good work. Since someone must be assigned the responsibility for systems maintenance, an in-house group is more convenient and will always be around when needed."

The business manager listened carefully to the discussion and then stated that it might be wise to investigate what other firms in their industry had done in this area.

QUESTIONS

1. What arguments in favor of an in-house systems group were overlooked?
2. What is the most important cost argument for using outside consultants?

FOOTNOTES

1. See *Fundamentals of Data Communications,* by Jerry FitzGerald and Tom Eason (New York: John Wiley and Sons, Inc., 1978).
2. Cited by Dick H. Brandon, "Contracting for Software," *Infosystems,* September, 1979, p. 72.

CHAPTER 19

RESEARCH NEEDS OF THE ANALYST

The systems analyst often will need information on the system under study. While much of this information may be internal to the area under study, much more may be located via the library. This chapter will discuss the type of materials the analyst can expect to find in the corporate library, how the materials are indexed, and how to retrieve citations to these materials, especially by computerized methods.

THE ANALYST'S MOST OVERLOOKED TOOL: THE LIBRARY

Among the many tools available to the systems analyst, the library is probably the most overlooked and underestimated of them all. The reasons are many, but they can be summarized in one: few of us have been shown the vast amount of useful information a library contains, much less how to utilize this very important resource in an effective and efficient manner. Today, however, most larger corporations recognize that the library is an expensive but vital resource.

Your own library needs may vary considerably from project to project, but if you become acquainted with this tool early, you will be prepared when the need arises. Corporate libraries today vary greatly in size and content, but most firms have some type of library facility available for their employees' use. Smaller firms in highly specialized technical fields may have the basic books relating to their field and some special reports, but the bulk of the collection might be current technical periodicals. These smaller libraries tend to depend on their local public and college libraries for information of a more general nature.

On the other hand, some large firms have libraries comparable to a public library, the only difference being in emphasis on the type of materials. These materials may include not only books, periodicals, and reports, but other highly specialized items such as manufacturers' catalogs and brochures or specifications and standards (these will be discussed later in the chapter). Technical libraries such as these usually have highly trained librarians who also are subject specialists. Additionally, some of these larger libraries, particularly those in the technical fields, have literature searchers whose sole job is to find information in specialized fields for staff members, but this is the exception rather than the rule.

COMPUTERIZED LITERATURE SEARCHING

One of the primary methods of finding references that are needed during the course of a systems study is to utilize various indexes. These indexes may be to periodical articles, technical reports, government documents, newspaper citations, or to any combination of these.

There are now two means by which the analyst may go about this searching: manual and computerized. Going through paper copies of these indexes by hand is not only very time consuming, but it is expensive in terms of the analyst's labor.

During the mid-1970s the advent of computers helped all library users find more efficient methods of using library resources. Primary among these is the computerized method of literature searching. Many of the indexes that are available on library shelves are now available through computerized systems. In addition, many new indexes exist only in a computerized form and have no paper counterpart. These systems are accessed generally by highly trained librarians who not only have an in-depth knowledge of the contents of the various indexes, but who also have specialized training both in the use of the system software and also for the unique capabilities of each individual file within the system. That is, one must have special training in order to know how to effectively get the information out of the computer.

Analysts who work with trained literature searchers will be pleasantly surprised at how quickly and easily the information they need can be retrieved from the computer. In addition, this also is a relatively inexpensive way to proceed. Further, the real advantage to using a computerized index is that the computer can do far more than the human being can in a much shorter period of time, both in terms of the greater number of years of data that the computer can search as well as being able to find more retrieval points. Not only can the searcher search on specific subject categories, but in addition frequently can pinpoint specific terminology that is unique to a particular problem. For example, let us say that the analyst needs to find information on a specifically named system, and wants everything that can be located about that system. The information probably could not be found easily using a subject approach; however, by using what is referred to as "free-text" searching for the specific name of the system, it can be retrieved if there is anything relevant in the particular data base that is being searched.

HOW TO FIND YOUR LIBRARY MATERIALS

Once the searcher or analyst finds citations to specific articles, the analyst will need to retrieve some of these in order to get the facts that are needed. While the library of the corporation or agency in which you work may have numerous publications available at hand, there may be many more that are not readily available. Just because the materials are not available in the same facility, however, does not mean that they are totally unavailable.

Today most corporate libraries, and indeed many other types of libraries as well, belong to cooperative networks. These networks tend to emanate from the Library of Congress in Washington, through the various state libraries (such as the California State Library in Sacramento), and down to smaller local level cooperatives in which member libraries are those from the corporate, educational, and public library communities. In practical terms, these cooperatives mean that the systems analyst can ask the corporate librarian to procure material that previously may have been unavailable because no one knew where it might be located. In many cases federal funds to library systems have enabled cooperating libraries to publish joint lists of holdings (called "union lists") so that materials can be located more effectively. Some libraries now have cooperative cataloging, and some even purchase costly items jointly that none could afford alone.

Cooperative borrowing is done through a formal Interlibrary Loan Request from one library to another (not by individuals). While there is a specific routine to follow in requesting these materials through interlibrary loan, in exceptional cases local librarians cooperate to the extent of providing materials on a same-day or next-day basis when feasible. This, of course, depends on the workload of the various libraries involved. You may be sure, however, that your corporate librarian will do everything possible to help you find the materials you need in order to complete your project.

WHAT YOUR LIBRARY CONTAINS . . . BASICS[1]

Books

The approach to the books in the library is generally through a card catalog or some computerized version of one. This catalog often is divided into author and title or subject categories, although some special libraries also may use a report number, corporate author (as opposed to personal author), or contract numbers. Each catalog card will have a "call number" by which each book in the library is individually identified. Books are placed on the shelves in the call number order. Libraries in the United States tend to use one of two cataloging systems with which to classify their books.

1. The *Dewey Decimal System* is probably the most widely used classification system because it has been in popular use the longest (developed in 1876), and it is the simplest to learn and use. It divides all knowledge into ten subject classes, each of which is further subdivided. These ten classes are arranged in numerical sequence.

   000–099   General Works
   100–199   Philosophy
   200–299   Religion
   300–399   Social Sciences
   400–499   Linguistics
   500–599   Pure Science
   600–699   Technology (Applied Science)
   700–799   Arts and Recreation
   800–899   Literature
   900–999   History, Geography, and Biography

   Further subdivision is by a well developed and logical decimal system. Until recently the Dewey system was considered flexible enough to encompass all knowledge.

2. The *Library of Congress System* is taking over quickly as the classification system for academic and technical libraries. Because it is an alphanumeric system, it provides much greater flexibility than the Dewey method. With Library of Congress, all knowledge is classified into twenty-one subject areas organized on an A to Z basis (the letters I, O, W, X, and Y are omitted). Further subdivision is through the use of alphanumeric designations. See Figure 19-1 for a brief comparison of the two systems.

Neither system has just one number on which the user can rely for a specific subject. It is for this reason that one must learn to use the card catalog or some alternate means of indexing. For example, the systems analyst may think all books on systems analysis are in the 600's if using Dewey or the "T's" when using the Library of Congress system. In reality, if the book's main emphasis is on mathematical methods of systems, it may be in the 500's or the QA's. If it is on systems analysis in libraries, it may be in the 000's or Z's. A business systems text such as this may be located in the 658's or the HF's.

Users of the Library of Congress system must be particularly wary of being trapped into a favorite shelf area. Incidentally, users seldom can go from one library to another expecting a certain book to have the same classification, or call number. Each library is distinct from others and so are its catalogers! As libraries get more mechanized methods, however, they tend toward shared cataloging which means more consistency from the viewpoint of the users and librarians alike.

Periodicals

In today's fast changing world, periodicals often make up the bulk of a special library's collection; for it is through them (not books) that the worker keeps up with what is happening currently in a given field. This is especially true in high-technology industries.

As the systems analyst, you will no doubt want to read each issue of the *Journal of Systems Management* or perhaps computer-oriented publications such as *Computerworld* if you deal primarily with computerized systems.

When you are looking for articles on a specific subject for a project on which you are working, however, it is inefficient and costly to go through periodicals on an issue-by-issue basis. Indexes arranged by subject have been designed to save you time. Perhaps the best known to the business systems analyst is *Business Periodicals Index,* which indexes articles from most of the major business periodicals. Another index that is available only in computerized format is *Abstracted Business Information* (also sometimes called *ABI/INFORM*). Others you may find useful are *Applied Science and Technology Index, Computer and Control Abstracts,* and *Engineering Index* to name just a few of the many that are available. (The latter two have computerized versions called INSPEC and COMPENDEX, respectively.)

DEWEY	LIBRARY OF CONGRESS		
000	A	General Books	
	B	Philosophy—Religion	
100		B–BL	Philosophy
200		BL–BX	Religion
900	C	History—Auxiliary Science	
900	D	History (except American)	
900	E–F	America	
		E	America (General) and U.S. (General)
		F	U.S. (Local) and America, except U.S.
	G	Geography and Anthropology	
900		G–GF	Geography
500		GN	Anthropology
300		GV	Physical Education
300	H	Social Science	
		HA	Statistics
		HB–HJ	Economics
		HM–HX	Sociology
300	J	Political Science	
300	K	Law	
300	L	Education	
700	M	Music	
700	N	Fine Arts	
		NA	Architecture
		ND	Painting
	P	Language and Literature	
400		PB–PH	Modern European Languages
800		PN–PZ	Literature
500	Q	Science	
		QA	Mathematics
		QC	Physics
		QD	Chemistry
		QH–QR	Biological Sciences
600	R	Medicine	
600	S	Agriculture	
600	T	Technology	
		TA	Civil Engineering
		TJ	Mechanical Engineering
		TK	Electrical Engineering
		TL	Aeronautics
		TP	Chemical Engineering
		TX	Home Economics
300	U	Military Science	
300	V	Naval Science	
000	Z	Bibliography and Library Science	

FIGURE 19-1 Comparison of Dewey Decimal and the Library of Congress classification systems.

It should be noted here that all subjects mentioned in this text are thoroughly covered by these periodical indexes. Books are often emphasized, however, since they generally give a more in-depth coverage (the user should recognize that books can never be as current as periodical articles). Since few students are familiar with such indexes,[2] descriptive information is provided in Figures 19-2 through 19-4 to aid the student in understanding this important systems tool.

Newspapers

Depending upon the nature of the information you are trying to locate, newspapers may be beneficial. Large dailies such as the *New York Times* have their own indexes; most have none. For the businessperson, the *Wall Street Journal* and its *Index* (divided by corporate name and subject) may be indispensable. See Figure 19-5 for a sample page of the *Wall Street Journal Index.* The advent of computerized typesetting has enabled newspaper publishers to index their newspapers more adequately, so the number of newspaper indexes has proliferated in recent years.

Manufacturers' Literature

These come in a wide variety of formats and are handled in many ways. They may be simple descriptive brochures on anything from a ball bearing to a computer. They may be general or highly technical. They are often the only up-to-date, factual source of information on a given type of equipment or product. Some of the materials of this type will be so important to the library that they will be placed in its reference section and may not be checked out. Other firms maintain all manufacturers' literature in their purchasing department rather than the library.

Pamphlets

Pamphlets are an ill-defined portion of any library's collection. Most libraries have pamphlets, but each handles them in a different way, depending upon the needs of their users and how important the librarians think each individual pamphlet is to the users. Often they are cataloged, but maintained in upright file cabinets, which librarians refer to as "vertical files." In other libraries, pamphlets are considered to be minor materials and are not considered important enough to catalog. When this happens, they are placed either in vertical files or in boxes on shelves and arranged in broad subject categories. Pamphlet files generally reflect current high-interest topics in which the users of that particular library are interested.

Maps and Atlases

Maps appear in many forms in libraries. They may be local highway maps, topographic maps, and either sea or air navigation charts. Because of their specialized nature, these are sometimes cataloged in a library, but unless there is a very large collection that is a

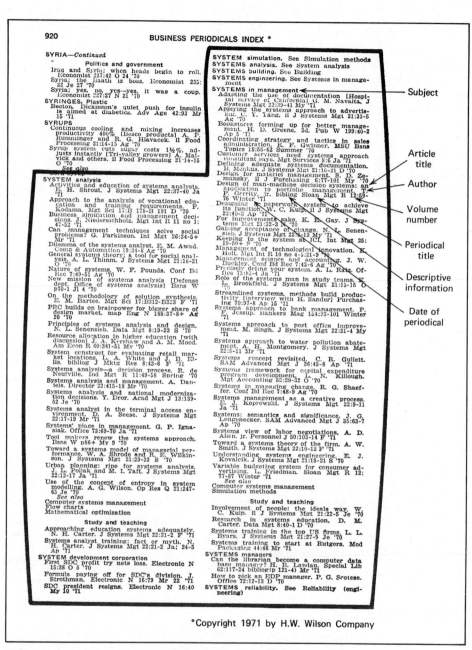

FIGURE 19-2   Sample page from **Business Periodicals Index**.

FIGURE 19-3   Sample page from **Computer and Control Abstracts.**

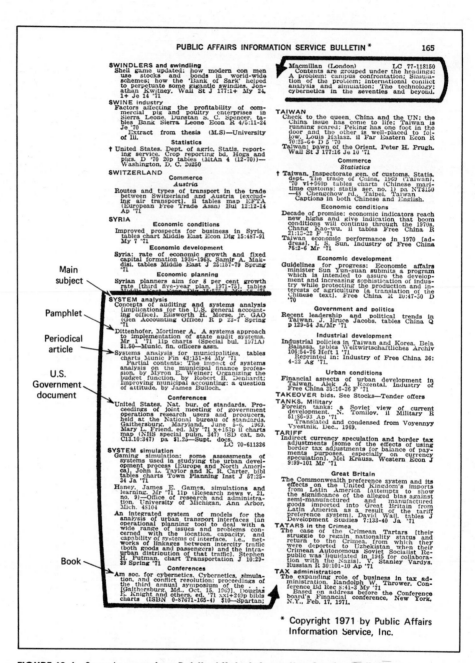

FIGURE 19-4   Sample page from **Public Affairs Information Service Bulletin.**

GENERAL NEWS   **THE WALL STREET JOURNAL INDEX**[*]   February, 1972

to CWA locals; members will vote by mail; one union officer sees small gain in new offer, urges rejection. 2/7-14;1

Federal judge issued an order to bar New York Telephone Co. temporarily from continuing to collect increased service rates that were put into effect Feb. 3. 2/10-36;2

Colorado Public Utilities Commission sets hearings on discounts by Mountain States Telephone & Telegraph Co.; asks justification for cut rates granted employes, favored customers. 2/15-9;1

Dean Burch, chairman of FCC, viewed with skepticism, a Defense Department offer to aid FCC's investigation of American Telephone & Telegraph Co.'s long-distance telephone rate structure. 2/17-4;4

New York strike over, phone men return at less than $2 a week above July 1971 offer. 2/18-6;2

Federal Communications Commission said there was 'significant overall improvement' in AT&T service during first 10 months of 1971, but pains remain. 2/25-20;1

New Jersey Bell Telephone Co. asked New Jersey Public Utilities Commission to approve rate increases for $137 million in additional

COMPUTERS ← — General subject

Three major computer companies in Europe agreed to form joint venture; plan close cooperation in field. 2/2-4;2

'Artificial brain' gives movement to monkeys paralyzed by 'strokes;' potential human use is seen for computerized, implanted electrodes that guide limbs. 2/9-20;3

Humans replace the computer at California's Commission for Teacher Preparation and Licensing. 2/15-1;5

Federal Reserve stated computer error caused its puzzling money-market steps. 2/18-19;1

Patenting of programming for computers to ← — Subject of article
get High Court ruling, probably in fall. 2/23-4;2

New York Stock Exchange trading halted 24 minutes by computer snarl; day's data incomplete. 2/25-5;2

International Computers (Holdings) Ltd. said talks with Britain on aid are making 'satisfactory progress.' 2/25-21;4

New York Stock Exchange's computer woes again briefly halted ticker. 2/29-33;3

} Date—February 23
Page—4
Column—2

U.S. economy to grow faster in 1970s, 1980s than in past two decades, Conference Board study forecasted. 2/4-13;1

Record demands on credit markets in 1972, interest-rate rises forecast at board's meeting. 2/24-8;2

CONFLICT OF INTEREST

Neil H. Jacoby, public member of President Nixon's Pay Board resigned as chairman of executive committee of Occidental Petroleum Corp.; labor vexed over corporate appointment. 2/9-11;2

CONGLOMERATES
(see Mergers)

CONGRESS (UNITED STATES)
(See also House and Senate)

Congressional attempts to legislate higher grain prices for 1973 harvests apparently have collapsed. 2/1-19;2

President Nixon sent Congress plan for developing job opportunities in rural America that included $1 billion in additional federally backed credit for industries, communities. 2/2-3;2

President Nixon shortly to ask Congress to devalue dollar by lifting price of gold $3 an ounce. 2/7-4;2

Bid to end western dock strike will be started Feb. 7; despite Nixon Administration's strong pressure, legislation unlikely to get far in week beginning Feb. 7. 2/7-9;1

Dollar devaluation formally asked by President Nixon through gold-price boost; increase

in official quote to $38 an ounce from $35 sought from Congress. 2/10-3;1

Editorial page article on some of the reasons a number of very senior Republicans are quitting Congress at the end of 1972. 2/10-16;3

Health-insurance bill sent to Congress; new questions, controversy seen certain. 2/11-6;2

SEC dropped opposition to utilities entering low-cost home field; W. Casey, however, asked Congress not to permit firms to build various commercial properties. 2/17-22;1

Congress' economizers move to cut costly concentrations of noncombat troops. 2/18-1;5

Monetary reform for long pull will start moving soon after Congress acts on dollar devaluation. 2/25-1;5

Faces week of busing battles; five senators to interrupt President drives. 2/28-8;2

CONSERVATIVE PARTY

Calming Conservatives; President Nixon takes conservative speechwriter Pat Buchanan along to China. 2/18-1;5

adjusted annual rate of $115.1 billion in December, up 1% from November's upward-revised $113.6 billion. 2/2-3;1

Canadian seasonally adjusted annual rate of housing starts totaled 217,500 units in January, down from 231,000 units in December. 2/10-8;5

Housing starts in January reached record, climbing to a seasonally adjusted annual rate of 2,549,000 units. 2/17-2;2

Value of construction contracts for January set record $6.23 billion, lifting index to record of 165. 2/29-9;1

CONSTRUCTION INDUSTRY STABILIZATION COMM.

Editorial on significance of wage decisions of committee, Pay Board. 2/3-10;1

Editorial page article highlighting recent discussion of several Brookings Institution economists on panel's performance. 2/22-14;3

CONSUMER CREDIT

Pace of consumer credit surge slowed in December by personal, auto sectors. 2/4-28;2

CONSUMER PRICE INDEX
(see Prices)

CONSUMERS

Federal poll finds cut in car-buying plans but sees rise for housing, furniture. 2/3-4;3

Bank of America pares most rates on consumer loans. 2/9-7;1

Price Commission plans to modify its tight policy on how much of their increase labor

costs companies may pass through to customers may pass through to customers in higher prices. 2/23-6;2

CONTAINERS
(see Packaging)

CONTESTS

Questionable Breakthrough: Downtown St. Louis Inc. boasts it will stage the first beauty-talent contest in the nation using the title Ms. for the winner. 2/3-1;5

Cities hustle to win a contest that isn't that easy to lose; All-America City title is held by Newark, White Bear Lake and 242 other communities. 2/25-1;4

CONVENTIONS

Valentine's Day is used as an excuse by New York City to court 300 convention buyers from around the country. 2/10-1;5

COOPERATIVES

Land O'Lakes Inc., a major butter selling cooperative, said it will diversify a bit later in 1972; it plans to make margarine. 2/10-1;5

COPPER INDUSTRY

Braden Copper Co. sued Chile in New York federal court seeking more than $5 million it claimed in losses from loans to mining company nationalized in July 1971 by Chile. 2/7-10;6

Biting the Hand... Some mining states outdo U.S. standards in air-pollution war; Arizona and Montana stances alarm copper producers; smelter closings forecast. 2/10-1;6

Phelps Dodge Tube Co. raised prices 5% for copper water pipe and drainage and refrigeration tubing. 2/10-12;3

Anaconda Co. took a major charge from Chile loss; wrote off $302.5 million for expropriated mines. 2/11-32;1

Indiana Standard said Africa venture found bed of copper. 2/18-2;2

Copper prices are increased 2 1/4 cents a pound; higher quotes are posted by Phelps Dodge Corp., Anaconda Co. and Inspiration Consolidated Copper Co.; rises aren't under controls. 2/25-2;2

Six U.S. and Canadian firms followed Phelps Dodge Corp.'s Feb. 24 increase in copper. 2/28-6;1

Chile agreed to pay Kennecott Copper Corp. $84.6 million toward mine loan; Anaconda Co. to file suit. 2/28-6;3

Heard on the Street item on significance of rise in per share earnings of copper stocks. 2/28-27;3

Recent general price boosts in lead, copper and brass mill products were followed by other U.S. and Canadian companies. 2/29-7;3

Chile said it won't pay $171 million in outstanding promissory notes to Anaconda Co. 2/29-12;2

CORN NEWS

Agriculture Dept. said 1972 cotton plan stands despite textile pleas; but it added voluntary diversion payments for corn, sorghums in bid to win set-aside goal. 2/3-20;4

Agriculture Secy. E.L. Butz suspended further government purchases of corn, citing an unexpected upsurge in the quantity of 1971-crop corn that farmers are withholding from market under federal loan program. 2/16-24;4

CORPORATE CAPITAL

Venturing Out: Corporate giants now providing some capital for risky new business; General Electric Co., others seek technology rather than a fast buck; rejections by the hundreds. 2/1-1;6

CORPORATE INCOME

Onward & Upward: Corporate earnings continued to recover in fourth quarter; survey of 464 firms showed jump of 25.3%; excluding General Motors Corp., profit boosts is 10.2%. 2/4-1;6

Editorial on what to expect from corporate profits in period of price and wage controls. 2/9-14;1

Fourth quarter 1971 net for 82 Canadian firms climbed 4.1% from 1970. 2/14-12;2

72

FIGURE 19-5   Sample page from the **Wall Street Journal Index.**

major portion of the library, they often are not. Atlases, by contrast, are often what librarians call "oversize" materials. As such, they do not fit in the regular scheme of shelving, so they are kept in special atlas cases. Usually they are cataloged as part of the reference collection.

Directories

Directories as we mean them here are guides to business associations or people. They may be membership directories such as for the Association for Systems Management, the American Marketing Association, or similar organizations. Directories to business firms may be either commercially published such as the publication called *California Manufacturers*, or they may be trade association directories listing member corporations. Directories such as these usually are cataloged and frequently are a part of the reference collection. There are, of course, also other types of directories such as telephone directories for cities, government agencies, and so on. Sometimes these are cataloged, but more often they are not.

Annual Reports

Corporate annual reports can constitute an important part of a corporate library's collection if the need for information on other corporations exists. By having these materials available, the analyst has ready access to financial and product information on other corporate entities.

Two other important areas we will explore in the next section are those of government documents and government reports.

THE TOPSY-TURVY WORLD OF GOVERNMENT DOCUMENTS

The United States Government is perhaps the largest and most complex system on the Earth. Its activities and publications have influence on almost every formal organization in the country. Thus, the systems analyst needs to be familiar with the government's system of information dissemination.

To the beginning systems analyst who may suddenly be in the position of needing government documents, one can only advise: take a deep breath and plunge! Even to the experienced practicing librarian who may have worked with these materials for years, the whole procedure often seems impossible. The reason is that, like Topsy, they "just grew that way."

Government Documents

First of all, what is meant by "document"? In this chapter we mean any publication published by any level of government, whether it is international, national (domestic or foreign), state, county, or municipal. Government publications constitute an increasingly major proportion of the world's publishing output. One reason is government's increasing participation in projects too costly on the private level. Another is the need for documentation that can assist others in learning from past mistakes.

Because of the many improvements in information retrieval, such as computerized indexing systems, we also have been made more *aware* of the availability of these publications.

It is essential when working with government document indexes to understand that none of these indexes is totally independent of the others. Or, to put it another way, no agency announces just its own publications in its index; it may announce publications of many other agencies. To complicate matters further, many of these government publications also are announced in commercial indexing services such as *Engineering Index*. To use these indexes intelligently, one should learn and understand the types of information found in the various indexes.

Because so many indexes today are actually printed versions of something that is computerized, one needs to understand some of the odd twists a computerized index can take. If the programmer had little knowledge of library filing procedures, he or she might utilize what could be called the "something before nothing" principle. For example, suppose you have only a National Aeronautics and Space Administration Technical Note number NASA-TN-55. You have seen a reference to this document and you want to find out more about it. Since you do not have NASA's index available, you decide to use the Commerce Department's *Government Reports Announcements Index*. You would expect to find TN-55 between TN-54 and TN-56. Instead (especially in the early computerized index), you may find it between TN-543 and TN-553, thus

NASA-TN-53

NASA-TN-54

NASA-TN-543

NASA-TN-55

NASA-TN-553

NASA-TN-56

In this example, all 54's, regardless of the number of digits, have to be printed before the computer can go on to the 55's.

Another confusing sequence sometimes programmed into indexes might be called alphanumeric reversal. We would expect to find alphanumeric report numbers with the "smallest" alpha designation first; for example,

L-6594

LA-4245

LA-TR-70-15

But, if the programmer or systems designer is unaware of library practices, you may instead have the following mixed-up sequence[3]:

LA-TR-70-15
LA-4245
LCI-4
LMEC-10
LMEC-8
L-6594

If these numbers are close to one another, you may be able to find the report number you want; however, if they are separated by pages you may not be able to locate the reference you need.

To further complicate matters, unnoticed and uncorrected input errors also cause problems which you should understand. An extra hyphen, a missed digit, a skipped space, or any other of a dozen possible errors can hamper your use of the index. Using the Technical Note 54 example, assume the keypuncher accidentally hit the space bar when typing TN-54. Depending on how the program is written, you might then have

NASA-TN-53
NASA-TN-543
NASA-TN-55
NASA-TN-553
NASA-TN-56
NASA-T N-54

The alphabetical sequence may also be broken if the people writing the inputs are not consistent: for example, Viet Nam, Vietnam, VietNam, Viet-nam; or Journal, Jnl, Jl, J, and Jour. Remember that the computer only prints what is put into it; and if no one catches the human errors, you, the user, have problems! Any of these seemingly small errors can cause you to be off by many pages in your searching of paper indexes.

The U.S. Government Printing Office. The Printing Act of 1895 established that the Superintendent of Documents would publish on the first day of each month a list of government publications printed during the preceding month. Today we know this list as the *Monthly Catalog of U.S. Government Publications.* [4] Its title has varied somewhat over the years; but essentially it announces what the Government Printing Office has printed, instructions for obtaining the items, and the prices. It has often happened that changes in administration or growth of field agencies caused omissions in the *Monthly Catalog,* so it can by no stretch of the imagination be considered complete.

Publications are listed under the issuing agencies, which appear alphabetically. Each publication is assigned an entry number beginning with 1 each January and working consecutively through the December issue. In addition to the title, information given includes date of issuance, number of pages, price, the Superintendent of Documents classification number (by which some libraries file the publications when received), the Library of Congress card number, and other useful information.

Each monthly issue contains indexes by author, title, and report number. Use of the entry number simplifies retrieval. In June and December of each year the index is cumulated for easy use. When using the *Monthly Catalog,* it is useful to note the symbols appearing with each entry. Items marked with an asterisk (*) are available for sale by the Superintendent of Documents. A single dagger (†) indicates the item is distributed by the issuing office. A double dagger (‡) means for official use only (but they often are available anyway). Items marked with a phi (ϕ) symbol are for sale by the Department of Commerce. A black dot (●) indicates the item is sent automatically to Depository Libraries (a list of which appears in each September issue). See Figure 19-6 for an example of a *Monthly Catalog* page.

Publications announced in the *Monthly Catalog* may be handled in any manner once they arrive in the library. Sometimes they are placed in a separate section and filed by the Government Printing Office designation; sometimes they are cataloged and treated like a book. Some may arrive in microform, and these will be treated in a different manner. Still other libraries have their own unique system for handling these documents, often a system dependent on the knowledge of their first documents librarian.

The *Monthly Catalog* is useful in the identification of most "official" United States Government publications. These include census information, statistics for most industries, many Congressional documents, and documents from agencies that have their own announcing media such as the National Aeronautics and Space Administration. In addition, lists of serials and periodicals currently being published by the government are announced.

Technical Reports

The next major section within the government documents sector is a group of publications referred to as technical reports. These publications are written by corporations, primarily under contract to the various government agencies, both state and federal. These contractor reports frequently are the method by which new techniques or new technologies are transmitted from the private sector to the public sector and vice versa. These reports frequently are part of a systems approach to a problem that needs to be solved. The contractor examines the problem using the various steps

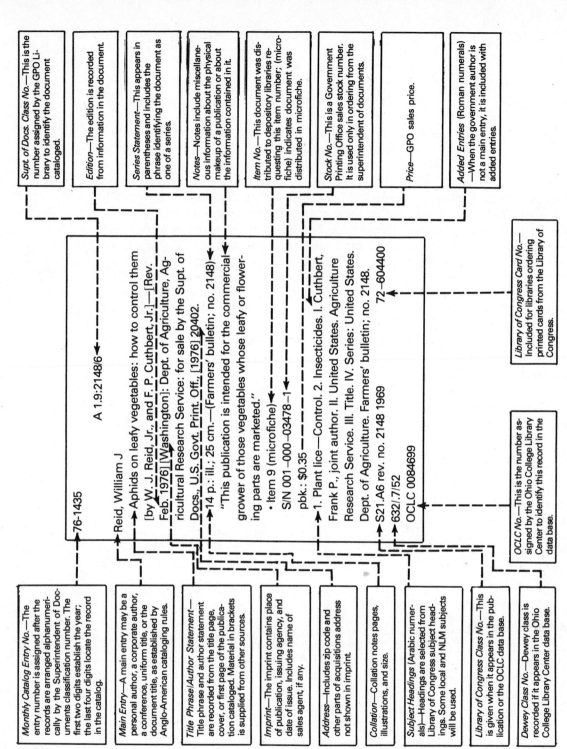

FIGURE 19-6 Example of an entry from the **Monthly Catalog of U.S. Government Publications.**

478

within the system analysis framework, and then provides in the contract report recommendations for action that should be taken.

Following is a discussion of the primary government agencies that have contractors perform work on their behalf, and which also provide indexes to these contractor reports. Since most firms today perform government research work, and since the amount of tax money spent on contract research is significant, these reports become essential to business and industry. They are used to determine the state of the art which can lead to further development. In addition, they provide important leads to what the competition is doing.

The Department of Defense. The dissemination of Department of Defense (DOD) documents began in 1945. At that time, what was then called the Air Documents Division of the Intelligence Department loosened wartime classification of research reports. Some of these reports retained their classified status; for example, only those people with the proper security clearance level and "need to know" were permitted access to these classified documents. These classified documents formed one-half of the dissemination program. The second half was composed of declassified reports and captured enemy research documents. These unclassified research reports were announced in the *Technical Information Pilot (TIP)*, originally begun by the Office of Naval Research (ONR).

In 1947, the *Air Technical Index* took over the announcement of unclassified research reports. Each announced report was identified by a number using an ATI prefix. In the 1948–1950 period there occurred some shifting around of responsibility in the announcement of these reports. Finally, in 1951 the Armed Services Technical Information Agency (ASTIA) was established to handle dissemination of *all* defense-oriented reports. By this time, the unclassified/classified approach was firmly entrenched. One of ASTIA's functions, therefore, was to act as the central document releasing agency for government contractors. ASTIA also continued publication of the *Air Technical Index.*

In 1953 the title was changed to *Title Announcement Bulletin (TAB)*. This new *TAB* was one of the first indexes to be printed using computerized methods. For approximately a year there were actually three *TAB*'s published: an unclassified version announcing unclassified reports, a classified version announcing to military contractors the available classified reports, and a military version. In 1954 the classified version was discontinued and that part which had been classified was "sanitized" to make it available to the public. At this point the unclassified and classified versions were merged.

The previously unclassified reports comprised the first half, a white section. The second half was buff colored, and it announced the still classified or restricted dissemination documents in a "sanitized" style; for example, instead of printing a classified title, it would simply note "Classified Title." Identification of needed reports was by subject arrangement and some descriptors. The new *TAB* used an AD prefix (*ASTIA*

*D*ocument) with six digits, for example, AD-349 856. This numbering system with slight modification is still used today. A new title in 1957 reflected the inclusion of abstracts: *Technical Abstract Bulletin.*

In the meantime, it was ironic that the *TAB* was not truly available to the public at large, primarily because of a statement in small print to the effect that it was not to be given to nationals of a foreign country. For this reason, its use was restricted largely to the military and its contractors. Any library having a *TAB* set was generally associated with a military contractor (such as some university libraries) and kept it in a closed area. For this reason, in 1961 the Department of Commerce began duplicating the *TAB*'s white section with its *U.S. Government Research Reports* (see next section for details).

In 1963 ASTIA's name was changed to Defense Documentation Center (DDC) to more accurately reflect its military orientation. Eventually, in 1967, since the Department of Commerce was entirely duplicating the *TAB* white section, the white section's announcement was transferred totally to the Department of Commerce. This was an economical move which should have been taken ten years earlier. Concurrently, the buff portion was once again classified "confidential" and it officially became available only to military contractors. This move supposedly made the *TAB* more useful to contractors. Since the need to "sanitize" was removed, the contractor could more readily identify the needed reports and, at least theoretically, eliminate the ordering of reports not relevant to the contract.

In mid-1978 DDC was permitted to once again publish the *TAB* in an unclassified format. It is still available, however, only to "authorized users" (e.g., Department of Defense contractors). In 1979, the *TAB* again incorporated the unclassified, unlimited distribution reports that were delegated to the Department of Commerce in 1967. DDC also changed its name to Defense Technical Information Center (DTIC) during 1979.

The later *TAB*'s are well indexed (including biweekly, quarterly, and annual cumulations) by personal author, corporate author, title, report number of the company writing the report, report number of the sponsoring military agency, contract number, subject, and AD number. Since 1965 each newly acquired report has been put on microfiche to eliminate the unnecessary cost of paper copies and their accompanying space problem. The reports may be purchased either in paper or microfiche.

To qualify for receipt of classified reports from DDC, one must

- Have a military monitor to sponsor the activity and to verify the need for the information.
- Have the military monitor specify in writing the "need to know," for example, this contractor needs information on this subject for use on this contract until this date.

- Send the written need-to-know to the proper agency (DTIC) requesting registration to receive documents.
- Signify the exact numerically identified subject areas of interest on the "Field of Interest Register" (FOIR) DTIC sends to you.
- Send the FOIR to military monitor for approval.

After submitting the FOIR, the following takes place.

- Military monitor approves all, some, or none of the subject areas on the FOIR and returns the form to the disseminating agency.
- The agency sets up a file and sends you a copy of the FOIR as it was approved so you may begin ordering on that contract.
- If the military monitor has crossed out a subject area you had indicated, you may not order documents falling into that area unless the monitor agrees to submit a change, following the same procedure as with the original FOIR.
- The FOIR procedure often takes six weeks, making it virtually useless for short-term contracts. In that case, the military monitor generally provides directly the documents he or she thinks will be needed to fulfill the terms of the contract.

Generally speaking, classified materials

- Are restricted by law from the public at large under terms of the Internal Security Act of 1950, and the Espionage and Sabotage Act of 1954.
- When received must be kept in "secure" containers of a type and with locks as specified by the Department of Defense.
- May be handled only by persons having a proper "security clearance" (which takes from three to six months to obtain) and with a definite "need to know."[5]

The systems analyst designing for classified materials should know that specific paperwork routines *must* be followed in transferral of classified documents from one person to another; therefore, a document-receiving facility must be prepared for new paperwork routines as well as special storage equipment.

DTIC estimated that in 1970 alone they processed 42,717 reports. The subject areas are wide-ranging, including aeronautics, agriculture, social sciences, missile technology, and physics, to name a few.

The Department of Commerce. In 1945 the President of the United States established the Office of the Publications Board (PB). Its purpose was to announce research documents available to the public, a function parallel to the Intelligence Department's Air Documents Division. In January of 1946 the Publications Board began its first announcement service, calling it *Bibliography of Scientific and Industrial Reports (BSIR)*. Each report announced was given a Publications Board identifying number, beginning with PB-1.

The following year the Publications Board was superseded by the Office of Technical Services (OTS), which continued publishing the *Bibliography*, retaining the PB- prefix. The Library of Congress, however, retained the "Publications Board Project" for many years and you often may find PB entries in the *Monthly Catalog* (see the previous section on the U.S. Government Printing Office).

Over the years a number of changes took place in the title and occasionally indexing changed. Also, in 1961 the *U.S. Government Research Reports (USGRR)*, as it was then known, included the previously mentioned Department of Defense's *Technical Abstract Bulletin*, including use of the AD- prefix for accession numbers.

In 1965 the OTS was superseded by the Clearinghouse for Federal Scientific and Technical Information (CFSTI); then CFSTI was superseded in 1970 by the National Technical Information Service (NTIS), which still exists today. In the meantime users were confused by numerous title changes and problems with the index itself. While the new index had subject, personal author, corporate author, and accession number indexing, it also used *derived* titles rather than *actual* titles, which caused many ordering problems. As was stated in the Foreword of the first issue, the combined index was produced entirely by computer manipulation of data records prepared for other purposes by the four contributing agencies. Since these agencies used different rules for indexing and for creating machine-readable records, it was inevitable that format errors, inconsistencies, or duplications would occur. The situation, although improved, has never been fully resolved. This tends to deter all but the most serious user from this very important data base. The NTIS index is computerized so retrieval has been somewhat simplified.

When the publishing agency became known as NTIS, the new agency was given broader functions. Its current title, *Government Reports Announcements*, reflects this. Today NTIS publishes not only for the Department of Commerce, but also reports of many other government agencies and their contractors. In addition, it includes some journal articles (usually translations), symposia, patents, and theses to aid the Department of Commerce in its mission of transferring technology from the government to the private sector. See Figure 19-7 for an example of a *Government Reports Announcements* entry.

The National Aeronautics and Space Administration. As far back as 1915 the United States had made a formal commitment to aviation even though by the end of that year only one-hundred aircraft (both civilian and military) had been built in this country. It was in 1915 that the National Advisory Committee for Aeronautics

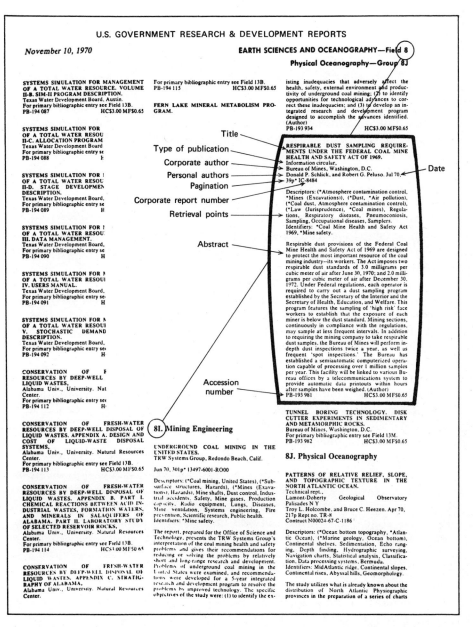

FIGURE 19-7   Reproduction of abstract entry page from **U.S. Government Research and Development Reports** (now titled **Government Reports Announcements**).

(NACA) was created by Congress. NACA published numerous reports and memoranda, most of which were announced in the Government Printing Office's *Monthly Catalog.* It was not until 1948 that an index was published. At that time the *Index of NACA Publications, 1915–47* appeared. This was superseded in 1949 by the *Index of NACA Publications, 1915–49.* The reason given for republication was the large number of omissions and the many new reports that had been added. The *Index* continued on an almost-annual basis until 1958. Identification of each report was by its internal NACA number, for example, MR for Memorandum Report, TM for Technical Memorandum, TN for Technical Note, and so forth.[6]

In 1958, when NACA was absorbed into the newly created National Aeronautics and Space Administration (NASA), the *Index* continued to be published. It was eventually absorbed by the *Technical Publications Announcements (TPA),* which began as a computerized index. This new index was formatted to reflect NASA's accession number system for computer recognition. Each entry now had not only an internal NASA number, but as primary identification an accession number, for example, N62-100 001.

Use of the accession number eventually facilitated entry into the system of contractor reports which had internal company numbers but no NASA number. Even to the casual observer, it is clear that the people who organized NASA's document dissemination program were not only well organized, but also knew how to set up a user-oriented logical system (as compared with the Departments of Defense and Commerce mentioned previously).

Concurrently the National Science Foundation and the Air Force Office of Scientific Research instituted the *International Aerospace Abstracts (IAA)* to announce and index periodical articles, conference proceedings, and translations in the field of aerospace. The *IAA* followed the same subject arrangement as NASA's *TPA,* and the format is similar to those mentioned above.

In 1963 NASA took over the publication of the *IAA.* The only format change was the addition of accession numbers to identify each entry, for example, A63-100 001. At the same time NASA's own *TPA* became the *Scientific and Technical Aerospace Reports (STAR).* In reality it became two *STAR*'s. The second, know as the *C-STAR* was a less well-known classified version, which utilized an X prefix in its accession numbering, for example, X63-100 001. As the reader can see, if one has a publication with any of the A, N, or X prefix accession numbers, one automatically knows where that publication is indexed, for example, in the *IAA, STAR* or *C-STAR.*

In 1970 the *C-STAR* ceased publication, but its place was taken in 1972 by a quarterly, unclassified version called the *Limited STAR (L-STAR).* The *L-STAR* announces both unclassified and classified limited distribution NASA work that must be announced under the Freedom of Information Act.

The Department of Energy. The Department of Energy's earliest predecessor, the Atomic Energy Commission (AEC) was created by the Atomic Energy Act of 1946.

The Commission's purpose was to administer programs on atomic energy and special nuclear materials in such a manner as to protect the health and safety of the public. Part of this program included dissemination of scientific and technical reports related to nuclear energy. The AEC's activities overlapped into many other government departments since they oversaw procurement and production of fissionable materials, testing of DOD's nuclear weapons, the development of NASA's nuclear rocket propellants, and development of nuclear reactors. A major purpose of the program was to encourage the development of peaceful uses for nuclear energy.

In 1948 the AEC began publication of the *Nuclear Science Abstracts,* which was an outgrowth of the older *Abstracts of Declassified Documents.* The *NSA* was instituted to announce availability on an international basis of reports, books, translations, proceedings, patents, and journal articles on nuclear science. An accession number system was used to identify each entry, beginning with the number 1 each January. *NSA* had excellent cumulative indexes for subject, corporate and personal authors, and report numbers. Like NASA, the AEC had a network of depository libraries through which many of the reports could be obtained. Many also were available through the National Technical Information Service mentioned earlier.

The AEC was dissolved in 1975 and replaced by the Energy Research and Development Agency (ERDA) which in turn was replaced by the Department of Energy in 1977. Since the AEC was dissolved, the indexing situation has been very confused. The last issue of *NSA* was in June of 1976 and it was replaced by an ERDA index that changed names several times. The name was changed again in 1977 when DOE took over the ERDA functions. They new *Energy Research Abstracts* is considerably broader than the old *NSA* because it includes all forms of energy including solar and geothermal. The index cites reports, journal articles, conference proceedings and papers, books, patents, theses, and monographs originated by ERDA/DOE laboratories, energy centers, and contractors.

Researchers who wish to study only nuclear energy tend now to use the *INIS Atomindex,* which is published by the International Nuclear Information Service of the International Atomic Energy Agency in Vienna. As with the old *NSA,* this index covers the literature on the peaceful uses of nuclear energy.

The Department of Health, Education and Welfare. The U.S. Office of Education (USOE) also has become interested in the field of systems analysis, so you should be aware of their publications. They too have a distinctive accession numbering process. The Office of Education has a system of clearinghouses, each of which is concerned with a particular aspect of educational research. This clearinghouse system began in 1966 and was called the Educational Research Information Center, now called the Educational Resources Information Center (ERIC).

The ERIC reports are abstracted, computerized, and announced by the Office of Education in the monthly *Resources in Education (RIE).* Its accession numbers have an ED (*ERIC D*ocument) number, for example, ED-010 001. Some entries have an EP

(*ERIC PROJECT*) number. Indexing is thorough and is by personal author, institution, subject, contract or grant number, and Office of Education project number.

In 1968 a secondary accession number was added to each entry, although arrangement is still by ED number. The new number indicates which clearinghouse processed the entry. For example,

AC = Adult Education

TE = Teaching of English

HE = Higher Education

and so on.

A private publisher in a cooperative venture with USOE began another indexing service in 1969. This one, *Current Index to Journals in Education (CIJE)* complements *Resources in Education,* which tends to announce reports. Each periodical notation has an EJ accession number prefix, and indexing is by personal author, journal title, and subject. Although some entries do not have abstracts, a number of subject terms (called descriptors) are included to help the user decide whether to read the actual periodical article.

Some of the ERIC clearinghouses also publish special abstract journals, such as the *Exceptional Child Education Abstracts,* which include only their own entries processed for ERIC.

Miscellaneous Documents

Although there are many other types of documents you may have use for as a systems analyst, only a few of the most likely will be included here.

Standards. Although both civilian and military agencies publish standards, the main concept in all standards is comparison. A standard may define a unit of length or define the purity of a piece of metal. The military has two standardization publications; a Military Specification (MIL-SPEC) is intended to establish the necessary characteristics of an item in terms of its expected performance, while a Military Standard (MIL-STD) actually defines dimensions of a particular item for purposes of interchangeability. A Military Handbook (MIL-HDBK) is not a standard but is designed to be used in conjunction with MIL-SPEC's and MIL-STD's to aid in their application and interpretation. These specifications and standards are indexed in the *Department of Defense Index of Specifications and Standards.* (See Figure 19-8 for other codes.)

Federal Specifications (FED-SPEC) and Federal Standards (FED-STD) are similar in format to those of the military. They are published, however, by the General Services Administration which is the "watch-dog" agency of the federal government. The *Index of Federal Specifications, Standards and Handbooks* partially duplicates the *DOD Index,* but it includes many nonmilitary items since it is intended for civilian usage.

The National Bureau of Standards (NBS) was founded by Congress in 1901 to

MS	Military Sheet Standard	Army FM	Army Field Manual
MIL-STD	Military Standard	AMCP	Army Materiel Command Pamphlet
MIL-SPEC	Military Specification		
FED-SPEC	Federal Specification	AMCR	Army Materiel Command Regulation
FED-STD	Federal Standard		
QPL	Qualified Products List	MPD	Missile Purchase Description
AN	Air Force Navy Aeronautical Standard		
		ORD M	Ordnance Manual
AND	Air Force Navy Aeronautical Design Standard	ORD P	Ordnance Pamphlet
		DOD	Department of Defense
		DOD DSAM	Defense Supply Agency Manual
DP	Description Pattern		
MIL-HDBK	Military Handbook	DOD DSAH	Defense Supply Agency Handbook
AFL	Air Force Letter		
AFM	Air Force Manual	DOD DSAR	Defense Supply Agency Regulation
AFR	Air Force Regulation		
AFP	Air Force Pamphlet	NBS	National Bureau of Standards
AFSCM	Air Force Systems Command Manual		
		FAA Regs	Federal Aviation Regulation
AFSCR	Air Force Systems Command Regulation		
		CS (GPO)	Commercial Standard
AFSCP	Air Force Systems Command Pamphlet	USASI	United States of America Standards Institute
AFPC	Air Force Procurement Circular	NAS	National Aerospace Standard
AFTO	Air Force Technical Order	ASTM	American Society for Testing and Materials
AFPI 71 Series	Air Force Procurement Instruction		
		AMS	Aerospace Material Specification
AF Exhibits	Air Force Exhibit		
USAF Spec Bul	United States Air Force Specification Bulletin	ARP	Aerospace Recommended Practice
ANA Bulletins	Air Force Navy Aeronautical Bulletin	AIR	Aerospace Information Report
		AS	Aerospace Standard
NavInst	Navy Instruction	AWS	American Welding Society
Navy WR	Weapons Requirement	AGMA	American Gear Manufacturing Association
Navy WS	Weapons System		
Navy OD	Ordinance Document		
NAVEXOS P	Navy Office Executive Office Pamphlet	AISI	American Iron & Steel Institute
AR	Army Regulation	NEMA	National Electrical Manufacturers Association
NAVDOCKS DM	Navy-Docks Design Manual		
		AFBMA	Anti-Friction Bearing Manufacturers Association
NAFVAC DM	Navy-Factory Design Manual		
Army TM	Army Technical Manual		

FIGURE 19-8   Some common document designations.

develop and preserve standards of physical measurement, with effective calibration methods, for use by American industry. The Bureau publishes *NBS Handbooks*, which are recommended codes of industrial and engineering practice. *NBS Handbook 28, "Screwthread Standards for Federal Services,"* is perhaps the most widely used of all the Handbooks. Business systems analysts have been concerned with recent NBS activity in setting standards for secure computer systems through the design of complex algorithms. NBS publications are listed in the *Monthly Catalog of U.S. Government Publications* and are available from the Government Printing Office.

Other standards are published by the American Standards Association, American Iron and Steel Institute, American Society for Testing and Materials, the Society for Automotive Engineers, the National Electrical Manufacturers Association, and the Underwriters' Laboratories, to name a few. Each has its unique numbering system. Codes, such as the *Uniform Building Code* and the *National Fire Codes*, also may be put in the standards category since they may place a limitation on some part of a proposed system.

Patents. Each country has patents to protect an inventor from having an invention stolen and credited to someone else. All U.S. patents are registered in the Patent Office and announced in the *Official Gazette of United States Patent and Trademark Office.* Since the *Official Gazette* only provides an abstract describing the invention, it is often only a starting point with the services of a patent attorney being needed for a full-scale patent search. Some patents also may be located through NTIS, *Chemical Abstracts,* Derwent's *World Patents Index,* or the cumputerized index to U.S. patents called *CLAIMS.* Your librarian should be able to assist with these indexes.

There are some publications that may be considered as supplementing the Printing Office's *Monthly Catalog.* These include the *Monthly Checklist of State Publications* issued by the Library of Congress. It includes official publications of the various states (if they are sent to the Library of Congress) and some quasiofficial publications of associations of state officials, statistical reports, and so forth.

Many states also have their own announcement services such as *California State Publications* and the *Checklist of Official Pennsylvania Publications.*

The *United Nations Documents Index* began in 1950. It lists publications of the U.N. and its immediate agencies. Many of these U.N., state, and federal documents also are listed in the commercially published *Public Affairs Information Service Bulletin (PAIS),* as well as other indexing services.

ADVICE TO THE NOVICE

As you have seen from the preceding discussion, there are so many types of publications that might help a systems analyst, that it is difficult to find a starting point. A hospital systems analyst, for instance, might find *Index Medicus* or the *Hospital*

Literature Index indispensable, while the corporate systems analyst may be quite dependent on *Business Periodicals Index, Abstracted Business Information (ABI/INFORM),* or *Computer and Control Abstracts.* Each reader will find a favorite library resource. In the meantime, the following hints are provided.

1. Read the history of your field. Try to find out the mistakes and accomplishments of others in the field so that you can profit by them. Find out what made the successes.

2. Get to know *your* librarian(s). Find out who knows what is happening in your field. Librarians consider keeping up on various fields to be an essential part of their job (in fact, some companies include "keeping up with the literature" in their performance evaluations). Librarians also like to feel wanted, and a liberal sprinkling of thank you's can mean a librarian who thinks of *you* when a good article or book is found.

3. While it is not practical for the analyst to learn how to perform literature searching in a detailed way (that is not what you are being paid for!), it is useful to learn what indexes are available and the type of materials they contain. To get the most out of these indexes, ask your librarian to explain them. Only people who use the indexes frequently can keep abreast of changes, so important features may be lost to a novice.

4. Understand that computerization has had a significant impact both upon libraries, librarians, and users. With about 250 indexes readily available to librarians in a computerized form, you can assume a certain percentage of overlap. The key is to select the two or three that will help you solve the problem at hand. Consult with your librarian as you begin a project in a new subject area. This will save frustrations caused by wasting time and in getting caught at the last minute without a plan of action. Most important, it will allow an orderly search of the best resources and still allow time to obtain materials via interlibrary loan.

5. In dealing with company and occasionally some government reports, it is useful to remember that there are often restrictions imposed by the company to protect its own designs or processes, or by the government to protect the nation's security.

SELECTIONS FOR FURTHER STUDY

1. ALDRICH, ELLA V. *Using Books and Libraries, 5th edition.* Englewood Cliffs, N.J.: Prentice-Hall, Inc., 1967.
2. *Business Periodicals Index.* New York: H.W. Wilson Company, 1958–

3. *Computer and Control Abstracts.* New York: Institute of Electrical and Electronics Engineers, 1969– .

4. *Computing Reviews.* New York: Association for Computing Machinery, 1960– .

5. *Current Index to Journals in Education.* New York: CCM-Information Corporation, 1969– .

6. *Department of Defense Index of Specifications and Standards.* Washington, D.C.: Defense Supply Agency, 1970– .

7. *Department of Defense Industrial Security Manual for Safeguarding Classified Information.* Washington, D.C.: Defense Supply Agency, 1970.

8. *Energy Research Abstracts.* Washington, D.C.: Government Printing Office, 1975– .

9. GATES, JEAN K. *Guide to the Use of Books and Libraries, 3rd edition.* New York: McGraw-Hill Book Company, 1973.

10. *Government Reports Announcements* (title varies). Springfield, Va.: National Technical Information Service, 1946– .

11. *Index of Federal Specifications and Standards.* Washington, D.C.: Government Printing Office, 1971.

12. *INIS Atomindex.* Vienna: International Atomic Energy Agency, 1970– .

13. *International Aerospace Abstracts.* New York: National Aeronautics and Space Administration and Institute of Aeronautics and Astronautics, 1961– .

14. JOHNSON, H. WEBSTER. *How To Use the Business Library, 4th edition.* Cincinnati, Ohio: South-Western Publishing Company, 1972.

15. *Journal of Systems Management* (formerly *Systems and Procedures*). Cleveland, Ohio: Association for Systems Management, 1950– .

16. *Monthly Catalog of United States Government Publications* (title varies). Washington, D.C.: Government Printing Office, 1895– .

17. *Monthly Checklist of State Publications.* Washington, D.C.: Government Printing Office, 1910– .

18. *New York Times Index.* New York: The New York Times Company, 1851– .

19. *Official Gazette of United States Patent and Trademark Office.* Washington, D.C.: Government Printing Office, 1872– .

20. *Public Affairs Information Service Bulletin.* New York: Public Affairs Information Service, 1915– .

21. *Quarterly Bibliography of Computers and Data Processing.* Phoenix, Ariz.: Applied Computer Research, 1971–   .

22. *Resources in Education.* Washington, D.C.: U.S. Office of Education, 1966–

23. RIVERS, WILLIAM L. *Finding Facts: Interviewing, Observing, Using Reference Sources.* Englewood Cliffs, N.J.: Prentice-Hall, Inc., 1975.

24. *Scientific and Technical Aerospace Abstracts* (title varies). Washington, D.C.: National Aeronautics and Space Administration, 1958–   .

25. TALLMAN, JOHANNA. "History and Importance of Technical Report Literature, Part II," *Sci-Tech News,* Winter 1962, pp. 164–172.

26. *Technical Abstract Bulletin* (title varies). Alexandria, Va.: Defense Technical Information Center, 1953–   .

27. *United Nations Documents Index.* New York: United Nations Publications, 1950–   .

28. *United States Government Organization Manual.* Washington, D.C.: Government Printing Office, annual.

29. *Wall Street Journal Index.* New York: Dow Jones and Company, Inc., 1958–   .

QUESTIONS

1. Describe some of the types of materials a typical corporate or business library might contain. Name at least six. Can you name more?

2. How do most libraries arrange their books?

3. Why do you need to use the card catalog in a library?

4. By what means are journal articles located?

5. Name at least five indexes you might find in a corporate library.

6. Discuss how you might utilize the following tools in a systems study: books, periodicals, newspapers, manufacturers' literature, maps, directories, annual reports, government documents, technical reports, specifications, standards.

7. Why might a systems analyst find it important to know about classified documents if the analyst does not need to use these materials in a study?

8. Why should the systems analyst know how to use government documents? (Name two reasons.)

9. Why should the systems analyst know how to use technical reports? (Name three reasons.)

10. Name four government agencies discussed in this chapter that disseminate documents and describe the types of materials they handle.

11. Why are accession numbers important in indexing schemes?

12. What is a MIL-SPEC?

13. What is a FED-STD?

14. Name four reasons why computerized literature searching might be used rather than manual searching.

15. How do libraries cooperate to help the analyst do a better job?

FOOTNOTES

1. If you are totally unfamiliar with libraries, there are many books available to guide you. Among them are Gates' *Guide to the Use of Books and Libraries,* Aldrichs' *Using Books and Libraries,* and Johnson's *How To Use the Business Library.*

2. Documented in a study by Bruner and Lee, "Student Knowledge, Attitude, and Use of the Business Library," *Collegiate News and Views,* volume 25, Winter 1971, pp. 1–2.

3. This example, in shortened form, was taken from *U.S. Government Research and Development Reports Index,* December 25, 1970, pp. AR11-12.

4. For an in-depth historical analysis, see Schmeckebier and Eastin's *Government Publications and Their Use* (1969), or Kling's *Government Printing Office* (1970).

5. Specifics are outlined in *Industrial Security Manual for Safeguarding Classified Information.*

6. This method of numbering was found to be so serviceable that many companies and other government agencies took it over. Today, you will see many numbering sequences like this but with an additional prefix for the name of the corporation or government agency, such as LMSC-TR-78-861 or NASA-TN-5903.

GLOSSARY

Accession number The number or alphanumeric designation assigned to any individual piece of information or document by which it becomes machine retrievable.

Active life Those records that are used in the day-to-day operation of the firm.

Activity In PERT/CPM, an activity links two successive events and represents the work required between these two events. It must be accomplished before the following event can take place.

Aperture card A type of microform in which a frame of 16-mm or 35-mm film is placed on an 80-column tab card.

Area cost sheet Documents the formal organization and economic data pertaining to the area under study and fits each area under study into its proper context.

Area, data base A specifically named section of contiguous data base pages.

Area under study That part of the firm for which a new system is to be designed. It may be a small functional area of one department or the whole firm. See also: Level

Audit trail The documentation that provides for traceability among various reports. It may trace either from the source document to the end report or from the end report to the original source document.

Audit trail, computerized Used by the auditor to test the reliability of a computerized system and the dependability of the data it generates. Each transaction may have an identifying number to print out the record upon request.

Audit trail, manual The documentation provided by journals, ledgers, or reports which enable an auditor to trace an original transaction either forward or backward through a manual system.

Ballot box design A type of form designed in such a way that all the user has to do is check the applicable box.

Batching A procedure involving grouping like documents to be submitted to the computer as input. Control totals are usually established on document count and one or more significant data fields.

Bench mark approach A bench mark is a surveyor's mark that is used as a reference point during subsequent measurements; analogously, the analyst's understanding of the current system is a bench mark in determining how much improvement can be made with the new system.

Bench mark test The test run on a recommended computer to see how long it takes to run one of the firm's selected applications.

Benefit list A list prepared for each key person who attends the verbal presentation of the proposed new system. It lists the items within the proposed system that will benefit each of the these people the most.

Box location number One of the most important pieces of information put on the storage box label form since it identifies exactly where the box will be physically located.

Boxed design A type of form on which each item is clearly in its own box so there is no question where the information is to be filled in.

Caption and line design A type of form in which the items are not boxed in. See also: Boxed design.

Changeover, one-for-one The existing system is completely replaced with some other system; the old system is stopped and the new begins.

Changeover, parallel The old system and the new system are operated simultaneously until it is proven the new system is reliable and can do the job for which it was designed.

Chart, structure A chart that defines and illustrates the organization of a system on a hierarchical basis in terms of modules and submodules.

Charting A graphic or pictorial means of presenting data.

Charting, activity A type of chart in which the analyst is pictorially summarizing the flow of work through the various operations of a system.

Charting, layout A type of chart in which the physical area under study is pictured. Work areas and equipment are shown both before and after the new system.

Charting, personal relationship A type of chart that depicts lines of authority, job responsibilities, or job duties.

Charting, statistical data A type of chart that converts statistical data into meaningful statistical information by graphic portrayal.

Charts, flow See: Flowchart

Charts, Gantt Used for scheduling, Gantt charts portray output performance against a time requirement.

Charts, linear responsibility A chart that relates the degree of responsibility of key individuals to their various job duties.

Charts, operations analysis Charts used to analyze manufacturing operations.

Charts, organization A chart that shows the official structure of the organization in terms of functional units or in terms of superior-subordinate relationships.

Charts, right-hand and left-hand Charts used in motion and time studies by industrial engineers.

Charts, Simo Charts used in motion and time studies by industrial engineers.

Charts, work distribution These provide an analysis of what jobs are being performed, by whom, how the work is divided, and the approximate time required to perform each job.

Choice-set The alternatives that the systems designer has to work with.

Coding Writing the program in whatever programming language was chosen.

Company background Characteristics of the firm, including management ideas, attitudes, and opinions, and the company's goals and style.

Component The moving part of a system: personnel, computers, facilities, and paperwork.

Computer hardware The major electronic components in a computer system including the central processing unit, disk drives, and tape drives.

Computer output microfilm Microfilm produced as computer output, either directly or first on magnetic tape to be put onto microfilm later (COM).

Conversion, parallel See: Changeover, parallel

Cost analysis That phase of the system design in which the cost of the proposed system is determined in order to decide whether implementation is justified.

Cost analysis, budgeting The method of cost analysis based on the cash-flow concept, which refers to the amount of money that will be required for a particular project and the dates when that money will be needed. The firm then budgets the required funds so they will be available when needed.

Cost analysis, planning A method of cost analysis based on the analysis of the opportunity costs of using a resource for one purpose rather than another. In other words, how much more can be gained by using a resource in one area rather than in its second best alternative area.

Costs, implementation A one-time outlay to create new capability.

Costs, investment Nonrecurring outlays to acquire new equipment.

Costs, operating Recurring outlays required to operate the system.

Court decision The result of a legal controversy which has been interpreted in a court of law and based upon the statutes under the court's jurisdiction.

CPU The Central Processing Unit of a computer.

Critical path The longest path through a PERT network.

Critical path method (CPM) See: PERT

CRT terminal A Cathode Ray Tube terminal that displays the input and output characters on a television-like screen.

CSMP A complete simulation language oriented toward the solution of problems that are stated as nonlinear equations.

Cycle One month's business, one quarter, or whatever period of time is thought necessary in conversion from one system to another.

Data 1. Specific individual facts, or a list of such items. 2. Facts from which conclusions can be drawn.

Data base Also called *data bank*. A set of logically connected files that have a common access. They are the sum total of all the data that exist within an organization.

Data communications Part of the overall system which permits one or more users to access a remotely located computer.

Data flow diagram A diagram used to define the changes or transformations data undergo during its flow through a system.

Data item, data base The smallest unit of named data in a data base.

Data processing The conversion of data to information.

Debugging Removing errors in a computer program to get the program to run correctly.

Decision making The process of choosing the best solution to a problem.

Decision point A point in a system where some person or automatic mechanism must react to input data and make a decision.

Decision table 1. A tabular format showing all possible criteria that might be involved in an operation, along with the action to be taken in each situation. 2. A method of describing the logic of a computer program that tells what action must be taken when a given condition is either met or not met.

Defense Technical Information Center The primary distributor of classified government documents, sponsored by the Department of Defense.

Degradation A slowdown in computer response time usually due to a large number of simultaneous requests for processing.

Dependence, random When a procedure is required because of some other procedure.

Dependence, sequential When one procedure *must* proceed or follow another procedure.

Dependence, time When a procedure is required at a set time with regard to another procedure.

Descriptor An identifying word assigned to a report so that potential users can determine whether it is within their field of interest. Also, the word that the indexer uses to identify a report within a computerized index.

Design A creative process that plans or arranges the parts into a whole that satisfies the objectives involved.

Desk checking Consists of checking for keypunch errors and checking the keypunched program code against the program flowchart.

Destruction card A tickler file, arranged by date, which tells the records storage personnel which boxes are ready for destruction in any given month.

Dewey Decimal System A subject classification system that divides all knowledge into ten subject classes arranged in numeric sequence and further subdivided by a decimal system. Used in libraries.

Document 1. A verb meaning to outline a program so others know how it operates. 2. A noun meaning any publication published by any level of government or government contractors.

Documentation A thorough written description of all the component parts and operations of the system. Includes forms, personnel, equipment, and input/output sequence. Communicates to other people the program characteristics. Both written and charted explanations are used.

DYNAMO A simulation language oriented toward expressing microeconomic models of firms by means of difference equations.

Empathy Mentally entering into the feeling or spirit of other people in order to understand them better.

Empire builder The type of person who seeks personal power, prestige, and recognition through authority and control.

Emulation A method of protecting the firm against loss of data during a one-for-one changeover. An emulator processes the old program by the old procedure using current data. It enables one computer to process computer instructions from another computer.

Environment, external The sources outside of the firm which may influence the firm's operations, such as unions, customers, competition, and so forth.

Environment, internal Sources within the firm which may influence the firm's operations, such as auditors, budgets, the informal orginization, and so forth.

Equipment sheet A documentation form used to describe needed equipment for various operations in the system.

Estimating The art of predicting by utilizing all available information in a systematic and informed way.

Estimating, comparison By meeting with individuals, inside or outside the firm, the analyst is able to evaluate comparable operations and make an estimate.

Estimating, conglomerate Representatives from each functional area within the area under study confer to develop estimates based on past experience.

Estimating, detailed The analyst makes a detailed study of the costs, times required, and any other pertinent factors for each step of each procedure within the system.

Estimation error An error involving time or money, which may be avoided through a feasibility study.

Evaluation criteria Performance standards through which management can have a valid measurement to evaluate the new system's performance.

Event In PERT/CPM, the beginning or ending of an activity; a milestone.

Family, data base A data base family consists of one parent record type and at least one child record type. Sometimes called master and slave, or mother and daughter, records.

Feas-view The level of detail (lying between prob-view and sys-view) where all decision points, most volumes or quantities, and important events are documented. See also: Prob-view and Sys-view.

Feasibility study A study undertaken to determine the possibility or probability of improving the existing system within a reasonable cost.

Feasibility study report The written documentation of the feasibility study that tells management what the problem is, what its causes are, and makes recommendations for solving the problem.

Feasibility study – Type A A feasibility study that recommends for or against a full systems study. It is generally reserved for a Level-I or Level-II problem involving the entire firm or a division of the firm.

Feasibility study – Type B A feasibility study to determine the practicality of upgrading to a larger computer, or the evaluation of a first-time computer rental or purchase. It is always computer oriented.

Field justification, left/right When data is to be entered into fixed length fields, and there is less data than space, a decision must be made as to whether the spaces left over will by convention be on the right or left side of the field. Alphabetic data is usually left justified (extra spaces on the right). Numeric data is usually right justified (extra spaces on the left).

File integrity A term generally used in records retention to mean that all records are in their proper location; thus, file integrity is lost when records are misfiled.

File sheet A documentation form describing a collection of information in the system.

Files, external Files maintained outside the area under study but which affect the area's systems in various ways.

Files, index-sequential Computer files are stored in order by their key, but an index is also created so specific data can be accessed directly without searching sequentially through the file.

Files, internal Files maintained by the area under study, usually containing up-to-date forms and data that are used by the area.

Files, partitioned Computer files in which various

496

areas are partitioned into unique file areas for some specific data only.

Files, random access Computer files stored in an order (prescribed by a mathematical formula) which can be accessed directly.

Files, sequential Computer-based files stored in order by their key; the key may be part number, or name, or whatever the data are filed or accessed by.

Flow approach A method of analyzing the system in which the analyst studies the flow of physical entities and builds a model that simulates the flow of these items through the business organization.

Flowchart A graphic picture of the logical steps and sequence involved in a procedure or a program.

Flowchart segment A form of documentation that describes in flowchart format a specific operation.

Flowchart, documented This chart traces the flow of a single activity through its sequence of operations. A number in each box of the flowchart is used to cross reference each flowchart box back to the documentation section.

Flowchart, layout This type of flowchart shows the floor plan of an area, including file cabinets and the storage areas, and indicates the flow of paperwork or goods.

Flowchart, paperwork A type of flowchart tracing the flow of written data through the system.

Flowchart, process A type of flowchart used by industrial engineers to break down and analyze successive steps in a procedure or system. Five special symbols are used.

Flowchart, program This flowchart may be derived from a systems flowchart. It is a detailed explanation of how the program or procedure is actually performed. It tells *how* the work is being done.

Flowchart, schematic This systems flowchart presents an overview of a system or an overview of a procedure. Pictures of the equipment are used in place of standard flowchart symbols.

Flowchart, systems A flowchart showing the overall work flow of a system. It is a pictorial description of the sequence of the combined procedures making up the system. It tells *what* is being done.

Follow-up After implementation, the analyst returns to observe actual operation of the installed system to find out if the system is really working as planned.

Forms control The coordination of forms design and usage among all users of all forms in the firm.

Forms design Designing the format of the form so that it performs its function simply and efficiently, and is easily completed, legible, uncomplicated, and economically feasible.

Forms, continuous strip A type of form in which the original and carbons are joined together in a continuous strip with perforations between each form.

Forms, flat A single-copy form; that is, there are no carbon copies.

Forms, tab card A type of form in which the bottom copy is a tab card and top copies are paper with carbon interleaved between each copy.

Forms, unit-set/snapout A type of form which has an original copy and several copies with carbon paper interleaved between each copy. The entire set is glued together.

Functional approach A method of analyzing the system in which there are no observable flowing entities through the system, but the analyst studies the sequence of events and builds a model that simulates the sequence of events as they happen throughout the business organization.

GASP A set of subroutines in FORTRAN which perform functions useful in simulations.

General Purpose Simulation System (GPSS) The most popular and often the most useful simulation language by IBM. It is a complete language oriented toward problems in which items pass through a series of processing and/or storage functions.

Handbook A manual used in conjunction with specifications and standards to aid in their application and interpretation.

Hardware The machines utilized in a system. Generally refers to the computer and its peripheral equipment within a system.

HIPO Hierarchy plus Input-Process-Output. A documentation technique that graphically represents functions in charts from a general level down to the detailed level. The technique was developed by IBM.

HIPO detail design package A HIPO documentation package prepared by a development group using the initial design package. The analysts and programmers specify in detail more levels of HIPO diagrams, and use the resultant package for implementation and comparison with the initial design package to ensure that all requirements have been satisfied.

HIPO detail diagrams Lover-level HIPO diagrams that describe the specific functions, show specific input and output items, and refer to other detail diagrams.

HIPO initial design package A HIPO documentation package prepared by a design group at the start of a project. It describes the overall functional design of the project and is used as a design aid.

HIPO overview diagrams High-level HIPO diagrams that describe the major functions of a system and reference the detail HIPO diagrams needed to expand the function to the described level of detail.

HIPO visual table of contents A HIPO diagram that contains the names and identification numbers of all the overview and detail HIPO diagrams in a documentation package, and shows the structure of the diagram package and relationship of the functions in a hierarchical fashion.

Implementation Consists of the installation of the new system and the removal of the current system.

Index An essential ingredient in any information storage and retrieval system; the index is a filter to let through to the user the wanted information, while keeping back the unwanted information. A listing that is: (a) a guide to primary information (such as that at the end of a book); (b) a guide to periodicals by subject, author, title, and so forth; (c) a guide to a specific page or frame within a microform system.

Industry background That part of the systems study which places the firm in perspective within its environment, or looks at how this firm performs in comparison with firms of like nature.

Information 1. A meaningful assembly of data telling something about the data relationships. 2. A meaningful aggregation of data.

Information storage and retrieval A field concerned with the structure, analysis, organization, storage, searching, and retrieval of information.

Input element boundary A line on a data flow diagram which represents the point at which the input data stream is most processed but still considered as input.

Input/output cycle That part of the system which consists of inputs, processing, and outputs.

Input/output sheet Describes inputs and outputs that will be different from those used in the existing system. It contains a functional description of the input or output and describes its purpose and use. A sketch of the input or output may be included.

Inputs The raw data, raw materials, paperwork, processed computer files, reports, or semifinished products which are used to make up the finished product, or output. The energizing element that puts the system into operation. May be either verbal or written. See also: Outputs.

Interactions The relationships to be studied between employees, departments, management personnel, or any combination of these elements.

Interview The most important tool for gathering data about the area under study is to talk with people in an organized, systematic way.

Iterative technique The creative process by which the analyst moves through the various activities or job procedures, one at a time, mentally tracing through the entire system.

Jackets A jacket is usually made of mylar film and is manufactured in such a way that strips of microfilm can be placed in rows, one frame at a time if desired. It looks similar to microfiche, but the film is removable.

Job shop simulator A simulation language that can be set up to represent a variety of job shops.

Key verification A procedure for checking the accuracy of keypunched data which involves rekeying the data and comparing it to the original data entered.

"Lateral thinking" The method of thinking described by de Bono which explores all the different ways of looking at the system. It often begins with the end of the system (outputs) and works backward through the system.

Ledger A summary of transactions to document an audit trail.

Legal life The retention period of a record that is required by law.

Legal requirements The laws of local, state, or federal government which affect company operations. Some help the company, some restrict the company, while others affect the firm's recordkeeping practices.

Letterpress The type of printing press that prints the form from a raised surface of type covered with ink.

Level of detail The level of detail involved in a study. See also: Prob-view, Feas-view, Sys-view.

Level-I study The systems study that involves the whole firm and in which management has the greatest stake.

Level-II study The systems study that involves one division of the firm, and one in which management will be vitally interested.

Level-III study The systems study that involves departmental interaction; the "middle management" area that involves exacting detail.

Level-IV study The systems study that involves functions within a specific department.

Level-V study The smallest systems study, generally dealing with a specific problem within a specific area of the department.

Library of Congress system A subject classification system that classifies all knowledge into twenty-one subject areas. Subdivision is by alphanumeric designations. Used in libraries.

Line printer (computer) A high-speed printing device that is attached to a computer.

Long-range plans The plans for the future of the firm. These plans may affect or obsolete a proposed system.

Management information system A system in which management or others having an established need to know are provided with historical information, information on current status, and projected information appropriately summarized. A decision-making tool. Also called an MIS.

Manuals A form of documentation to guide employees in doing their assigned tasks. See also: Procedure.

Marginal efficiency of investment The rate of return that a potential new system is expected to earn after all of its costs are covered, excluding interest.

Maximize To get the highest possible degree of use out of the system without regard to other systems.

Microcard A term generally used for the type of microform which is opaque as opposed to transparent. These vary a great deal in size and reduction ratio.

Microfiche A unitized type of microform with a number of images arranged in rows on a transparent card.

Microfilm A continuous strip of film, usually either 16-mm or 35-mm, with all the "pages" placed in order. The film is usually placed on some type of reel for easy wind and rewind.

Microform/micrographics A miniaturized record in which the original record has been photographically reduced in order to save space.

Model Any representation of reality that embodies those features that are of particular interest to the user of the model. See also: Simulation.

Modularity Applies to computer equipment, and is the ability to add to the basic computer system to increase its power either through more data storage or more input and output devices.

Narrative A description written in story form. A verbal model in which the analyst details the sequences of steps involved in necessary operations.

National Technical Information Service (NTIS) The primary distributor of unclassified government documents, sponsored by the Department of Commerce.

Need-to-know A term originated by the military to identify a person who has a legitimate need to use a specific bit of information for a specific purpose. Also used within a firm to indicate who may have access to sensitive corporate information.

Network Events and activities in PERT/CPM.

Objectives As used in systems, what one intends to accomplish in the problem definition phase or during a full systems study.

Occurrence, data base An occurrence of a specific record type is said to exist when a value for each data element within it exists within the data base.

Offset The type of printing press which photographs the finished drawing of the form onto a printing surface from which the form is then printed.

On-line Pertaining to equipment or devices under the control of the central processing unit, or pertaining to a user's ability to interact with a computer.

On-line system A system in which the input data enters the computer directly from the point of origin or in which output data is transmitted directly to where it is used.

Operations The procedure or activity that must take place to transform inputs into outputs.

Optimize To get the most favorable degree of use out of a system, taking into account all other systems.

Organization, formal The organization that is built around the firm's goals, upper management's policy statements, and the written procedures that carry out the policies. Visually shown on the firm's organization chart.

Organization, informal The organization that is built around the job at hand as the employees see it, as opposed to the manner in which management wishes the job to be performed.

Outline The detailed plan of action by which the systems study will be carried out.

Outputs The end product of the area under study. Outputs can be completed paperwork, processed computer files, reports, semifinished products, or finished products. See also: Inputs.

Output element boundary A line on a data flow diagram which represents the point at which the output data stream is least processed but still able to be considered output.

Page, data base A block is the physical unit of data transfer between a data base and the CPU, and is referred to as a page.

Paper, bond High-quality paper. Types include rag, sulfite, and duplicating.

Paper, index bristol Paper that is heavier than ledger or bond; also known as card stock.

Paper, ledger A heavyweight bond paper used for maching posted ledgers.

Paper, manifold A lightweight bond paper; sometimes called onionskin.

Paper, safety A special type of paper which cannot be erased without leaving a mark.

Parameter Elements of an activity or job procedure that are almost always constant. Defines the limits of a system or its phases.

Payback period A criterion used to judge the profitability of a system. The number of years required to accumulate earnings sufficient to cover the cost of the proposed new system.

Peopleware The operating personnel of a system, and the most essential ingredient in the workability of a system.

Periodical A regularly issued publication, usually oriented toward one specific subject, through which the specialist keeps abreast of new developments. Often used synonymously with magazine or journal.

Personnel sheet A documentation form specifying the job description, job title, and approximate pay range for each needed position in the system.

PERT Programmed Evaluation Review Technique is a planning and control tool for defining and controlling the efforts necessary to accomplish project objectives on schedule.

Phase out By some preplanned date the old system should no longer be operating. Phasing out can pertain to parallel conversion where the old and new system are operated simultaneously until the new system has been proven. *The old is generally then phased out.*

Policy Management guidelines for regulating progress toward the firm's goals.

Politics, company The maneuvering of personalities within the informal organization, which is the individual's strategy to achieve success.

Prob-view The lowest level of detail in which only general-purpose items are documented, such as major decision points, key volumes or quantities, and critical events. See also: Feas-view and Sys-view.

Problem A question proposed for solution or consideration.

Problem definition report A short report that sets the stage for an advanced feasibility study or a major systems study. It includes subject, scope, and objectives.

Problem flow The directional flow of a problem from either the internal or external environment to the systems analysis department.

Problem report form A form to be filled out which formally reports the existence of a problem and the circumstances in which the problem occurs.

Problem reporting machinery An expression used to describe the method by which the systems analyst learns of problems. It may be either a written or verbal message.

Problem solving The process of recognizing a problem, specifying exactly what the problem is, determining what is causing the problem, and providing a solution to the problem.

Procedure, formal A precise series of step-by-step instructions that explain what is to be done, who will do it, when it will be done, and how it will be done.

Procedure, informal The tasks performed by an individual which are not in writing but which the individual performs to "get the job done."

Procedure, narrative A procedure composed of words, sentences, and paragraphs; a story form.

Procedure, playscript A procedure that uses sequence numbers, actors, action verbs, and a straight chronological sequence of who does what in the procedure.

Procedure, step-by-step outline A procedure in which the reader sees item-by-item what each step contains.

Processing The activity in the input/output cycle which transforms the input into an output.

Product The output of a system or the output of the firm.

Program run book Operating instructions for the benefit of others who may run the program; included are special restart procedures in case of failure prior to normal program end.

Program test Trying out the program on the computer to see whether it will compile and run.

Project assignment sheet A form indicating not only who will perform a systems study and its priority, but its subject, scope, objectives, phases, schedule, followup notes, and records of time spent. An integral part of project control.

Project control The primary function of the systems analysis manager is to assign projects to each analyst and to ensure that the assigned projects are completed in the specified length of time.

Project schedule Each phase of a systems project is put on a Gantt chart to indicate the length of time allotted to each phase.

Rapport A meeting of the minds, or absence of friction.

Record retention schedule A timetable governing the retirement and destruction of all company records.

Records inventory The starting point in initiating a records control program when the analyst makes an inventory of the active records being maintained by each department.

Records retention A systems technique that provides control over records.

Records, vital Those records that would enable the firm to reconstruct its operations after a disaster.

Re-evaluation After implementation and followup, the analyst makes whatever changes are needed for the refinement and improvement of the new system. Some portions may be redesigned and others may be revised.

References The titles of any documents that have a governing or otherwise vital bearing upon the procedure.

Report analysis A function similar to forms control, in which the analyst examines the reports being generated to determine which can be eliminated, improved, or added.

Report, final The most important report of the systems study. It summarizes the work to date, presents the new system, specifies cost comparisons, describes the existing system, and makes recommendations.

Reports, action A report that initiates or controls a necessary procedure or operation.

Reports, exception A report that is generated only when certain parameters are out of line with what is expected of these parameters.

Reports, feeder A report that consists of bits of data to be used later in conjunction with other data, another report, or for accumulating data for a decision.

Reports, informational A report that provides data or information for further analysis and control.

Reports, on-demand A report generated upon the user's request.

Reports, progress A report that delineates the work done during the interval since the last progress report.

Reports, reference A report that keeps the manager informed on operations.

Reports, scheduled A report that is prepared and distributed at a fixed time on a regular basis.

Requirements The objectives as set during the problem definition phase. These are determined in terms of outputs, inputs, operations, and resources.

Resources Items used in the day-to-day operation to convert inputs to outputs.

Resources, facilities The resources of the area under study, consisting of land, buildings, data processing equipment, or other capital equipment.

Resources, financial The assets of the area under study, consisting of the budget for the area and the area manager's ability to get financial backing for new projects and systems.

Resources, inventory 1. The "stock-in-trade" resources such as raw materials, parts, supplies, semifinished products, and finished products. 2. The "files of information" or data collected over the years.

Resources, personnel The assets of the area under study in terms of key managers and other skilled and able personnel. It includes personalities and talents.

Response time The elapsed time between the last CRT terminal operator entry and the display of the response from the computer on the screen.

Retrieval, data A type of information storage and retrieval system in which the user is provided with data displayed as words or numbers. Commonly known as "lookup" and not truly information storage and retrieval.

Retrieval, document A type of information storage and retrieval system in which the system ultimately provides the user with the full text of the document.

Retrieval, reference A type of information storage and retrieval system in which the user is given citations to document locations.

Sampling The collection of a limited quantity of data from the total data available, for the purpose of studying that fraction to infer things about the total.

Satisfice To choose a particular level of performance for which to strive and for which management is willing to settle.

Schema The complete description of all the elements in a data base is referred to as a schema.

Scope The area or range that the systems study will encompass.

Sensitivity The quality of being readily affected by other people, responding to their feelings.

SIMSCRIPT A complete language similar to FORTRAN, oriented toward event-to-event simulation in which discrete logical processes are common.

Simulation A mathematical model which is programmed into a computer and which represents the system through all its phases.

Software The programs that control the operation of the computer.

Sources, external Secondary sources of information which aid the analyst in understanding the area under study, such as periodical articles, or special industry reports.

Sources, internal Primary sources of information which aid the analyst in understanding the area under study, such as long-range plans or employee handbooks.

Specification Establishes the necessary characteristics of an item in terms of its expected performance.

Standard Defines the dimensions of a particular item for purposes of interchangeability.

State-of-change approach A method of analyzing the system in which no specific sequence can be observed in a large number of interdependent relationships. Various checkpoints of the model are monitored to determine what, if any, changes occur as the inputs to the subsystem are varied.

Statute A law passed by Congress or a state legislature.

Storage box label The form used to identify each box of records put into storage.

Storage life The time during which records that are inactive or no longer in current use are stored where they can be located if needed.

Subject The topic or central theme of a systems study or problem definition study.

Subschema A subschema is a consistent and logical subset of the schema from which it was obtained.

Summation The verbal summary of an interview with a person. It mentions the points covered and verifies any agreements reached on important or controversial points.

Sys-view The most detailed level of documentation. It includes all decision points, all volumes or quantities, and all events. See also: Prob-view and Feas-view.

System A network of interrelated procedures that are joined together to perform an activity.

System requirements model A model illustrating the relationships between inputs, operations, resources, and outputs.

System, closed A system that automatically controls or modifies its own operation by responding to data generated by the system itself.

System, conceptual A system existing only in thought.

System, empirical A working system.

System, existing The current system with all its faults and inadequacies.

System, open A system that does not provide for its own control or modification. It needs to be supervised by people.

System, proposed The newly developed system designed to replace the existing (current) system.

Systems analysis The approach to a system that is the opposite of trial and error. All influences and constraints are identified and evaluated in terms of their impact on the various parts of the system.

Systems analyst A methods person who can start with a complex problem, break it down, and identify the solutions.

Systems department A staff activity that renders service to all other departments. It advises and assists, rather than directs.

Systems design The art of developing a new system. The ten steps outlined in Appendix I portray the systems design cycle. It is concerned with the coordination of activities, job procedures, and equipment utilization in order to achieve organizational objectives.

Tact The ability to say or do the right thing without offending the other person.

Theory "X" A form of managerial style explained by McGregor which assumes that humans naturally avoid work, are irresponsible, desire security above most other things, and that tight controls are needed to keep humans working properly.

Theory "Y" A form of managerial control explained by McGregor which assumes that humans actually seek work and responsibility, that humans are self-motivated, and that the human being has a capacity for a high degree of ingenuity and creativity, under proper conditions.

Time, calendar The overall time in terms of days, months, or years that it takes to complete a new system from start to finish.

Time, chargeable The actual number of hours to be spent in developing a new system. Each activity or phase fits into a timetable.

Time, expected activity (PERT) The time in weeks calculated for an activity from the time estimates given.

Time, expected event (PERT) The sum of all expected activity times along the longest path leading to an event.

Time, latest allowable (PERT) The latest time by which an event can be accomplished without affecting the date scheduled for completion of the entire project.

Time, most likely (PERT) The time that responsible managers or analysts think will be required for the job.

Time, optimistic (PERT) The time estimate for an

activity, assuming everything goes better than expected.

Time, pessimistic (PERT) The time required if many adverse conditions are encountered, not including acts of God, strikes, or power failures.

Time, scheduled (PERT) The time in weeks from the starting event to the planned-for or contractually-obligated completion date.

Time, slack (PERT) The difference between the latest allowable time and the earliest expected time for an activity to be completed.

To Do list 1. A list of all the tasks that you plan to accomplish tomorrow. 2. An outline of how you plan to carry out the problem definition project. It notes the tasks to be accomplished and how you will go about them.

Ultrafiche One of the newer types of photoreduction onto microform. Similar to microfiche in that it is transparent; but because of the method of reduction, many more frames can be "packed" onto a single 4 × 6 inch or 3 × 6 inch transparency.

User The personnel in various parts of an organization who prepare data for input to the computer and also those personnel who receive and use the output.

Variable Those elements of an activity or job procedure that are subject to change or variation.

"Vertical thinking" The method of thinking described by de Bono where thinking begins with the most promising method of approaching the problem and proceeds from that point to a solution.

Visual aid Any aid the analyst uses during the verbal presentation which helps demonstrate what the analyst is talking about.

Work sampling Consists of a large number of observations taken at random intervals, noting what the employee does and recording it into a predefined category. The information gained from work sampling can be used to evaluate the existing system or as a bench mark for comparison with the new system design.

APPENDIXES

APPENDIX 1

HOW TO EVALUATE FILES

The following checklist can be used when appraising filing operations. It is taken from U. S. National Archives and Records Service, General Services Administration, "Checklist for Appraising Files Operations in Your Office," Washington: Government Printing Office, 1968. It is subdivided into the following areas.

A. Reduction of quantities being filed.

B. Classification and filing system.

C. Classifying practices.

D. Finding aids.

E. Filing practices.

F. Reference service.

G. Workload.

H. Documentation.

I. Equipment and supplies.

J. Space and work flow.

K. Training.

A. Reduction of quantities being filed.

   1. Does someone in my office determine whether or not given types of papers being created or received must be filed?

   General rule: In every office where files are maintained someone should make this determination; otherwise many papers will be filed which are not worth filing.

   2. Does my office have a policy that the following materials are not to be filed?
      a. Envelopes.
      b. Route slips on which there are no significant notations.
      c. Superseded drafts that show no important substantive changes.

    d. Duplicates of correspondence and reports other than those needed for cross references.

General rule: Every office should have a firm rule that such obviously unneeded papers will not be filed. This prevents wasted effort, filing space, and equipment, and makes needed papers easier to file and find.

3. Does my office have a policy that file copies of form letters will not be made in instances when (a) no retained record is necessary or (b) a notation on incoming correspondence showing the form reply used and the date will suffice or (c) one copy showing distribution will do?

General rule: Such copies serve no purpose, increase the filing workload, clutter files, and so should not be made.

4. Does my office have policy governing which publications will or will not be filed?

5. Are originators of publications requested to discontinue sending those no longer needed?

General rule: These practices should be followed, because publications rapidly consume filing space and equipment, sometimes require indexing.

6. Does my office, in responding to purely routine correspondence (such as requests for publications, applications, stereotyped inquiries, etc.) reply (a) on the incoming letter, which is returned to the sender or (b) by form letter, printed slip or other readymade answer, returning or discarding the incoming letter?

General rule: Unless policy prohibits, one or more of these practices should be followed. They eliminate preparation of file copies of replies and filing of incoming letters.

B. Classification and filing system.
  1. Have I listed the subjects and types of records in my office?
  2. Have I compared my list with the subjects and types of records provided for in the filing manual prescribed for my use?
  3. Have I defined all subjects on my list which do not appear in the filing manual prescribed for my use?

General rule: These actions should be taken as first, essential steps, if files are not already arranged according to authorized system.

4. Does the filing manual prescribed for my use adequately provide for the records of my office?

   General rule: When a filing manual developed for general use does not satisfactorily provide for the records of an office, that office should see that the manual is supplemented to fit local needs by contacting the records management office.

5. Are the manual's subjects logically arranged in relation to the way my office operates?

   General rule: Subject outlines work best when they conform to the functions to which they pertain.

6. Do I usually find it rather easy to select the subject under which a paper should be filed (i.e., there are not too many subjects under which a given paper can go)?

   General rule: A good filing manual avoids providing an excessive number of subjects from which to choose in deciding where to file a paper. This in turn improves likelihood that all papers on the same subject will be consistently filed and found together.

7. Are the manual's instructions and definitions complete and clear?

   General rule: A good filing manual provides clear instructions, definitions, and references not only as general guidance on how to use the manual but also wherever they are needed in connection with subjects.

8. Are the coding symbols that represent subjects short, simple, and easy to remember?

   General rule: They should be, to make the marking, sorting, filing, and finding of papers fast, easy, accurate.

9. Are my files actually arranged in accordance with the manual prescribed for my use?

   General rule: They should be, unless the manual is believed unsuitable for the office's files, in which event the records management office should be contacted.

C. Classifying practices.

1. Do I assemble directly related papers (e.g., incoming letter and copy of outgoing reply) before determining their file designation?

   General rule: By assembling directly related papers, more information is available on which to make a sound decision on the correct file designation.

2. Do I *mark* papers with their file designations (e.g., underlining or checkmarking name, writing file code in corner, etc.)?

   General rule: Papers should be so marked to make it unnecessary to re-read a paper when filing it or returning it to file.

3. When I am in doubt regarding the right file designation for a paper do I refer to an index to the files, to papers already filed to verify or reject a tentative choice, or ask the opinion of people who are acquainted with the subject, case, or project?

   General rule: These steps, in the order given should be taken, rather than to guess.

4. Do my superiors refrain from marking file designations on papers before sending them to me to file?

   General rule: If you are held responsible for finding papers when they are asked for, you should at least participate in deciding how to file them. This is true even if (a) the filing system was devised by a superior and he or she understands it best; (b) he or she is intimately acquainted with the technical content of papers and so feels best qualified to mark them for filing; or (c) he or she feels that certain papers are so important that he or she wants to be certain of their location in the files so he or she can produce them quickly.

D. Finding aids.

1. Are all of the indexes and other finding aids I have really worth the time, effort, and cost of preparing and maintaining them?

   General rule: All finding aids are costly and, therefore, should be held to a minimum. Several types of indexes to one kind of file system is a sign of weakness in that system.

2. When a paper covers more than one subject, name, and so forth, do I

provide cross references only for the additional subjects or names by which I feel the paper is likely to be requested?

General rule: Cross referencing should be restricted to just those which experience has proven are useful. Resist the temptation to cross reference every subject, name, and so forth, in a paper.

3. When a cross reference is needed, do I use an extra carbon copy or obtain a "quick-copy" of the paper involved?

General rule: A copy of the paper is preferable to the preparation of a cross-reference form, because it provides the full text of the paper and is usually faster and cheaper to obtain.

4. Do I mark each copy to show (a) that it is a cross reference, (b) where it should be filed, and (c) where the paper from which it was copied is filed?

General rule: This should be done to clearly identify the nature and purpose of the copy and location of the paper copied.

E. Filing practices.
1. When others release papers to me for filing, do they initial or otherwise mark them to show that their filing is authorized?

General rule: This should always be done; otherwise there is no assurance that a paper has been seen or acted on.

2. Do I sort papers that are ready to be filed, into the same sequence as the files in which they will be placed?

General rule: This should be done. It prevents backtracking and thus saves filing time and effort.

3. Do I fasten papers together which will be asked for as a group?
4. Do I leave papers unfastened which will be asked for singly?

General rule: Whether or not to fasten papers should be governed by the way papers are asked for. The perforating of papers, placing them on fasteners, opening and closing fasteners, and so forth, are tedious, time consuming operations. For these reasons, papers should be fastened only when entire folders are requested.

5. Are my files arranged according to the way they are asked for?

General rule: This should be so, as far as it can be carried, because it makes finding much easier.

6. Have I arranged as many papers as possible into case or project files?
7. Do I clearly understand what constitutes the essential papers that belong in each kind of case or project file, so I can tell when such a file is complete?
8. When a case or project is closed, do I remove the file from among those of still active cases or projects and place it with the files of other closed cases or projects?

General rule: Papers should be arranged into case or project files, if possible, as this is the simplest way to file and find information. Recurrent, repetitive kinds of information that belong in a case or project file should be known to those who maintain such files. Closed files should be separated from active ones.

9. Is file material separated or identified in some way to
   a. Show its age, so that it will be easy to dispose of or retire at scheduled times?
   b. Distinguish that of permanent or long-term value from that of transitory value?
   c. Keep heavily used material from being mixed with and encumbered by seldom used material?
   d. Keep files designated as "official files" apart from those which have not been so designated?
10. If so, is this done in one or more of the following ways?
   a. By maintaining material in (a) separate filing cabinets, drawers, or sections of drawers or (b) separate shelves or sections of shelves, with inclusive dates shown on drawer, shelf, or guide card labels.
   b. By maintaining material in separate file folders, with inclusive dates or values shown on folder tabs.
   c. Color coding to show periods of time or values of papers; for example, yellow label for the current period or for permanent papers, green label for the preceding period or temporary papers, and so forth.
   d. By dividing material within file folders to separate permanent from temporary or to separate heavily used from seldom used material.
   e. By affixing clip-on signals to file folders, using different colors or positions to indicate time periods.
   f. By using staggering positions of file folder tabs to indicate retention periods and methods of disposition.

General rule: Separation of papers for one or more of the purposes listed above should be done, as far as practicable. The method used must depend on the amount, kind, and so forth, of papers, and the advice of the records management person should be requested to help arrive at the right selection in each instance.

11. Are my files neat and orderly in appearance, with file folder and guide card tabs aligned in simple patterns which are easy to scan when locating files?

    General rule: Uncluttered, simple arrangement of folder and guide tabs and clear, standardized labeling of such tabs are definite aids to filing and finding papers.

12. Are the contents of my file folders and/or containers limited in volume to avoid overloading them?

    General Rule: Overloading should be avoided, as it makes filing and finding difficult and can damage papers

13. Are my files virtually free of empty or nearly empty file folders?

    General rule: Only folders for which there is a present or expected need should be established.

14. Is there the right number of file guide cards (dividers) in my files?

    General rule: The number of guide cards needed will vary somewhat, due to the number and thickness of file folders, whether the files are subject, name, or number files, and so forth. It is better to have too few than too many, as too many actually slow filing and finding.

15. Do all file drawer or shelf file labels clearly identify the files involved?

    General rule: When such labels identify contents of cabinet drawers or shelving sections, filing and finding is expedited.

16. If I have an alphabetical name file of persons or organizations, do I provide for name changes by refiling the papers involved under the new name?

    General rule: This should be done, as requests will most likely mention new name. (See also next question.)

17. If name file papers are refiled under a new name is a cross reference placed under the old name referring to the new one?

General rule: This should be done, since some requests may mention only old name.

18. When I have bulky or oversized file material that cannot be suitably placed with my regular file material, do I (a) mark it with the appropriate file designation and identification of the particular letter, report, and so forth, to which it relates and (b) place it in other equipment suited to its size?

General rule: These steps should be taken, for reasons that are obvious. (See also next question.)

19. If I place bulky material apart from my regular files, do I indicate on the related letter, report, and so forth, in the regular files where the bulky material has been placed?

General rule: This should be done, for reasons that are obvious.

20. When I must remove papers from an earlier group of files to combine them with current papers, do I replace the earlier papers with a cross reference showing their new location?

General rule: This should be done; otherwise whereabouts of earlier papers is uncertain.

21. Do I maintain a suspense (tickler, reminder) file on (a) correspondence to which replies are due or on which action should be taken by a given date or (b) files needed by someone on a predetermined date?

General rule: Such a file is extremely useful, and should be established if supervisor approves.

22. Do I regularly straighten and tamp down papers in folders, crease expansion folds (scoring) on bottom of folders to keep papers from hiding labels, and check for misfiles?

General rule: These practices should, of course, be followed to make files easier to use and to ensure that papers are where they belong.

23. Do I remove paperclips, rubber bands, spring clips, and pins from papers, and staple those which should be stapled before filing?

General rule: These actions should be taken to prevent papers being inadvertently attached to others or separated as clips slip off, and to reduce bulk.

24. Do I keep my current files free of records that should be disposed of according to authorized schedules?

    General rule: This should be done to the fullest practicable extent, to save filing space and equipment, making filing and finding easier.

25. Do I periodically dispose of records (by retirement or destruction as authorized) according to schedules provided by my records management office?

    General rule: These things should be done, to save filing space and equipment and make filing and finding easier.

26. Do I file security-classified papers in separate file containers from papers not security-classified or papers marked "For Official Use Only"?

    General rule: This practice should be followed. However, in most firms, security regulations permit filing unclassified papers with classified ones when they are needed together for reference purposes. Such interfiling should be restricted to papers which directly support, explain, or document a decision or transaction. Be sure to check this point with your records management office or your security officer.

F. Reference service.
   1. When records are removed from my files for use are they replaced by a charge-out form?

      General rule: This should be done, unless the users are in the same room or within a very limited distance from your files, so that location of records is known.

   2. When additional papers arrive for inclusion in a file that is charged out, do I take these additional papers to the person who has the file?

      General rule: This should be done, so that he or she will have the benefit of the additional information.

   3. Do I periodically contact persons to whom files are charged, after a reasonable period of time, to see if files can be returned?

      General rule: This should be done, to make these files available to other persons and lessen chance of their being misplaced.

4. Over a period of, say, one year, would the total "can't finds" in my file be less than 3%?

General rule: Three percent is regarded as the break-off point between efficient and inefficient reference service.

G. Workload.

1. Do I keep my classifying and filing up-to-date?

General rule: Classifying and filing should, of course, be kept up-to-date; that is, done daily, so that backlogs do not accumulate. If this is not possible because the volume of papers is too great or other duties are given priority, and so forth, this should be discussed with supervisor.

---

2. Am I able to attend to requests for files or information from files as such requests are received?

General rule: Requests for files service should be handled when they are received, not backlogged. If this is not happening, determine cause and discuss with supervisor.

H. Documentation.

1. Are my files complete, free of information gaps?
2. Do other offices or organizations always supply information due my office (such as periodic reports, requested data, and so forth)?
3. Do other persons in my office always turn papers over to me for my files?
4. Do other persons in my office always tell me when they remove papers from my files during my absence?

General rule: Files should completely document, as far as possible, the office's role in a transaction, decision, project, and so forth. This is not possible if the answer to questions 2, 3, or 4 is "no," in which case you should consult your supervisor.

---

5. Do I fully understand regulations and procedures for the protection of security-classified files?

General rule: All such policy and instructions must be thoroughly understood.

I. Equipment and supplies.

    1. Am I using legal-size equipment and supplies only when the amount of legal-size papers is 20% or more?

General rule: Because legal-size equipment and supplies cost more and take up more space, their use should be held to a practical minimum.

    2. Am I using five-drawer filing cabinets?

General rule: These should be used when available and when modern shelving should not be used, because of their greater capacity and saving of floor space.

    3. Have I explored the advantages of using modern shelving instead of filing cabinets?

General rule: Unless there are strong reasons why they should not be used, such shelves offer benefits, particularly in floor space savings. Their use should be considered.

    4. Am I using fire resistant insulated file equipment and security type file equipment only for records that require this protection?

General rule: Such equipment costs more than regular equipment, occupies more floor space, and is a greater floor load. It should be used only for records that warrant the degree of protection it affords.

    5. Am I keeping filing cabinets free of stocks of blank forms, office supplies, stocks of publications, and so forth?

General rule: Filing cabinet space should not be wasted on such items. Materials of this kind should be stored in nearby supply cabinet, shelving, or other suitable housing.

    6. Am I using the right kind of file folders?

General rule: Choice of folders should be governed by the kinds of papers, frequency of use, kind of container, and so on, involved.

    7. Am I using the right kind of file guide cards (dividers)?

General rule: Here, too, choice should be based on the kind of file, the tab position prescribed, whether color coding is to be used, and so on.

8. Am I using the right kind of file folder labels?

General rule: Pressure-sensitive (self-adhesive) labels are easiest to apply. Their size, color, and other features should be chosen according to the kind of file, amount of information on label, and so on.

---

9. Do I use such aids as a sorter, hook-on shelves to hold papers as I file into containers, filing stool, tiered desk tray, and so on.

General rule: Such devices make classifying, sorting, and filing easier. They should be used unless volume of papers is quite small.

J. Space and work flow.
   1. Are my filing aids and file containers and their contents so arranged that steps are saved and filing moves progressively forward (e.g., from top to bottom of containers, left to right, etc.)?

General rule: They should be so arranged, so that filing can be accomplished with least effort and in least time.

---

   2. Are file containers placed so that I can get to them easily?
   3. Are they placed so they do not interfere with the flow of other work and movement of other personnel?
   4. Are they placed so they do not expose files to damage?
   5. Are they placed so they do not unnecessarily expose my files to unauthorized access?
   6. Are they placed so they are not a safety hazard?
   7. Are they placed so they are in an area of good light?

General rule: Naturally, as many of these objectives should be realized as possible.

K. Training.
   1. Does my firm (a) present files workshops or other files training courses or (b) encourage attendance at such courses when presented by other sources?
   2. If so, have I attended one recently (within the past two years)?

General rule: Such courses should be given or supported, and be attended by all who maintain files. Only in this way can skills be improved, and the latest techniques, equipment, and supplies be introduced.

APPENDIX 2

CONTROLS FOR INPUTS, DATA COMMUNICATIONS, PROGRAMMING, AND OUTPUTS

HOW TO USE THE CONTROL MATRIX APPROACH

In order to effectively utilize the control matrix review approach, the reader should become generally familiar with the overall data processing functions and the various components of each system such as the on-line terminals, the data communication network, and the data bases. Figure 9-1 identified the nine components of a data processing system (see Chapter 9, p. 194).

After choosing the appropriate control matrix (only the input, data communications, programming, and output matrices are included here), use it in conjunction with whatever internal control review methodology is used by the organization. In other words, the control matrices are a tool that will interrelate the several hundred controls described to each component of a data processing system.

To utilize the control matrices, look at the sample matrix in Figure 9-2 (see Chapter 9, p. 196). The various concerns/exposures* to which a data communication network may be subjected are listed across the top of the matrix. The various resources/assets† that should be reviewed are listed on the left vertical axis of the matrix.

In conducting the control review, any resource/asset (left vertical column) that might be subject to a concern/exposure (top horizontal row) should have some type of control/safeguard that must be evaluated and reviewed. These controls/safeguards are represented by the numbers within the cells of the matrix, which correspond to the list of controls. Therefore, referring to Figure 9-2, if the concern/exposure was "message loss or change" and if the resource/asset being reviewed was "the front-end communication processor," then the appropriate controls that should be evaluated and reviewed

---

* A *concern/exposure* is defined as a potential threat area or an adverse occurrence that could be injurious to the organization.

† A *resource/asset* is defined as the item that is being reviewed, such as computer hardware, software programs, operational policies or procedures, personnel, and other tangible or intangible assets.

519

are numerically listed at the intersecting cell of the matrix. There are numerous numbers listed in this cell. The specific controls that correspond to these numbers are defined within the chapter. Our example in Figure 9-2 highlights one of the controls (control 89) in order to exemplify the specific type of control to which we are referring.

As a further example of the versatility of the matrices, they may be utilized in any of four major approaches, as follows:

1. *The resource/asset approach*
   - Identify the appropriate resource/asset on the left vertical axis of the matrix. This would correspond to a physical asset or some other identifiable function or procedure that was being reviewed.
   - Read across the row identifying each potential concern/exposure to which the resource/asset may be subjected.
   - Evaluate and review each potential control/safeguard (these are listed by number within the cells of the matrix) for its applicability with respect to protecting the resource/asset from each concern/exposure.

2. *The concern/exposure approach*
   - Identify the appropriate concern/exposure along the top row of the matrix. This would correspond to the specific concern or exposure that was being reviewed.
   - Read down the column identifying each resource/asset upon which this concern/exposure has an effect.
   - Evaluate and review each potential control/safeguard (these are listed by number within the cells of the matrix) for its applicability with respect to protecting each resource/asset from the concern/exposure.

3. *All controls for a resource/asset*
   - Identify the specific resource/asset that requires some controls/safeguards.
   - Read across the row copying the numbers (disregard duplicates) for the controls/safeguards. In this way *all* the controls that relate to a specific resource or asset will be covered.
   - Evaluate and review each potential control/safeguard for its applicability with respect to protecting the resource/asset.

4. *All controls for a concern/exposure*
   - Identify the specific concern/exposure that requires some controls/safeguards.
   - Read down the column copying the numbers (disregard duplicates) for the

controls/safeguards. In this way *all* the controls that relate to a specific concern or exposure will be covered.

- Evaluate and review each potential control/safeguard for its applicability with respect to mitigating the effects of the concern or exposure.

This appendix is an excerpt from a larger book that contains nine matrices. It is entitled *Internal Controls for Computerized Systems* and it contains over 650 controls. It is published by Jerry FitzGerald & Associates, 506 Barkentine Lane, Redwood City, California 94065.

INPUT CONTROL MATRIX

THE MATRIX APPROACH

The internal control area to be reviewed using this matrix covers inputs to the computer system. These inputs may involve transaction origination or transaction entry and will be somewhat oriented toward batch computer systems. When reviewing the input controls, match each resource/asset with its corresponding concern/exposure as listed in Figure 1: Input Control Matrix. This matrix lists the resources in relation to the potential exposures and cross-relates these with the various controls/safeguards that should be considered when reviewing inputs to computer systems.

Immediately following the matrix is a definition of each of the concerns/exposures that are listed across the top of the matrix and each of the resources/assets that are listed down the left vertical column of the matrix. Following these definitions is a complete numerical listing and description of each of the controls/safeguards that are listed numerically in the cells of the matrix.

CONCERNS/EXPOSURES

The following concerns/exposures are those that are directly applicable to the inputs (transaction origination and transaction entry) of a computer-based system. The definition for each of these exposures, listed across the top of the matrix, is as follows:

- Authorization—The proper authorization either prior to the input (source document) or during input into the computer system.
- Source Document Origination—The procedures and methods used to insure the proper and timely recording of data. The data may be recorded directly in a machine-readable form, or it may be recorded initially on a human-readable document.
- Source Document Error Handling—The procedures and methods used to insure that all manual transactions rejected at any point in the system are corrected and reentered in a timely manner.
- Source Document Retention—The procedures and methods used to insure the proper retention of source documents, including the adequate backup of source data maintained to provide audit trails or to be used for recovery should the computer data be inadvertently destroyed.

CONCERNS / EXPOSURES

RESOURCES / ASSETS

	AUTHOR-IZATION	SOURCE DOCUMENT ORIGINATION	SOURCE DOCUMENT ERROR HANDLING	SOURCE DOCUMENT RETENTION	DATA ENTRY	TRANS-ACTION DATA VALIDATION	TRANS-ACTION ENTRY ERROR HANDLING	NEGOTIABLE DOCUMENT CONTROL	PRIVACY
INPUT DEVICES		6, 10			10, 43, 41	10	39, 43		57
OPERATIONAL PROCEDURES	5, 7, 9, 11-19, 25, 58	2,4,5,7,8,10, 13, 14, 20-25, 31-34, 41, 44, 55, 57, 59-62	4,5,15,21,27,34 38,40-42,46,49, 50, 54, 56, 59, 64	15, 16, 29, 47, 48, 61, 63	1-8, 14, 20, 23, 24,27,28,30,31, 33,34,40,42,44, 45,54,55,57,59 60, 62, 67, 68	1-3, 6, 8-10, 12, 25, 28, 31, 33, 35-38, 40 53, 56, 67, 68	21, 24, 28, 34 39, 40, 42, 49-56, 64	15-18, 26, 57	4, 9, 11, 15, 18, 20, 32, 46, 57, 58, 60, 61
PEOPLE	5, 7, 9, 11-19, 25, 28	4,5,9,10,12-14, 20, 44, 45, 57, 59-62, 66	4,15,45,46, 49, 50, 56, 59, 64, 65	15, 16, 29, 47, 48, 63	2,4,5,7,14,20, 23, 24, 27, 28, 40,44,45,54,55, 57, 59, 60, 62, 66	9, 10, 12, 25, 53, 56	24, 28, 49-56, 64, 65	15-18, 26, 57	4, 9, 11, 15, 18, 20, 32, 46, 57, 58, 60, 61
FILES (MANUAL OR MAGNETIC MEDIA)		3, 4, 8, 9		15, 16, 29, 47, 48, 61, 63	3		51		
RECORDS STORAGE			56	15, 16, 29, 47, 48, 61, 63					
FORMS		22, 23, 30, 62		15, 16, 61	26, 62			15, 17, 18, 26	

INPUT CONTROL MATRIX

523

- Data Entry—The procedures and methods used to insure the proper collection of data for recording as well as the transcription of that data to a machine-readable form for input to the computer system. This begins after the source documents have been originated and authorized. At this point it is necessary to prepare the data further for data processing input.
- Transaction Data Validation—The validation of data as it enters the computerized system.
- Transaction Entry Error Handling—The methodologies and controls for handling errors after they have been entered into the system, correcting these errors, and reentering the corrected data into the system.
- Negotiable Document Control—The manual handling and use of negotiable document forms, either before, during, or after the use of these negotiable documents within the data processing department.
- Privacy—The accidental or intentional release of data about an individual, assuming that the release of this personal information was improper to the normal conduct of the business at the organization.

RESOURCES/ASSETS

The following resources/assets are those that should be reviewed during the input control review. The definition for each of these assets, listed down the left vertical column of the matrix, is as follows:

- Input Devices—Any or all of the input devices used to interconnect with the computer system. This would specifically include (without excluding other devices) key punches, keytape/disk units, card readers, tape and disk units, optical readers, local terminals, and the like.
- Operational Procedures—The written procedures to be followed during the origination of, preparation of, and inputting of data to the computerized system.
- People—The individuals responsible for preparing and inputting data, operating and maintaining the equipment, following the operational procedures, and performing any other operations during the input of data to the computerized system.
- Files (Manual or Magnetic Media)—The manual files of source documents and the data once it is stored upon magnetic devices such as magnetic tapes or disks.

- Records Storage—The long-term storage of various source documents or other types of data or programs that may be needed at some time in the future. This resource also includes the long-term storage of data that has been written onto microforms.
- Forms—Any of the specially designed and preprinted forms that might be utilized prior to or during the inputting of data to the computerized system.

CONTROLS/SAFEGUARDS

The following controls/safeguards should be considered when reviewing the inputs (transaction origination and transaction entry) of a computer-based system. This numerical listing describes each control.

It should be noted that implementation of various controls can be both costly and time consuming. It is of great importance that a realistic and pragmatic evaluation be made with regard to the probability of a specific exposure affecting a specific asset. Only then can the control for safeguarding the asset be evaluated in a cost-effective manner.

The controls, as numerically listed in the cells of the matrix, are as follows.

1. When keypunching or keytaping, use verification techniques to insure minimum errors. The operators might verify the entire data record or just critical fields.

2. Consider inputting critical fields twice when entering the data. In this way the computer system can match these two fields to insure correctness.

3. When using magnetic tape or magnetic disk as an input device, consider using both internal and external labels on the tape and disk media devices.

4. When source documents are passed between various departments for manual processing, log in these documents as to time received and from whom received as they move between these manual operations.

5. Perform a manual check of source documents for items such as control figures, prior editing, signature authorization, and the like.

6. Whenever self-checking numbers are used, consider building hardware into keypunch or keytape equipment to automatically verify these self-checking numbers during the input function.

7. Review the current operating procedures and verify that the handling procedures, signature authorization, and items similar to these are being followed during the manual handling of source documents or data input.

525

8. Log all inputs in sequence for on-line systems.

9. Use passwords for people and lockwords for files to protect from unauthorized data entry.

10. Restrict the access to various input devices.

11. Whenever feasible, segregate the functions of the generation of the transaction, the recording of the transaction, and the custody of the assets.

12. Consider establishing an independent control group to verify the authorization of transactions.

13. Insure that the function responsible for inputting the transaction verifies authorization signatures by comparing them to an authorized signature list.

14. Establish batch controls close to the point of input preparation to prevent introduction of unauthorized input between the source and the entry into the computerized system.

15. Store source documents in a locked cabinet to prevent unauthorized modifications or unauthorized use of the data prior to its entry into the system.

16. Restrict access to blank input forms and especially to negotiable documents.

17. Keep negotiable documents under lock and key and control them through the use of prior serial numbering.

18. Control sensitive documents and especially negotiable documents by using a dual custody method where two people must be present when the documents are being used. These two people should also perform a manual count, which would be reconciled by a third person back to the serial numbers that were preprinted on the negotiable documents.

19. Insure that the authorized individual does, in fact, sign source documents where this authorization is required.

20. Separate the computer operations functions from the transaction-generation and the transaction-recording functions.

21. Clearly describe the coding requirements, the batching requirements, and the scheduling requirements in the operations procedures manual.

22. Whenever possible, record data on a preprinted form to insure against errors and omissions.

23. When designing forms, insure that the data is recorded in a predetermined and uniform format in order to minimize errors and omissions.

24. Clearly describe the input keying requirements, response/error checking interpretation, and any special requirements in the operations procedures manual.

25. Insure that the personnel who input data either initial, sign, or identify the data that they prepare.

26. Design source documents with preprinted sequential serial numbers for control and to insure against lost documents.

27. Impose restrictions on batch size to allow ease of correction and control.

28. Record and/or maintain the identification number of the source document on the transaction to be processed.

29. Maintain the file of source documents by identification number or subject so it is readily retrievable. The identification number can be a preprinted sequential number or an assigned number such as employee number, part number, and the like.

30. Design and utilize precoded forms which contain information common to all given transaction types in order to reduce continual re-keying of the same data.

31. Include either the business date or processing date as a field in the input transaction.

32. Verify correctness of source documents prior to their conversion to a machine-readable form.

33. Include the Julian date as part of the transaction number.

34. Match transactions by processing cycle, and maintain uniqueness of batch and transaction numbers.

35. Assign transaction types to reflect the update process and desired update sequence.

36. Publish processing cycles to allow users to control cut-off dates.

37. Identify business or processing cut-off dates with specific logical accounting cycles.

38. Clearly define in the operations procedures the control points at which batch controls should be reconciled in the movement of data between and/or through various departments.

39. Establish backup processing procedures in case of a total and lengthy computer failure. These procedures should encompass alternative computers and/or manual data handling and processing procedures.

40. Insure that every department that handles data reconciles its batch controls, verifies the processing schedule to insure meeting cut-off points, and maintains a log of transactions passed between its department and other departments.

41. Try to centralize key-tape or key-disk operations close to the information source in order to insure against lost source documents.

42. Compare computer-produced batch totals, hash totals, transaction totals, sequence numbers, and the like to predetermined manually prepared totals.

43. Incorporate the batch-balancing comparison process into the key-tape or key-disk equipment.

44. Stamp the source document at the time of inputting to insure against inputting the same source documents twice.

45. Consider having a data control clerk visually verify all transactions prior to input, anticipate input, research missing input, and the like.

46. Consider having a data control clerk maintain a log of all source documents returned to the user for correction. This log should be reviewed frequently to insure that the corrected document has been returned and reentered into the system.

47. Store source documents in a safe place. Filing of these documents should provide for rapid access to the documents.

48. Assign a retention date to each source document, and insure that it is placed in long-term storage which can be accessed at any time in the future.

49. Clearly outline the correction procedures in the operations manual. These procedures should include the types of errors that occur, correction procedures for all errors, and recycling of input and balancing of output reports.

50. Establish a central data control group responsible for error detection, correction, and resubmission.

51. Suspend the rejects on a file. Remove the file when the reentered corrected version is verified and accepted for processing.

52. In addition to the original error message, regularly print the overdue suspended items and their messages so follow-up can be done on errors that have not been corrected in a timely manner.

53. When reentering new transactions, edit the correction with the same module used to edit the original transaction.

54. Provide adequate user manuals and operations procedure manuals that cover items such as how to prepare the documents, regulate document flow, adhere to cut-off schedules, describe input keying requirements, and the like.

55. Insure that each transaction to be entered into the computer system either has its own transaction identification or the batch identification from which it came.

56. Whenever possible, utilize a cross-reference field, in which the source document number might be part of the transaction identification. That will provide a cross-reference useful in tracing information to and from the source document.

57. Insure that there is adequate separation of duties during the data preparation and data inputting to the computer system.

58. Insure that there is evidence of approval (stamp the document) with regard to written authorizations and/or signatures.

59. Develop a control desk function to monitor the timely receipt of transactions and/or batches and to maintain proper source transaction schedules and compliance with requirements to the cut-offs.

60. Utilize a transmittal document to control the movement of paperwork between various users and the data entry function.

61. Physically secure input during the transportation of source documents and/or source data between work areas.

62. Utilize source turnaround documents to insure proper turnaround times. The turnaround portion of the document contains prerecorded data that can be used as the input medium for computer processing.

63. Develop a source document retention schedule and maintain a cross-index system so as to know how long to keep source documents and when to dispose of them. This can also be used for source document retrieval and for changing the retention dates, should it become necessary.

64. Develop written error-handling procedures to provide user personnel with comprehensive instructions for source document error detection, error correction, and corrected data resubmission.

65. Insure that the operations personnel receive adequate job training.

66. Periodically rotate the duties between the operations personnel to reduce boredom and to give new job duties to different personnel.

67. Insure that there are run-to-run totals so as not to lose data or information between jobs.

68. Cross-reference data input to record counts, control totals, hash totals, batch totals, and the like.

DATA COMMUNICATION CONTROL MATRIX

THE MATRIX APPROACH

The internal control area to be reviewed using this matrix covers the data communication links between the computer and the input/output terminals. These data-communication-oriented controls may involve hardware controls, software controls, and personnel controls. When reviewing the data communication controls, match each resource/asset with its corresponding concern/exposure as listed in Figure 2: Data Communication Control Matrix. This matrix lists the resources in relation to the potential exposures and cross-relates these with the various controls/safeguards that should be considered when reviewing the data communication controls.

Immediately following the matrix is a definition of each of the concerns/exposures that are listed across the top of the matrix and each of the resources/assets that are listed down the left vertical column of the matrix. Following these definitions is a complete numerical listing and description of each of the controls/safeguards that are listed numerically in the cells of the matrix.

CONCERNS/EXPOSURES

The following concerns/exposures are those that are directly applicable to the data communication network review of an on-line system. The definition for each of these exposures, listed across the top of the matrix, is as follows:

- Errors and Omissions—The accidental or intentional transmission of data that is in error, including the accidental or intentional omission of data that should have been entered or transmitted on the on-line system. This type of exposure includes, but is not limited to, inaccurate data, incomplete data, malfunctioning hardware, and the like.
- Message Loss or Change—The loss of messages as they are transmitted throughout the data communication system, or the accidental/intentional changing of messages during transmission.
- Disasters and Disruptions (natural and man-made)—The temporary or long-term disruption of normal data communication capabilities. This exposure renders the organization's normal data communication on-line system inoperative.
- Privacy—The accidental or intentional release of data about an individual, as-

CONCERNS / EXPOSURES

DATA COMMUNICATION CONTROL MATRIX

RESOURCES / ASSETS	ERRORS AND OMISSIONS	MESSAGE LOSS OR CHANGE	DISASTERS AND DISRUPTIONS	PRIVACY	SECURITY/ THEFT	RELIABILITY (UP-TIME)	RECOVERY AND RESTART	ERROR HANDLING	DATA VALIDATION AND CHECKING
CENTRAL SYSTEM	1-4, 7, 39, 41-43, 47, 48	1-5, 7, 37, 39, 48, 49, 89	1, 8, 11, 13, 16, 29, 40, 48, 50, 51, 54, 57, 58, 64, 65, 79, 85	6, 8, 24, 35, 53, 56, 60, 62, 70, 72-74, 78-80	6, 8, 24, 35, 53, 56, 60, 62, 68, 70, 72-74, 77-80	1, 13, 16, 29, 38, 40, 50, 51, 63-65, 68, 81, 88	50, 51, 63-65, 68	48, 85, 89	6, 24, 39, 41, 47, 88
SOFTWARE	1-5, 7, 39, 41, 44, 46, 47	1-5, 7, 37, 39, 41, 42, 48, 49, 52, 54, 89	1, 8, 16, 40, 48, 50-54, 57-59, 63, 85	6, 8, 24, 35, 53, 56, 60, 62, 68, 70, 72-74, 78-80	6, 8, 24, 35, 39, 53, 56, 60, 62, 68, 70, 72-74, 78-80	1, 38, 40, 50, 51, 56-59, 61, 63, 68, 88	50-52, 61, 63, 64, 68	48, 61, 85, 89	6, 24, 39, 41, 47-49, 52, 53, 55, 60, 88
FRONT-END COMMUNICATION PROCESSOR	**1-4, 7, 34, 39, 41-44, 46-48**	1-5, 7, 34, 37, 39, 41, 42, 49, **89**	1, 8, 13, 16, 29, 40, 44, 48, 50, 51, 54, 57, 58, 64, 65, 79, 85	6, 8, 24, 35, 37, 45, 60, 62, 68, 70, 72-74, 78-80	6, 8, 24, 29, 35, 37, 39, 45, 60, 62, 68, 70, 72-74, 78-80	1, 13, 16, 29, 30, 34, 36, 40, 43, 44, 50, 51, 63-65, 81, 88	37, 50, 51, 63-65	43, 48, 85, 89	6, 24, 39, 41, 45, 47, 48, 88
MULTIPLEXER, CONCENTRATOR SWITCH	1-4, 7, 37, 39, 41, 44, 46, 47	1-5, 7, 37, 39, 41, 42, 49, 89	1, 8, 13, 16, 29, 30, 32, 33, 40, 44, 48, 50, 51, 54, 57, 58, 65, 79, 85	6, 3, 24, 35, 37, 45, 60, 62, 68, 70, 72-74, 78-80	6, 8, 24, 29, 35, 37, 39, 45, 60, 62, 68, 70, 72-74, 78-80	1, 13, 16, 29, 30, 32-34, 36, 40, 44, 50, 51, 63-65, 81, 88	37, 50, 51, 63, 64	48, 85, 89	6, 24, 39, 41, 45, 47, 48, 88
COMMUNICATION CIRCUITS (LINES)	12, 26	28, 70, 91	10, 15, 16, 18, 26, 63, 64, 66, 75, 76, 79, 91	25, 28, 68, 70, 75, 76, 78-80, 91	25, 28, 68, 70, 75, 76, 78-80, 91	15, 16, 20, 21, 23, 26, 27, 63, 64, 66-68, 88	63, 64, 66, 68	85	
LOCAL LOOP	12	25	25, 75, 85	25, 76	25, 29, 75, 76	68, 88	63, 64, 68	85	
MODEMS	12, 18	18, 24	8-11, 13-16, 18	24	24, 29	9-11, 13-18, 20, 21, 23, 36, 88	9-11, 14, 15, 63, 64	18-20, 22, 23	
PEOPLE	5, 39	5, 7, 31, 39, 70	79-87	6, 8, 24, 53, 69-71, 74, 77, 79, 80	6, 8, 24, 29, 53, 69-71, 74, 77, 79, 80	81, 82, 85-87	50, 51, 86, 87	49, 86, 87, 89, 90	6, 88
TERMINALS/ DISTRIBUTED INTELLIGENCE		2	SEE CHAPTER 7 FOR MOST OF THESE CONTROLS	6, 8, 24, 45, 53, 56, 62, 70	6, 8, 24, 29, 45, 53, 56, 62, 70	1, 40, 88	63, 64		6, 24, 45

531

suming that the release of this personal information was improper to the normal conduct of the business at the organization.

- Security/Theft—The security or theft of information that should have been kept confidential because of its proprietary nature. In a way, this is a form of privacy, but the information removed from the organization does not pertain to an individual. The information might be inadvertently (accidentally) released, or it might be the subject of an outright theft. This exposure also includes the theft of assets such as might be experienced in embezzlement, fraud, or defalcation.

- Reliability (Up-Time)—The reliability of the data communication network and its "up-time." This includes the organization's ability to keep the data communication network operating and the mean time between failures (MTBF) as well as the time to repair equipment when it malfunctions. Reliability of hardware, reliability of software, and the maintenance of these two items are chief concerns here.

- Recovery and Restart—The recovery and restart capabilities of the data communication network, should it fail. In other words, How does the software operate in a failure mode? How long does it take to recover from a failure? This recovery and restart concern also includes backup for key portions of the data communication network and the contingency planning for backup, should there be a failure at any point of the data communication network.

- Error Handling—The methodologies and controls for handling errors at a remote distributed site or at the centralized computer site. This may also involve the error handling procedures of a distributed data processing system (at the distributed site). The object here is to insure that when errors are discovered they are promptly corrected and reentered into the system for processing.

- Data Validation and Checking—The validation of data either at the time of transmission or during transmission. The validation may take place at a remote site (intelligent terminal), at the central site (front-end communication processor), or at a distributed intelligence site (concentrator or remote front-end communication processor).

RESOURCES/ASSETS

The following resources/assets are those that should be reviewed during the data communication control review. The definition for each of these assets, listed down the left vertical column of the matrix, is as follows:

- Central System—Most prevalent in the form of a central computer to which the data communication network transmits and from which it receives information. In a distributed system, with equal processing at each distributed node, there might not be an identifiable central system (just some other equal-sized distributed computer).

- Software—The software programs that operate the data communication network. These programs may reside in the central computer, a distributed-system computer, the front-end communication processor, a remote concentrator or statistical multiplexer, and/or a remote intelligent terminal. This software may include the telecommunications access methods, an overall teleprocessing monitor, programs that reside in the front-end processors, and/or programs that reside in the intelligent terminals.

- Front-End Communication Processor—A hardware device that interconnects all the data communication circuits (lines) to the central computer or distributed computers and performs a subset of the following functions: code and speed conversion, protocol, error detection and correction, format checking, authentication, data validation, statistical data gathering, polling/addressing, insertion/deletion of line control codes, and the like.

- Multiplexer, Concentrator, Switch—Hardware devices that enable the data communication network to operate in the most efficient manner. The multiplexer is a device that combines, in one data stream, several simultaneous data signals from independent stations. The concentrator performs the same functions as a multiplexer except it is intelligent and therefore can perform some of the functions of a front-end communication processor. A *switch* is a device that allows the interconnection between any two circuits (lines) connected to the switch. There might be two distinct types of switch: a switch that performs message switching between stations (terminals) might be located within the data communication network facilities that are owned and operated by the organization; a circuit or line switching switch that interconnects various circuits might be located at (and owned by) the telephone company central office. For example, organizations perform message switching and the telephone company performs circuit switching.

- Communication Circuits (Lines)—The common carrier facilities used as links (a link is the interconnection of any two stations/terminals) to interconnect the organization's stations/terminals. These communication circuits include, not to the exclusion of others, satellite facilities, public switched dial-up facilities, point-to-point private lines, multiplexed lines, multipoint or loop configured private lines, WATS services, and many others.

- Local Loop—The communication facility between the customer's premises and the telephone company's central office or the central office of any other special common carrier. The local loop is usually assumed to be metallic pairs of wires.

- Modems—A hardware device used for the conversion of data signals from terminals (digital signal) to an electrical form (analog signal) which is acceptable for transmission over the communication circuits that are owned and maintained by the telephone company or other special common carrier.

- People—The individuals responsible for inputting data, operating and maintaining the data communication network equipment, writing the software programs for the data communications, managing the overall data communication network, and those involved at the remote stations/terminals.

- Terminals/Distributed Intelligence—Any or all of the input or output devices used to interconnect with the on-line data communication network. This resource would specifically include, without excluding other devices, teleprinter terminals, video terminals, remote job entry terminals, transaction terminals, intelligent terminals, and any other devices used with distributed data communication networks. These may include microprocessors or minicomputers when they are input/output devices or if they are used to control portions of the data communication network.

CONTROLS/SAFEGUARDS

The following controls/safeguards should be considered when reviewing the data communication network review of an on-line system. This numerical listing describes each control.

It should be noted that implementation of various controls can be both costly and time consuming. It is of great importance that a realistic and pragmatic evaluation be made with regard to the probability of a specific exposure affecting a specific asset. Only then can the control for safeguarding the asset be evaluated in a cost-effective manner.

The controls, as numerically listed in the cells of the matrix, are as follows:

1. Insure that the system can switch messages destined for a down station/terminal to an alternate station/terminal.

2. Determine whether the system can perform message-switching to transmit messages between stations/terminals.

---

3. In order to avoid lost messages in a message-switching system, provide a store and forward capability. This is where a message destined for a busy station is stored at the central switch and then forwarded at a later time when the station is no longer busy.

4. Review the message or transaction logging capabilities to reduce lost messages, provide for an audit trail, restrict messages, prohibit illegal messages, and the like. These messages might be logged at the remote station (intelligent terminal), they might be logged at a remote concentrator/remote front-end processor, or they might be logged at the central front-end communication processor/central computer.

5. Transmit messages promptly to reduce risk of loss.

6. Identify each message by the individual user's password, the terminal, and the individual message sequence number.

7. Acknowledge the successful or unsuccessful receipt of all messages.

8. Utilize physical security controls throughout the data communication network (see Chapter 8 of the book *Internal Controls for Computerized Systems* for the Physical Security Control Matrix). This includes the use of locks, guards, badges, sensors, alarms, and administrative measures to protect the physical facilities, data communication networks, and related data communication equipment. These safeguards are required for access monitoring and control to protect data communication equipment and software from damage by accident, fire, and environmental hazard either intentional or unintentional.

9. Consider using modems that have either manual or remote actuated loopback switches for fault isolation to insure the prompt identification of malfunctioning equipment. These are extremely important in order to increase the uptime and to identify faults.

10. Use front panel lights on modems to indicate if the circuit/line is functioning properly (carrier signal is up). This may not be a viable alternative with organizations that have hundreds of modems.

11. Consider a modem with alternate voice capabilities for quick trouble-shooting between the central site and a major remote site.

12. When feasible, use digital data transmission, because it has a lower error rate than analog data transmission.

13. For data communication equipment, check the manufacturer's mean time between failures (MTBF) in order to insure that the data communication equipment has the largest MTBF.

14. Consider placing unused backup modems in critical areas of the data communication network.

15. Consider using modems that have an automatic or semiautomatic dial backup capability in case the leased line fails.

16. Review the maintenance contract and mean time to fix (MTTF) for all data communication equipment. Maintenance should be both fast and available. Determine from where the maintenance is dispatched, and determine if tests can be made from a remote site (for example, in many cases modems have remote loopback capabilities).

17. Increase data transmission efficiency. The faster the modem synchronization time, the lower will be the turnaround time and thus more throughput to the system.

18. Consider modems with automatic equilization (built in microprocessors for circuit equalization and balancing) in order to compensate for amplitude and phase distortions on the line. This will reduce the number of errors in transmission and may decrease the need for conditioned lines.

19. With regard to the efficiency of modems, review to see if they have multiple-speed switches so the transmission rate can be lowered when the line error rates are high.

20. Utilize four-wire circuits in a pseudo-full duplex transmission mode. In other words, keep the carrier wave up in each direction on alternate pairs of wires in order to reduce turnaround time and gain efficiency during transmission.

21. If needed, use full duplex transmission on two-wire circuits with special modems that split the frequencies and thus achieve full duplex transmission.

22. Increase the speed of transmission. The faster the speed of transmission by the modem, the more cost effective are the data communications, but error rates may increase with speed, and therefore you may need more error detection and correction facilities.

23. Utilize a reverse channel capability for control signals (supervisory) and to keep the carrier wave up in both directions.

24. Consider the following special controls on dial-up modems when the data communication network allows incoming dial-up connections: change the phone numbers at regular intervals; keep the phone numbers confidential; remove the phone numbers from the modems in the computer operations area; require that each "dial-up terminal" have an electronic identification circuit chip to transmit its unique identification to the front-end communication processor; do not allow automatic call receipt and connection (always have a per-

son intercept the call and make a verbal identification); have the central site call the various terminals that will be allowed connection to the system; utilize dial-out only where an incoming dialed call triggers an automatic dial-back to the caller (in this way the central system controls those phone numbers to which it will allow connection).

25. Physically trace out and, as best as possible, secure the local loop communication circuits/lines within the organization or facility. After these lines leave the facility and enter the public domain, they cannot be physically secured.

26. Consider conditioning the voice-grade circuits in order to reduce the number of errors during transmission (this may be unnecessary with the newer microprocessor-based modems that perform automatic equalization and balancing).

27. Use four-wire circuits in such a fashion that there is little to no turnaround time. This can be done by using two wires in each direction and keeping the carrier signal up.

28. Within an organizational facility, fiber optic (laser) communication circuits can be used so as to totally preclude the possibility of wiretapping.

29. Insure that there is adequate physical security at remote sites and especially for terminals, concentrators, multiplexers, and front-end communication processors.

30. Determine whether the multiplexer/concentrator/remote front-end hardware has redundant logic and backup power supplies with automatic fall-back capabilities in case the hardware fails. This will increase the up-time of the many stations/terminals that might be connected to this equipment.

31. Consider logging inbound and outbound messages at the remote site.

32. Consider uninterruptible power supplies at large multiplexer/concentrator type remote sites.

33. Consider multiplexer/concentrator equipment that has diagnostic lights, diagnostic capabilities, and the like.

34. If a concentrator is being used, is it performing some of the controls that are usually performed by the front-end communication processor and therefore increasing the efficiency and correctness of data transmissions?

35. See if the polling configuration list can be changed during the day in order to exclude or include specific terminals. This would allow the positive exclusion of a terminal as well as allowing various terminals to come on-line and off-line during the working day.

36. Can the front-ends, concentrators, modems, and the like handle the automatic answering and automatic outward dialing of calls? This would increase the efficiency and accuracy when it is preprogrammed into the system.

37. Insure that all inbound and outbound messages are logged by the central processor, the front-end, or remote concentrator in order to insure against lost messages, keep track of message sequence numbers (identify illegal messages), and to use for system restart should the entire system crash.

38. For efficiency, insure that the central system can address either a group of terminals (group address), several terminals at a time (multiple address), one terminal at a time (single address), or send a broadcast message simultaneously to all stations/terminals in the system.

39. See that each inbound and outbound message is serial numbered as well as time and date stamped at the time of logging.

40. Insure that there is a "time out" facility so the system does not get hung up trying to poll/address a station. Also, if a particular station "times out" four or five consecutive times, it should be removed from the network configuration polling list so time is not wasted on this station (improves communication efficiency).

41. Consider having concentrators and front-ends perform two levels of editing. In the first level the front-end may add items to a message, reroute the message, or rearrange the data for further transmission. It may also check a message address for accuracy and perform parity checks. In the second level of editing, the concentrator or front-end is programmed to perform specific edits of the different transactions that enter the system. This editing is an application system type of editing and deals with message content rather than form and is specific to each application program being executed.

42. Have the concentrators, front-ends, and central computers handle the message priority system, if one exists. A priority system is set up to permit a higher line utilization to certain areas of the network or to insure that certain transactions are handled before other transactions of lesser importance.

43. See that the front-end collects message traffic statistics and performs correlations of traffic density and circuit availability. These analyses are mandatory for the effective management of a large data communication network. Some of the items included in a traffic density report might be the number of messages handled per hour or per day on each link of the network, the number of errors encountered per hour or per day, the number of errors encountered per program or per program module, the terminals or stations that appear to have a higher than average error record, and the like.

44. Insure that the front-ends and concentrators can perform miscellaneous functions such as triggering remote alarms if certain parameters are exceeded, performing multiplexing operations internally, signaling abnormal occurrences to the central computer, slowing up input/output messages when the central computer is overburdened due to heavy traffic, and the like.

45. Insure that the concentrators and front-ends can validate electronic terminal identification.

46. Insure that there is a message intercept function for inoperable terminals or invalid terminal addresses.

47. See that messages are checked for valid destination address.

48. Insure adequate error detection and control capabilities. These might include echo-checking, where a message is transmitted to a remote site and the remote site echoes the message back for verification, or it might include forward error correction, where special hardware boxes can automatically correct some errors upon receipt of the message, or it might include detection with retransmission. Detection with retransmission is the most common and cost-effective form of error detection and correction. This may include identification of errors by reviewing the parity bit or utilizing a special code to identify errors in individual characters during transmission. A more prevalent form is to utilize a polynomial (mathematical algorithm) to detect errors in message blocks. Whichever way is used, when a message error is detected, it is retransmitted until it is received correctly.

49. When reviewing error detection in transmission, first determine whatever error rate rate can be tolerated, then determine the extent and pattern of errors on the communication links used by the organization, and then review the error detection and correction methodologies in use and determine if they are adequate for the application systems utilizing the data communication network. In other words, a purely administrative message network (no critical financial data) would not require error detection and correction capabilities equal to a network that transmits critical financial data.

50. Insure that there are adequate restart and recovery software routines to recover from items such as a trapped machine check, where instead of bringing down the entire data communication system, a quick recovery can be made and only the one transaction need be retransmitted.

51. Insure that there are adequate restart and recovery procedures to effect both a warm start and a cold start. In other words, a data communication system should never completely fail so the user has to perform a cold start (start up as if it is a new day, all message counters cleared). The system should go into a

warm start procedure, where only parts of the system are disabled and recovery can be made while the system is operating in a degraded mode.

52. Insure that there is an audit trail logging facility to assist in the reconstruction of data files and the reconstruction of transactions from the various stations. There should be the capability to trace back to the terminal and user.

53. Provide some tables for checking for access by terminals, people, data base, and programs. These tables should be in protected areas of memory.

54. Safe store all messages. All transactions/messages should be protected in case of a disastrous situation, such as power failure.

55. Protect against concurrent file updates. If the data management software does not provide this protection, the data communication software should.

56. For convenience, flexibility, and security, insure that terminals can be brought up or down dynamically while the system is running.

57. Make available a systems trace capability to assist in locating problems.

58. Insure that the documentation of the system software is comprehensive.

59. Provide adequate maintenance for the software programs.

60. Insure that the system supports password protection (multilevel password protection).

61. Identify all default options in the software and their impact if they do not operate properly.

62. For entering sensitive or critical systems commands, restrict these commands to one master input terminal and insure strict physical custody over this terminal. In other words, restrict those personnel who can use this terminal.

63. Insure that there are adequate recovery facilities and/or capabilities for a software failure, loss of key pieces of hardware, and loss of various communication circuit/lines.

64. Insure that there are adequate backup facilities (local and remote) to back up key pieces of hardware and communication circuits/lines.

65. Consider backup power capabilities for large facilities such as the central site and various remote concentrators.

66. Consider installing the capabilities to fall back to the public dial network from a leased line configuration.

67. When utilizing multidrop or loop circuits, review the up-time problems. These types of configurations are more cost-effective than point-to-point configurations, but when there is a circuit failure close to the central site, all terminals/stations downline are disconnected.

68. Review the physical security (local and remote) for circuits/lines (especially the local loop), hardware/software, physical facilities, storage media, and the like.

69. For personnel that work in critical or sensitive areas, consider enforcing the following policies: insist that they take at least five consecutive days of vacation per year, check with their previous employers, perform an annual credit check, and have them sign hiring agreements stating that they will not sell programs, and so forth.

70. With regard to data security, consider encrypting all messages transmitted.

71. Develop an overall organizational security policy for the data communication network. This policy should specifically cover the security and privacy of information.

72. Insure that all sensitive communication programs and data are stored in protected areas of memory or disk storage.

73. Insure that all communication programs or data, when they are off-line, are stored in areas with adequate physical security.

74. Insure that all communication programs and data are adequately controlled when they are transferred to microfiche.

75. Lock up phone equipment rooms and install alarms on the doors of those phone equipment rooms that contain the basic data communication circuits.

76. Do not put communication lines through the public switchboard unless it is a new electronic switchboard (ESS) and the intent is to gain verbal identification of incoming dial-up data communication calls.

77. Review the communication system's console log that shows "network supervisor terminal commands" such as: disable or enable a line or station for input or output, alternately route traffic from one station to another, change the order and/or frequency of line or terminal service (polling, calling, dialing), and the like.

78. Consider packet-switching networks that use alternate routes for different packets of information from the same message; this would offer a form of security in case someone were intercepting messages.

79. Insure that there is a policy for the use of test equipment. Modern-day test equipment may offer a new vulnerability to the organization. This test equipment is easily connected to communication lines, and all messages can be read in clear "English language." Test equipment should not be used for monitoring lines "for fun"; it should be locked up (key lock or locked hood) when it is not in use and after normal working hours when it is not needed for testing and debugging; programs written for programmable test equipment should be kept locked up and out of the hands of those who do not need these programs.

80. Review the operational procedures, for example, the administrative regulations, policies, and day-to-day activities supporting the security/safeguards of the data communication network. These procedures may include:

- Specifying the objectives of the EDP security for an organization, especially as they relate to data communications.
- Planning for contingencies of security "events," including recording of all exception conditions and activities.
- Assuring management that other safeguards are implemented, maintained, and audited, including background checks, security clearances and hiring of people with adequate security oriented characteristics; separation of duties; mandatory vacations.
- Developing effective safeguards for deterring, detecting, preventing, and correcting undesirable security events.
- Reviewing the cost-effectiveness of the system and the related benefits such as better efficiency, improved reliability, and economy.
- Looking for the existence of current administrative regulations, security plans, contingency plans, risk analysis, personnel understanding of management objectives, and then reviewing the adequacy and timeliness of the specified procedures in satisfying these.

81. Review the preventive maintenance and scheduled diagnostic testing such as cleaning, replacement, and inspection of equipment to evaluate its accuracy, reliability, and integrity. This may include schedules for testing and repair, adequate testing of software program changes submitted by the vendor, inventories of replacement parts (circuit boards), past maintenance records, and the like.

82. Determine whether there is a central site for reporting all problems encountered in the data communication network. This usually results in faster repair time.

83. Review the financial protection afforded from insurance for various hardware, software, and data stored on magnetic media.

84. Review the legal contracts with regard to the agreements for performing a specific service and a specific costing basis for the data communication network hardware and software. These might include bonding of employees, conflict of interest agreements, clearances, nondisclosure agreements, agreements establishing liability for specific security events by vendors,

agreements by vendors not to perform certain acts that would incur a penalty, and the like.

85. Review the organization's fault isolation/diagnostics, including the techniques used to ascertain the integrity of the various hardware/software components comprising the total data communication entity. These techniques are used to audit, review, and control the total data communication environment and to isolate the offending elements either on a periodic basis or upon detection of a failure. These techniques may include diagnostic software routines, electrical loopback, test message generation, administrative and personnel procedures, and the like.

86. Review the training and education of employees with regard to the data communication network. Employees must be adequately trained in this area because of the high technical competence required for data communication networks.

87. Insure that there is adequate documentation, including a precise description of programs, hardware, system configurations, and procedures intended to assist in the prevention of problems, identification of problems, and recovery from problems. The documentation should be sufficiently detailed to assist in reconstructing the system from its parts.

88. Review the techniques for testing used to validate the hardware and software operation to insure integrity. Testing, including that of personnel, should uncover departures from the specified operation.

89. Review error recording to reduce lost messages. All errors in transmission of messages in the system should be logged and this log should include the type of error, the time and date, the terminal, the circuit, the terminal operator, and the number of times the message was retransmitted before it was correctly received.

90. Review the error correction procedures. A user's manual should specify a cross-reference of error messages to the appropriate error code generated by the system. These messages help the user interpret the error that has occurred and suggest the corrective action to be taken. Insure that the errors are in fact corrected and the correct data reentered into the system.

91. Consider backing up key circuits/lines. This circuit backup may take the form of a second leased line, modems that have the ability to go to the public dial-up network when a leased line fails, or manual procedures where the remote stations can transmit verbal messages using the public dial-up network.

PROGRAM/COMPUTER PROCESSING CONTROL MATRIX

THE MATRIX APPROACH

The internal control area to be reviewed using this matrix covers program/computer processing controls. These program/computer processing controls involve those automated controls that can be built into computer programs and into computerized systems. When reviewing the program/computer processing controls, match each resource/asset with its corresponding concern/exposure as listed in Figure 3: Program/Computer Processing Control Matrix. This matrix lists the resources in relation to the potential exposures and cross-relates these with the various controls/safeguards that should be considered when reviewing the programmed controls in a system.

Immediately following the matrix is a definition of each of the concerns/exposures that are listed across the top of the matrix and each of the resources/assets that are listed down the left vertical column of the matrix. Following these definitions is a complete numerical listing and description of each of the controls/safeguards that are listed numerically in the cells of the matrix.

CONCERNS/EXPOSURES

The following concerns/exposures are those that are directly applicable to the program/computer processing of either on-line or batch systems. The definition for each of these exposures, listed across the top of the matrix, is as follows:

- Program Errors and Omissions—The accidental or intentional creation of an error during the processing of the data or the running of the application programs, including the accidental or intentional omission of data (loss) during the processing of a computer program. This type of exposure includes, but is not limited to, multiprogram code, trapped machine checks where programs just quit processing, loss of data during the running of a program, and the like.

- Unauthorized Program Changes—The temporary or permanent change of program code by individuals who are unauthorized to make these changes, as well as by individuals who are so authorized but who make illegal program changes for whatever reason.

- Security/Theft—The security or theft of information or programs that should have been kept confidential because of their proprietary nature. In a way, this is a form of privacy, but the information removed from the organization does not specifically pertain to an individual. The information or computer pro-

CONCERNS / EXPOSURES

RESOURCES / ASSETS	PROGRAM ERRORS AND OMISSIONS	UNAUTHORIZED PROGRAM CHANGES	SECURITY/ THEFT	DATA VALIDATION	HARDWARE ERRORS	RESTART AND RECOVERY	AUDIT TRAILS	COMPUTER PROGRAM GENERATED TRANSACTIONS	ERROR HANDLING
APPLICATION PROGRAMS AND SYSTEMS	1-9, 16-20, 63, 71-74, 86, 89, 90	13, 22, 23, 25, 30-33, 35, 38, 43, 44, 49-51, 70, 85, 88, 90, 92	24, 25, 30-35, 38, 43, 49, 62, 70	2-8, 16, 69, 72-74, 89	70, 72-74, 89	20, 76, 77	6, 16, 22, 23, 25-28, 30, 32, 33, 35, 43, 49	1, 5, 7, 43, 44, 51, 64, 65, 67, 68, 70, 71, 90	9-15, 45, 71, 78-81, 83
DATA RECORD INTEGRITY	1, 3-7, 16, 17, 27, 69, 71, 73, 74, 80, 81	30-33	24-26, 29, 30-35, 38, 39, 48, 49	1-5, 7, 10, 12-14, 16, 17, 24, 25, 27, 28, 63, 69	41, 46, 47, 72, 87	20, 21, 28	6, 16, 25-28, 31, 32, 49, 64, 66, 80-82	64-68	10-15, 45, 79-83
OUTPUT INTEGRITY	2-4, 6, 7, 16, 18-20, 27, 69, 71, 73, 74, 80, 81	39	24-26, 29-35, 38, 39, 48, 49	2, 4, 7, 9, 10, 16, 28, 69		21, 28	17	64, 65, 67, 81	15, 71, 75, 78-83
CENTRAL SYSTEM	16, 40-42, 46, 47, 69, 76, 77, 91	36-39, 48, 49, 51-53, 85, 86, 91	36-39, 42, 47-49, 51-53, 55, 62, 84, 91	16, 40, 69	46, 54-61, 87	42, 76, 77	30, 33, 42, 84, 91		91
SOFTWARE PROGRAMS	1, 86, 87, 90	13, 22, 23, 25, 30-33, 35, 38, 44, 49, 50, 51, 70, 85, 88-90, 92	24, 25, 30-35, 38, 49, 62, 70	89	70, 87, 89		SEE CHAPTER 10 FOR MOST OF THESE CONTROLS	51, 70, 90	

PROGRAM/COMPUTER PROCESSING CONTROL MATRIX

grams might be inadvertently (accidentally) removed from the organization or might be the subject of outright theft.

- Data Validation—The computer program editing of data prior to its processing and the preprogrammed specific actions that should be taken when erroneous data is discovered (this may also include the discovery of omissions in certain data that should have been included).

- Hardware Errors—The malfunctioning of the computer hardware so it appears that a program has made some sort of an error in processing. The concern here is that a hardware malfunction may cause erroneous data, data omissions, loss of specific data, and the like.

- Restart and Recovery—The restarting of computer programs that have failed during their normal course of processing and the recovery that should take place so no data is lost, erroneously processed, or processed twice because of the failure (the failure may have been caused by program failure or computer hardware failures).

- Audit Trails—Insurance that the processing of the data can be traced backward and forward through the entire computer processing cycle.

- Computer Program Generated Transactions—Insurance that any transactions that are automatically generated within an on-line system are adequately controlled. In other words, some on-line systems automatically create transactions during the time they are being run and these transactions should have adequate controls to prevent errors, erroneous transactions, and illegal transactions.

- Error Handling—The procedures and methods used to insure that all transactions or data that are rejected during the computer processing are, in fact, corrected and reentered into the system in a timely manner. This involves accounting for and detecting data errors, loss, or the nonprocessing of transactions, as well as the reporting of these errors, error correction, and the corrected data resubmission.

RESOURCES/ASSETS

The following resources/assets are those that should be reviewed during the program/computer processing control review. The definition for each of these assets, listed down the left vertical column of the matrix, is as follows:

- Application Programs and Systems—Any or all the computer programs that are utilized in the data processing operations. This resource should also be viewed as the overall macrosystems that operate within the organization (these systems may be made up of a group of computer programs). This is far and away the most valuable asset of the organization because, in the long run, the computer programs are more costly than the hardware upon which they operate.

- Data Record Integrity—The data that is stored in the computer files or data bases and is used in the everyday processing of the organization's computerized record-keeping system.

- Output Integrity—The believability and integrity of the output reports from the system. The auditor should review this resource to insure that the output reports are Consistent, Accurate, Timely, Economic, and Relevant to the intended purpose (reports that meet these criteria will CATER to the needs of the organization).

- Central System—Most prevalent in the form of a central computer in which the computer programs operate. This asset may be in the form of a central computer system, or it may be in the form of numerous computer systems spread around in a distributed network.

- Software Programs—The software programs that run the overall computerized systems. These may include the operating system software (usually supplied by the computer vendor) as well as the software programs utilized to maintain and operate the data communication network, or the data base system (data management software). These software programs usually operate at the "systems control level" because any controls that are built into, or programmed into, this level of software affect all application programs. For example, a control that is built into the operating system software, data communication control software, or data management software would have its effect upon any incoming transaction that passed through that level of software programs without regard to whether it was a payroll transaction, inventory control transaction, financial balancing transaction, or the like.

CONTROLS/SAFEGUARDS

The following controls/safeguards should be considered when reviewing the programs or computer processing of either on-line or batch systems. This numerical listing describes each control.

It should be noted that implementation of various controls can be both costly and time consuming. It is of great importance that a realistic and pragmatic evaluation be made with regard to the probability of a specific exposure affecting a specific asset. Only then can the control for safeguarding the asset be evaluated in a cost-effective manner.

The controls, as numerically listed in the cells of the matrix, are as follows:

1. Transactions that are consecutively numbered by the station transmitting (these might be computer generated transactions) to the computer should be sequence number checked by the computer programs. In other words, the computer programs should verify the unbroken sequence of input or output transactions and take corrective action, should there be a break in sequence. One form of corrective action would be to notify the terminal operator and to close down the transmitting station's ability to transmit data until the remote station takes some sort of corrective action.

2. Have the programs compare the total count of input transactions to a predetermined total count or to a count of output transactions.

3. Let the program perform automated and/or preprogrammed editing for all input after it gets into the computer. Some of the editing that the program can perform might be as follows:

   - Count the number of fields in a record and compare that with a predetermined number of fields.
   - Check for the reasonableness of the input data with regard to some set of preestablished boundaries.
   - Test the data for blanks, sign (plus or minus), numeric, or alphabetic, and compare that with a preestablished criteria.
   - Check for consistency between fields of an input transaction (this would be a specific control with regard to a specific application input).
   - Conduct a limit test, and reject data or take corrective action whenever the data falls outside of some limit or predetermined range.
   - Check for completeness of data, for example, the zip code field should be full, and it should contain numeric data only.
   - Conduct sequence checking in order to insure correct sequence.
   - Conduct date checking in order to insure that the dates are correct whenever this is applicable.
   - Use self-checking numbers that pinpoint erroneous entry of account numbers or whatever type of number the organization is using.

- Enter critical data twice on one transaction input and have the computer programs cross-check these two inputs to insure that, first it was entered correctly, and second there was no error during transmission.

4. Let the computer programs compare or crossfoot predetermined control figures such as:
   - Record counts
   - Control totals
   - Hash totals
   - Batch control totals

5. Have the program recompute various totals of significant financial or accounting figures and transmit these totals back to the original input station.

6. Have the programs prepare specific reports that will display the contents of batch controls, header controls, and any other types of control totals that can be sent back to the original station that inputted the data.

7. Have the programs compare the current data totals with historical totals in order to maintain a logical relationship over time.

8. Have the programs perform logical relationship tests. Logical relationship tests are solely dependent on a specific application because there may be logical relationships within a specific application system.

9. Have the programs look for duplicate entries of data. Whenever duplicate entries are suspected, the original station inputting the data should be immediately notified.

10. Design systems so upon the discovery of erroneous data during processing, the original entry station is immediately notified so correction can take place as soon as possible.

11. Insure that whenever the program edits an incoming record and it finds an error, it continues editing the entire record to see if there is more than one error. This is to avoid the possibility of the program rejecting an input transaction because of an error, and after correcting that error and reentering the correction into the system finding a second error in the original input. This control reduces the cycling of erroneous input messages.

12. Where feasible, incorporate the editing/validation routines into the remote areas of a distributed or data communication network to reduce transmitting erroneous data.

13. Do not provide users with the capability of overriding computer program edits.

14. Do not allow fall-through comparison tests when editing.

15. Have the system produce a report containing all erroneous transactions and identify the invalid data, out of balance data, and the like.

16. Maintain an opening day transaction count record and a closing day transaction count record. These should be equal and should be compared between the central system and the remote data input stations.

17. Have the programs compare transaction date to the cutoff date table. Transactions entering after the cutoff date should be suspended until after closing. The totals of significant fields of the suspended items are to be reported. These suspended items will not be included in the closing balance because they are for the following period.

18. Insure that the data management program (the scheduler) determines potential conflicts between two users attempting to access the same file and keeps these two separate.

19. See that the data management system can prohibit two programs from simultaneously updating the same record but not the same file.

20. Have the master file update program (especially data management—data base) log the after-image of each data base update. This may not be necessary if the data base is a small one that can be copied over each day.

21. Maintain a file that reflects all updated master records and that can be used to recover from a master file loss.

22. Develop and maintain a formal system to control program changes. This system should control and log the changes to any computer program (application software or operations software).

23. Maintain a program that will control the various computer program libraries and will show whenever modifications have been made to these computer program libraries that reside within the computer.

24. See that there are various tables within the computer programs to validate and verify individual user security codes (passwords), unique terminal identification codes, transaction security codes, and any other type of verification of input data that might be necessary. This verification involves verifying prior to allowing entry to the system.

25. See that there is a cross-correlation between the individual user's security code (password) and the transactions, computer programs, computer systems, or other areas that that individual user is allowed to access.

26. See that there are reporting programs that can be used to identify all transactions entered by a specific user (password).

27. Insure that all transactions are dated, time stamped, and logged immediately upon entry into the system. Output transactions should be handled in a similar way.

28. Retain all input and output transactions on an independent file that is backed up.

29. Whenever computer programs or data are to be copied over to microfiche, insure that the handling and storage procedures for the microfiche at least equal the security and control within the data processing operations.

30. Maintain a special log of all unscheduled or unusual interventions by computer operations personnel. This should include location, date, time, type of intervention, and action taken.

31. Maintain a file of all changes to any security tables. There should be a formal procedure to be followed when changing the "up front" security tables. These are the tables against which the individual users' passwords are compared. All resident tables that require periodic update should be created and maintained external to the system.

32. See that the update program that produces changes to tables also produces a report containing the changes and the corresponding table entry, both before and after the change.

33. Make sure that all table changes receive appropriate authorization external to the system before the change takes place. This should be a written form approval and it should be reviewed before the change takes place.

34. Restrict the master commands to the computer system to one physical terminal. These master commands should also be restricted to as few computer operators as possible.

35. Have the computer programs accumulate data for a periodic report showing any unauthorized attempts to access the system, special security type programs, the operating system software, restricted "up-front" tables, and the like.

36. Limit computer operator intervention to system and/or port start-up, terminal backup assignments, emergency message broadcasting, system shutdown, communications debugging, and the system or job status reporting.

37. Design application programs so they do not display messages on the system console or accept data from the system console.

38. Establish access restrictions on system utilities and other sensitive programs that might be utilized to manipulate the system.

39. Consider assigning security codes to files (lockwords) to restrict their access to only certain programs and/or users.

40. Utilize disk and tape labels (internal) and check them with the various programs or program modules.

41. Have programs check the position of various computer console switches if there are any.

42. Log all interruptions by computer operations personnel, and especially save the computer console printing log for review by other personnel.

43. Keep a count on the number of program instructions executed, run time, or any other data for sensitive programs. This can be compared periodically with similar data from a prior period to uncover irregularities.

44. Use a check-sum methodology to control or detect unauthorized program changes. A check-sum count on computer programs can be compared with similar data at a later date to determine if a change was made to a program.

45. Immediately write errors, scuttles, or suspense accounts to eliminate computer operator intervention.

46. Consider using a read-after-write option on magnetic recording devices (especially real-time data base updates).

47. Use program software protection keys to safeguard data in memory, on tape, or on disk.

48. Internally store data in the computer and on magnetic tape/disk using cryptographic techniques. This will protect the organization from a computer memory dump.

49. Whenever a "program control" is overridden or bypassed, note the event on an exception report.

50. Consider using a program that compares a controlled duplicate copy of the source/object program with the program currently being used. This is time consuming and probably should only be used on very sensitive programs whether they are application programs or system software programs.

51. Consider including a listing of the job control language with the output to insure that unauthorized programs have not been executed.

52. Consider having the computer operations personnel insert all job control so computer programs cannot illegally execute programs by manipulating the job control on valid jobs.

53. Consider cataloging the job control language cards so it is more difficult to

illegally enter unauthorized job control and also to insure fewer errors from erroneous job control.

54. Consider having hardware that carries the parity bit into memory and throughout the entire system to more readily detect hardware memory or parity errors.

55. Consider hardware that has memory protection areas that separate parts of memory through hardware controls rather than software controls.

56. Use tape units that have write-ring protection, parity, and read-after-write.

57. Use disk units that allow file protect areas and read-after-write.

58. Use card readers that read twice and compare before entering the data.

59. Use specially equipped key disk/tape/punches and verifying machines that automatically provide for self-checking numbers (if applicable).

60. Use hardware that has circuit diagnostic routines to detect errors.

61. When using optical readers, insure that they read twice and compare before transmitting the data to the central system.

62. Allow for lower privilege levels for users as contrasted with the operating system. This may be accomplished either through hardware or software, depending on the computer manufacturer.

63. Have the computer programs programmed so they look for a unique identifier such as a transaction code to direct the transaction to the proper portion of the application program for processing.

64. Have all computer-program-generated transactions printed out in a listing and sorted to whatever is appropriate for direct feedback to users. This control provides for the use of computer-program-generated transaction control at the user level.

65. Control computer-program-generated transactions by putting severe limits or other preprogrammed restrictions on them so they cannot exceed certain values. Any transactions that exceed these values are automatically rejected and are printed out on a special report.

66. Control computer-generated transactions by keeping a count of them, and on a daily basis graphing this and comparing it with other co-related variables. As time goes by, the operations personnel will get to know what is a reasonable quantity of computer-generated transactions.

67. Control computer-generated transactions by developing a daily report of the ten largest, ten smallest, average number of, average dollar value per transac-

tion, and any other averages or quantities that make sense with regard to the specific application being run. In this way the users get a feel for the types of and quantities of computer-program-generated transactions in their system.

68. Control computer-generated transactions by building in balance controls between program modules. These balance controls are automated internal cross-checks. Such controls may take the form of reasonableness checks, who entered the transactions, program-to-program control totals, and the like.

69. Have control totals passed between jobs and job steps during the running of on-line programs. A system of automated controls should include balancing of the entire system as well as balancing between different program modules and/or the final file (data base) control totals.

70. Consider developing default tables for computer programs that have numerous options. This is where a program has several decisions and there is a standardized action for most of or all of these decisions. When developing a default table, make sure that one of the options is to report that none of the options fits the specific situation at hand.

71. Consider anticipation controls. These controls are used as a method of insuring accountability of input. Sequence-checking is a common example of such a control, but the general thought here is to have the programs anticipate what might be coming next.

72. Consider having the computer hardware use techniques such as double arithmetic, arithmetic overflow checks, and reverse multiplication. These checks would only be used at very critical points in the application program calculations.

73. Insure that programs perform a file completion check to determine that the application file has been completely processed and includes all transactions and all items on a master file.

74. See that programs balance computer files and/or data base record changes. The number of records on the opening of a file should be balanced against the changes made during the day and the closing balance. On a regular basis the total of the detail records might be compared to the total in the control records.

75. See that erroneous data is written out to an error file and returned to the user as soon as possible (preferably immediately for immediate correction). Dummy records should never be used to hold erroneous data.

76. Have a specific set of computer operator instructions developed and in use for each application program.

77. Use computer program run books for the specifics of each application program. These would cover items such as console message instructions, error message instructions, program halts, rerun procedures, check-point and restart instructions, check-point control totals, job control setup, and the like.

78. See that the information on error reports indicates all data fields that are in error in a transaction or record.

79. Maintain an automated error suspense file that includes all rejected transactions. These files should be used for follow-up and to correct and reenter the transactions. These files should be aged so it is evident when certain data has not been corrected and reentered.

80. Issue special discrepancy reports to insure that the handling of errors results in their correction and reentry in a timely manner.

81. Automatically assign unique serial numbers to transactions entered into the automated error suspense files. Such serial numbers are used to control subsequent updating or correction of or reentry of the data that was in error.

82. Never allow destructive updates to correct error conditions without first logging both the before and after image of the update. Debit and credit type entries should never be deleted or erased. Debit and credit type of entries should be corrected by instituting the opposite debit and credit rather than deleting or erasing.

83. Insure that there is an automatic program to reenter erroneous data and to correct the error suspense listing so it is brought up to date in a timely fashion.

84. See that computer operations personnel are rotated between various job functions periodically, and that there is a segregation of duties, within reasonable limits, within large data processing operations facilities.

85. Have signature authorizations thoroughly checked with regard to the external formal paperwork approvals that are required for program changes and special runs and other items that are out of the ordinary.

86. Keep both source and object copies of programs (applications and systems) under secure custody so they cannot be easily removed from the organization. The same is true for various historical records whether they reside on magnetic tapes or in data base systems.

87. Test the default options of vendor-supplied operating system software to determine what would happen if the default option itself failed.

88. Be sure that all programs, application programs, and vendor-supplied operating system programs are adequately documented.

89. Have computer programs conduct an overflow check on all numeric fields for which there could be a data overflow.

90. Insure that any fictitious branches or other fictitious entities that are incorporated into the on-line programs for testing or other training purposes are adequately separated from the "real" records and/or computer programs. Insure that these systems cannot contribute to errors and omissions, fraud, or any other serious types of problems such as disasters, system crashes, and the like.

91. Insure that there are adequate manual controls to duly record whenever computer operations personnel are called in at odd times or for an emergency.

92. Downline load new programs to distributed computer sites periodically to insure program code integrity.

OUTPUT CONTROL MATRIX

THE MATRIX APPROACH

The internal control area to be reviewed using this matrix covers outputs from the computer system. These outputs may be directly affected by input controls, program/computer processing controls, and on-line terminal controls. When reviewing the output controls, match each resource/asset with its corresponding concern/exposure as listed in Figure 4: Output Control Matrix. This matrix lists the resources in relation to the potential exposures and cross-relates these with the various controls/safeguards that should be considered when reviewing the outputs from computer systems.

Immediately following the matrix is a definition of each of the concerns/exposures that are listed across the top of the matrix and each of the resources/assets that are listed down the left vertical column of the matrix. Following these definitions is a complete numerical listing and description of each of the controls/safeguards that are listed numerically in the cells of the matrix.

CONCERNS/EXPOSURES

The following concerns/exposures are those that are directly applicable to the outputs of an on-line or batch type of computerized system. The definition for each of these exposures, listed across the top of the matrix, is as follows:

- Output Balancing and Reconciliation—The responsibility for the monitoring of data processing related controls. This concern sometimes manifests itself as a quality assurance function within the data processing department, insuring that the integrity of the data has not been lost during the data processing cycle.

- Privacy—The accidental or intentional release of data about an individual, assuming that the release of this personal information was improper to the normal conduct of the business at the organization.

- Security/Theft—The security or theft of information that should have been kept confidential because of its proprietary nature. In a way, this is a form of privacy, but the information removed from the organization does not pertain to an individual. The information might be inadvertently (accidentally) released, or it might be the subject of an outright theft. This exposure also includes the theft of assets such as might be experienced in embezzlement, fraud, or defalcation.

CONCERNS / EXPOSURES

RESOURCES / ASSETS	OUTPUT BALANCING AND RECONCILIATION	PRIVACY	SECURITY/ THEFT	OUTPUT DISTRIBUTION	USER BALANCING AND RECONCILIATION	RECORDS RETENTION AND DESTRUCTION	NEGOTIABLE DOCUMENTS	OUTPUT ERROR HANDLING			
OUTPUT DEVICES	9, 31-34, 41	9	9, 31, 33	9, 25, 34	41			9, 27, 31, 41			
PEOPLE	1-3,8,9,11,12, 14, 16, 28, 30, 31, 36-38	4-6, 14	4-6, 14, 21, 37	1,4,6,10,12-15, 17, 21, 30, 34	1, 2, 4, 8, 11, 27, 28, 36, 37	5, 7, 10, 15, 19, 26, 35, 43	5, 7, 20-23, 43	3, 8, 16, 27-31, 36-38			
RECORDS STORAGE AND DISPOSAL		7, 43	7, 37, 43			5, 7, 15, 19, 26, 35, 43	5, 7, 43	35, 39			
OPERATIONAL PROCEDURES	1-3, 8,9, 11, 12, 14, 16, 28-34, 36-42	4-7, 14, 18, 19, 24	4-7, 14, 18, 19, 21, 24, 33, 37, 40	1,4,6,10,12-17, 21, 24, 30, 34	1, 2, 4, 8, 11, 27-29, 36, 37, 40, 41	7, 10, 19, 26, 35, 43	20-23, 43	3, 8, 16,27-30, 36-41			

OUTPUT CONTROL MATRIX

- Output Distribution—A review to insure the delivery of complete and accurate reports to the authorized recipients in a timely manner.
- User Balancing and Reconciliation—Insurance that the integrity of the data has not been lost during the processing by the application system. This concern involves the final users' reviewing, scanning, balancing, and reconciling the data that they have received.
- Records Retention and Destruction—The controls that are used to guide the retention and disposal of confidential or private computer system output. In this context, the disposal may be immediate, or it might come after some years of retaining computer output. This concern should include both paper documents, magnetic media, and microforms.
- Negotiable Documents—The handling of negotiable documents (blank check stock, etc.) and prevention of their loss. This concern may also involve accountable documents which are not negotiable, but whose loss would be injurious to the organization.
- Output Error Handling—The procedures and methods used to insure that all transactions rejected during the system processing are corrected and reentered. The object here is to insure that any output documents found to be in error are corrected.

RESOURCES/ASSETS

The following resources/assets are those that should be reviewed during the output control review. The definition for each of these assets, listed down the left vertical column of the matrix, is as follows:

- Output Devices—The peripheral devices connected to the central computer system (also including the computer itself) and the types of errors that might be made by this hardware. The output devices for any controls, their current usage, operational procedures, and any written documentation pertaining to them should be reviewed.
- People—The individuals responsible for reconciling the output data, operating and maintaining the output devices, writing operational procedures, and managing the output quality assurance function. This may also include user department personnel.
- Records Storage and Disposal—The storage of output data (hard copy or mi-

croform) as well as the disposal of this data. This includes reviewing the storage area and the disposal techniques.

- Operational Procedures—The written operational procedures on how to obtain data, verification, reconciliation, and those other aspects of delivering output data from data processing (whether it is centralized or distributed) to the various users.

CONTROLS/SAFEGUARDS

The following controls/safeguards should be considered when reviewing the outputs of an on-line or batch type of computerized system. This numerical listing describes each control.

It should be noted that implementation of various controls can be both costly and time consuming. It is of great importance that a realistic and pragmatic evaluation be made with regard to the probability of a specific exposure affecting a specific asset. Only then can the control for safeguarding the asset be evaluated in a cost-effective manner.

The controls, as numerically listed in the cells of the matrix, are as follows:

1. Visually scan output reports for completeness and proper formatting.
2. On a random basis, thoroughly check a specific output report for correctness.
3. Always review all errors and the reason for these errors in order to determine if there is a program bug or input problem.
4. Control the distribution of reports so they are sent only to the proper and authorized personnel.
5. Keep all sensitive reports in a secure area so unauthorized personnel cannot obtain copies.
6. Verify that the current handling procedures for output reports are being followed.
7. Immediately destroy aborted output runs (paper shredder) for sensitive outputs.
8. Print out computer-generated control totals and cross-relate these back to the manually inputted control totals, for example, verify record counts, hash totals, batch totals, and the like.
9. Periodically review computer console error messages and system output error messages to try to determine if there are program bugs.
10. Institute an overall report analysis program to determine whether any reports

can be eliminated, combined, rearranged, simplified, or whether new reports may be required.

11. Manually maintain an output log of expected results and compare the actual output to the expected output.

12. Consider developing an independent control group within the data processing function that is responsible for the quality control of output reports.

13. Maintain a job schedule desk with appropriate cutoff times so it can be determined when jobs are to be run and so the output control group can insure timely report delivery.

14. Determine the number of output documents received and ensure that it agrees with the number of documents the program reported as producing. This may manifest itself as a page count routine.

15. Try to include the following elements in the heading of each report: date prepared, processing period covered, a descriptive title of the report contents, the user department, the processing program's identification job number, how to dispose of the report, and its confidentiality.

16. Number the pages consecutively, and indicate that the report came to a normal ending with a positive statement as to such.

17. Label the cover of all reports to indicate the recipient's name and location.

18. Identify any confidential reports as being confidential and proprietary information.

19. Label any reports that must be positively and completely destroyed as to their disposal procedures. In other words, a report that should be returned to a central destruction area should be labeled as such, or it should be labeled to be destroyed in a paper shredder or whatever.

20. At regular intervals, take an inventory of any prenumbered negotiable documents.

21. Consider having a processing program print an additional sequence number on prenumbered forms. The differences between the beginning and ending preprinted number and the beginning and ending of the computer-generated number must agree. The output control group should reconcile these two numbers.

22. Maintain all negotiable documents in a secure location and control access to this location.

23. Maintain signature stamps (such as for payroll) in a different physical location from the negotiable documents themselves.

24. Produce only the required number of output reports.

25. Clearly describe in the operations procedures manual what type of paper stock and/or forms should be used in the print devices.

26. Consider filing output reports by date so at the end of a calendar year they can be discarded when their retention date has been reached.

27. When reviewing output reports that are in error, indicate all data fields in error on the report, and submit them to the programming department for possible evaluation and correction of programs.

28. Produce control totals for all rejects and pass these control totals back to the input area for correction and reentry.

29. Clearly define in the operations procedures manual the procedures for error correction and reentry responsibility.

30. The output control totals for each application should be reconciled with the input totals before the release of the output report from the data processing to the user department.

31. Compare any transaction log maintained by the computer system with a transaction log maintained at each output device. These totals should be verified against individual application control totals established at other steps in the processing stream to verify that everything has been properly processed to its final step.

32. In order to monitor process flow, maintain a record that indicates the average time between the input of user data and the actual starting of each application run. One of the sources that might be used for this purpose is the automatic job accounting routines in some vendors' system software.

33. Develop and review an automated job accounting system. This type of control can be utilized to review the time utilized to process jobs, which files and/or programs were used by the jobs, the time it took to print jobs, and other valuable pieces of information. Vendors' accounting record job routines (for example, IBM has Systems Management Facilities) can be utilized for many unique output control tasks.

34. Consider reviewing the job control used for specific jobs whenever there is a discrepancy in the output.

35. Insure that there are appropriate waste disposal procedures for the immediate disposal of certain paper products (sensitive output reports).

36. Consider obtaining a list of all transactions that went into a specific report whenever there are discrepancies.

---

37. Consider maintaining an independent history file of errors, which is independent of all processing files, and is regularly analyzed to report error trends and statistics by type, source, and frequency of errors within an application system.

38. Insure that all data rejected from a processing cycle of an application is entered in an error log by a control group. This log is then used to insure the correction and reentry of data.

39. Insure that error correction procedures are defined in the operations procedures manual.

40. Insure that header and trailer labels/record counts are printed out at the end of each output report.

41. Even though some on-line systems are individual transaction oriented, consider having the remote terminal operators enter their data in small batches of two to ten transactions.

42. Insure that lengthy output reports have checkpoint and restart capabilities so an entire report need not be printed if there was an error near its end. In this way, the outputting of this report can go back to a prior partition and all the processing time is not lost.

43. When decollating a sensitive job, insure that the carbon paper is disposed of in a secure manner.

APPENDIX 3

INFORMATION RETRIEVAL[1]

1. Other methods such as Uniterm Cards or Superimposable Card systems are explained in Dyke's book. Dyke, Freeman H. Jr. *A Practical Approach to Information and Data Retrieval.* Publication Number PEN-20. Industrial Education Institute, 221 Columbus Avenue, Boston, Massachusetts 02116

KEYSORT NOTCHING and SORTING MANUAL

A McBEE CUSTOMER SERVICE

Coding

The marginal holes on KEYSORT cards are known as *Code Positions, Code Fields,* or *Code Sections.* A *Code Position* is a single hole assigned to a number, letter or word. A *Code Field* contains one or more *Code Positions* relating to a single subject or classification.

Code Position   Code Field

A *Code Section* contains one or more fields relating to the same subject. For example, a numerical Code Section contains a Code Field for each digit of a number. Each of the digits is usually identified in its own Code Field with its position in the series. That is, the first Code Field of a three-digit number is identified as "hundreds," the second field as "tens," and the last field as "units."

Requirements of the particular job determine the coding to be applied to the KEYSORT card. The two primary types of KEYSORT codes are the *Direct Code* and the *Numerical Code.*

Direct Code

A specific classification, such as First or Second (Shift) is assigned to a hole.

Numerical Code

To save space on a card, and time in notching and sorting, KEYSORT systems use a special numerical code system. Only four holes are used for each set of numbers from 0 through 9. These four holes are assigned the values of 7, 4, 2 and 1. By notching either a single number or a combination of two numbers, any number from 1 through 9 may be expressed. Ciphers are not notched.

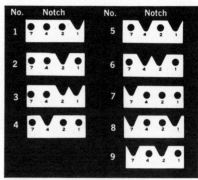

KEYSORT cards can also have alphabetical coding.

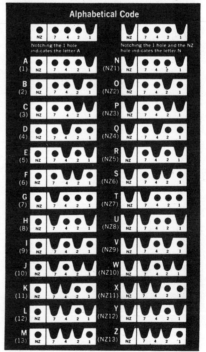

Alphabetical Code

Notching

Hand Notching

Code positions are notched individually when volume of cards is small, when speed is not an important factor—or to make a correction.

Model #5201 KEYSORT Hand Punch

This punch cuts a V shaped notch for single-row coding.

Technique

1. Hold card with thumb and forefinger of left hand, close to position to be notched.
2. Hold punch in right hand, rest chip receptacle on last three fingers of left hand.
3. Card edge must rest squarely against the guide back of the cutting blade.

Procedure

1. When learning, it is best to mark the code positions to be notched with a red pencil. As proficiency is gained, this can be eliminated.
2. Locate information and codes to be notched on the card.
3. Locate related code sections or fields along the edges of the card.
4. To notch top edge, turn card upside down.
5. Begin notching at upper left code field (now in lower right edge).
6. Notch codes from right to left, around the four edges of the card.

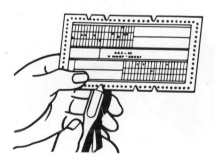

Machine Notching

Keypunching is used for high speed notching when codes vary from one card to another.

Keysort Keypunch
Model 6145 Manual; Model 6146 Electric

The KEYSORT Keypunch notches one entire side of a card at a time.

Keysort Groover — Model 6203

This desk model is for notching a group of cards having a common code. Capacity is 45 cards at a time.

Keysort Groover — Model 6210

This model can groove a maximum of 85 cards in one operation.

Sorting

Equipment

The KEYSORT Sorting Needle is a steel needle set in a plastic handle. One style is made with the needle permanently attached to the handle, while the other style is made with a screw collet which permits adjusting the length of the needle. When the latter is used, the length of the needle should be adjusted to provide the greatest ease and speed in sorting. As a rule, however, the needle should be shortened to handle unusually large or unusually small cards.

The Alignment Block increases the speed and ease of sorting. The drop front guide fits flush against the front edge of a desk. The vertical side guide on the right is used for aligning the cards.

General Guideposts in Sorting

There are several pertinent factors in KEYSORT sorting procedures with which all operators should be thoroughly familiar. They are most important in acquiring a smooth, efficient technique. These factors apply to any size or shape of KEYSORT record and should be constantly remembered by the operator.

1. **Always use an Alignment Block.** It is a most important part of the KEYSORT equipment and is essential to achieving speed and card control.

2. **Keep the KEYSORT Needle handle in the palm of the hand.** The fingers should never touch the needle while sorting.

3. **Keep wrist relaxed and flexible.** Tenseness slows the sorting operation and affects card control.

4. **Keep the hands and fingers relaxed.** This is essential for all steps of the sorting technique, but especially so in step #8 in Basic Technique. Too much tenseness or too tight a grip may cause a slight tearing out of the holes.

5. **Keep the KEYSORT Needle parallel with the tray of the Alignment Block.** This is essential to maintain the air space between the cards which allows the notched cards to fall out easily.

6. The recommended height for the notching or sorting desk is 26″ to 28″ from the floor. Proper desk and chair height is essential to the best notching and sorting technique.

With these factors in mind, sorting by the KEYSORT method is simple and easy for anyone to learn.

Basic Technique

A new operator can quickly attain a high degree of sorting efficiency without previous experience or mechanical aptitude. It is merely necessary to follow certain basic principles and techniques.

Sorting the Long Side

1. Place a convenient handful of cards (approximately 1″ or slightly larger) on the Alignment Block with the front of the cards facing the operator and the side to be sorted at the top.

2. Holding the cards loosely with the left hand, jog them against the guide of the Alignment Block.

3. Grasp the cards close to the position to be sorted.

4. Hold the handle of the KEYSORT Needle firmly with the right hand. Keep the fingers away from the needle at all times.

5. Insert the needle in the position to be sorted until the front card is approximately one inch from the handle.

9. Lift KEYSORT Needle slightly—approximately one-half inch from the Alignment Block. Strike the cards several times against the guide of the Alignment Block, at the same time gently raising the KEYSORT Needle away from the cards that are falling.

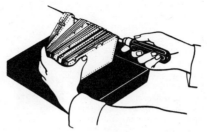

6. Slide the left hand to the left side of the cards. Hold them lightly with only slight pressure of the thumb and fingers against the cards.

7. Move the handle of the KEYSORT Needle to the left and at the same time move the cards to the center of the Alignment Block. Hold the cards with the left hand. Exert pressure with thumb in the lower left corner.

10. If some cards stick and do not fall, grasp all the cards, placing the thumb and fingers in the upper left-hand corner of the cards that are still on the needle. Move the cards towards the center of the Alignment Block. Lower right hand slightly; release the pressure of the left hand and strike the cards again to break loose those that do not fall in the first operation.

8. Swing the KEYSORT Needle to the right until resistance is felt. This will cause the cards to spread out on the needle. For sequence sorting (see following section), release the pressure of the left hand. Spread the fingers to balance the cards that will fall.

11. Life the KEYSORT Needle to the right over the guide. With the left hand open, slide the cards that have fallen against the right guide of the Alignment Block.

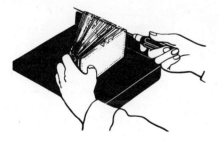

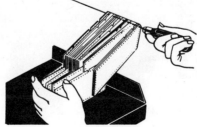

Never pull out the cards that are falling, but rather lift out the cards that are on the needle. Always keep the KEYSORT Needle parallel with the top of the desk; if tilted down, the cards will fall off the end of the needle, and if tilted up, the cards will bind together and the notched cards will not fall.

Sorting the Short Side

To sort the short side of a card the technique is slightly different from that described for the long side of the card.

1. Follow the same procedure as described in steps #1 through #5 in Sorting the Long Side of the card.

2. Slide the left hand to the bottom of the cards. Hold them lightly with only slight pressure of the thumb and fingers against the cards. Move the cards about one inch to the left away from the guide of the Alignment Block.

3. Lower the handle of the KEYSORT Needle and at the same time raise the batch of cards slightly off the floor of the Alignment Block so that the front cards are resting on the little finger of the left hand. Hold the cards with the left hand. Exert pressure with thumb in the lower left corner. The inside of the other three fingers should be flush against the beveled edge of the cards.

4. Raise the handle of the KEYSORT Needle until it is parallel to the top of the desk and resistance is felt. This will cause the cards to spread out on the needle.

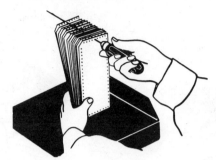

5. Release the pressure of the left hand. Tap the lower left corner of the cards several times on the Alignment Block, shaking gently two or three times.

6. Lift the cards that are on the needle to the right over the guide.

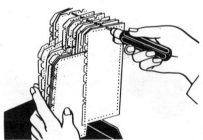

Direct Sorting

To sort in direct code positions, insert the needle in the proper hole and follow Basic Technique. For example, if the needle is inserted in the "First Shift" position, all cards punched in this position will drop out and all "Second Shift" cards will remain on the needle.

Sequence Sorting

Sequence sorting is used to sort a convenient handful of numerically coded cards into numerical sequence. To sequence sort a batch of cards, follow the operations described under Basic Technique.

Just four sorts in each 7-4-2-1 field will arrange the cards in numerical sequence. Always sort from right to left. First, sort in the 1 position of the units field and place the cards that drop to the rear of those that remain on the needle.

Before removing the KEYSORT Needle, jog all cards against the guide of the Alignment Block and let the needle fall into the grooves of the cards in the rear.

Then remove it.

In the same way, sort in the 2 position, then in the 4 position, and, finally, in the 7 position of the units field. Continue sorting all remaining digits (the tens digit, the hundreds digit, etc.) until the handful has been completely sorted into numerical sequence.

In sequence sorting, it is vitally important that all cards drop in the same relative position to one another that they originally held in the batch. Thus, if it is believed that some have not dropped, strike them against the guide of the Alignment Block as explained in Operation No. 9 of Basic Technique, prior to the removal of the needle. If a card unavoidably falls out of its proper sequence, place it to one side and insert it in its proper place when the sort is completed.

If sequence sorting is interrupted, the furthest clear groove in the rear of the batch being sorted will indicate the last position that was sorted. To resume sorting, start with the next position to the left of this clear groove.

How Keysort Works for Sequence Sorts

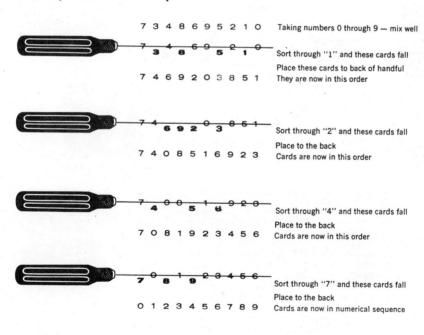

7 3 4 8 6 9 5 2 1 0   Taking numbers 0 through 9 — mix well

7 3 4 8 6 9 5 2 1 0   Sort through "1" and these cards fall

Place these cards to back of handful
7 4 6 9 2 0 3 8 5 1   They are now in this order

7 4 6 9 2 0 3 8 5 1   Sort through "2" and these cards fall

Place to the back
7 4 0 8 5 1 6 9 2 3   Cards are now in this order

7 4 0 8 5 1 6 9 2 3   Sort through "4" and these cards fall

Place to the back
7 0 8 1 9 2 3 4 5 6   Cards are now in this order

7 0 8 1 9 2 3 4 5 6   Sort through "7" and these cards fall

Place to the back
0 1 2 3 4 5 6 7 8 9   Cards are now in numerical sequence

Breakdown Sorting

This method of sorting is used to subdivide a large quantity of cards into convenient handfuls for sequence sorting.

Assume that thousands of cards numbered from 1 through 999 are to be sorted into numerical sequence. The numbers on the cards will be all mixed up. For example, reading from the top, the number on each succeeding card might be 256, 21, 875, 12, etc. Before this batch of cards can be sequence sorted, it must be divided or broken down into smaller batches of convenient handfuls.

To Break Down the Hundreds Field

Stack the cards to be sorted on the right side of the table. Take as large a handful of cards as possible. Usually twice as many cards can be handled as in sequence sorting, but the quantity will depend on the size and weight of the cards. *In breakdown sorting, always start sorting in the extreme left-hand position of the classification to be sorted.* In the example given, the 7 of the hundreds field is the starting position.

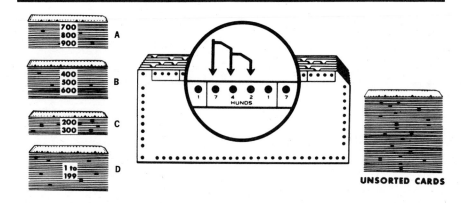

UNSORTED CARDS

Insert the needle in the 7 position of the hundreds field and follow Basic Technique. All 700, 800 and 900 cards will drop. Place these cards in Group A.

Sort the balance of the handful in the 4 position of the hundreds field. All 400, 500 and 600 cards will drop. Place these cards in Group B. Sort the balance of the handful in the 2 position of the hundreds field. All 200 and 300 cards will drop. Place these cards in Group C. Place the remaining cards, or those numbered from 1 through 199, in Stack D.

In this manner continue to sort large handfuls of cards in the hundreds field until all cards are arranged in the four groups.

To Break Down the 700—800—900 Stack

Take a handful of cards from the 700—800—900 group. Sort in the 2 position of the hundreds field. All 900 cards will drop. Place these in a stack.

Sort the balance of the handful in the 1 position of the hundreds field. All 800 cards will drop. Place these in a stack. Place the remaining 700 cards in another stack. In this manner continue to sort large handfuls of cards of the 700—800—900 group until they are arranged in the three stacks.

To Break Down the 900 Stack

Take a handful of cards from the 900 stack and sort in the 7 position of the tens field. All 970—980—990 cards will drop. Place these in a stack. In this manner continue to sort large handfuls of cards of the 900 stack.

Special Instructions

Upon completion of the initial Breakdown Sort—move the stacks of cards to the right of the alignment block.

Make sure they are in the sequence in which they will be sorted.

Complete sorting the 700—800—900's into final sequence.

Now—finish sorting the 400—500—600's into their final sequence. Then—sort the 200—300's and finally the 100's and less than 100.

Remember always to keep a well-organized sequence when moving cards from left to right of the Alignment Block so as not to mix the groups or stacks of cards. Breakdown sorting may be continued by applying the same principles just described. Merely continue to break down each stack until convenient handfuls are obtained. Then sequence sort from right to left up to the position where the last breakdown sort was made.

These two rules are extremely important: (1) Always sort each new handful of cards in the same positions as the previous handful, and (2) Whenever possible, combine small stacks to make a convenient handful for sequence sorting.

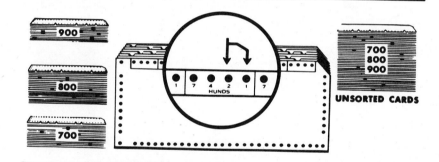

UNSORTED CARDS

AUTOMATED BUSINESS SYSTEMS [B]
DIVISION OF LITTON INDUSTRIES
600 WASHINGTON AVENUE, CARLSTADT, NEW JERSEY 07072

S-605

Printed in U.S.A.

573

INDEX